Forensic Analysis on the Cutting Edge

THE WILEY BICENTENNIAL—KNOWLEDGE FOR GENERATIONS

*E*ach generation has its unique needs and aspirations. When Charles Wiley first opened his small printing shop in lower Manhattan in 1807, it was a generation of boundless potential searching for an identity. And we were there, helping to define a new American literary tradition. Over half a century later, in the midst of the Second Industrial Revolution, it was a generation focused on building the future. Once again, we were there, supplying the critical scientific, technical, and engineering knowledge that helped frame the world. Throughout the 20th Century, and into the new millennium, nations began to reach out beyond their own borders and a new international community was born. Wiley was there, expanding its operations around the world to enable a global exchange of ideas, opinions, and know-how.

For 200 years, Wiley has been an integral part of each generation's journey, enabling the flow of information and understanding necessary to meet their needs and fulfill their aspirations. Today, bold new technologies are changing the way we live and learn. Wiley will be there, providing you the must-have knowledge you need to imagine new worlds, new possibilities, and new opportunities.

Generations come and go, but you can always count on Wiley to provide you the knowledge you need, when and where you need it!

WILLIAM J. PESCE
PRESIDENT AND CHIEF EXECUTIVE OFFICER

PETER BOOTH WILEY
CHAIRMAN OF THE BOARD

Forensic Analysis on the Cutting Edge

New Methods for Trace Evidence Analysis

Edited by
Robert D. Blackledge

WILEY-INTERSCIENCE

A JOHN WILEY & SONS, INC., PUBLICATION

Library of Congress Cataloging-in-Publication Data:

Forensic analysis on the cutting edge : new methods for trace evidence analysis /
Robert D. Blackledge, editor.
    p. cm.
  Includes index.
  ISBN 978-0-471-71644-0 (cloth)
1. Forensic sciences. 2. Criminal investigation. I. Blackledge, Robert D.
  HV8073.F557 2007
  363.25'62—dc22

                                       2007001700

Printed in the United States of America

10 9 8 7 6 5 4 3 2 1

To Sally, my beloved wife and constant source of support.

Contents

Preface xvii

Foreword xix

Contributors xxiii

1. All that Glitters *Is* Gold! 1

Robert D. Blackledge and Edwin L. Jones, Jr.

1.1   What Is Glitter? / 1
1.2   The Ideal Contact Trace / 2
    1.2.1   Nearly Invisible / 2
    1.2.2   High Probability of Transfer and Retention / 3
    1.2.3   Highly Individualistic / 3
    1.2.4   Quickly and Easily Collected, Separated, and
           Concentrated / 8
    1.2.5   Easily Characterized / 9
    1.2.6   Computerized Database Capability / 9
1.3   Characterization Methods / 10
    1.3.1   Color / 11
    1.3.2   Morphology / 12
    1.3.3   Shape / 13
    1.3.4   Size / 13
    1.3.5   Specific Gravity / 13
    1.3.6   Thickness / 14
    1.3.7   Cross Section / 15
    1.3.8   Infrared Spectra / 16

1.3.9   Raman Microspectroscopy / 19
1.3.10   Scanning Electron Microscopy/Energy Dispersive
         Spectroscopy / 22
1.4   Glitter as Evidence in Criminal Cases / 24
References / 31

**2. Forensic Analysis of Automotive Airbag Contact—Not Just
   a Bag of Hot Air**   **33**
Glenn D. Schubert

2.1   History of Airbags / 34
2.2   How Do Airbags Work? / 34
2.3   Types of Forensic Evidence to Look for / 35
2.4   Airbag Case Reports and Examples / 39
2.5   Changes that Are Occurring / 54
2.6   Final Discussion / 55
References / 55

3. Ink Analysis Using UV Laser Desorption Mass Spectrometry   **57**
John Allison

3.1   Introduction / 57
3.2   The Instrumentation / 58
3.3   The Analyte Target Molecules / 63
3.4   LDMS for the Analysis of Dyes in Pen Inks / 64
3.5   Related Applications / 74
3.6   LDMS Analyses that "Don't Work" / 75
3.7   Conclusions / 76
Acknowledgments / 77
References / 77

**4. Condom Trace Evidence in Sexual Assaults: Recovery and
   Characterization**   **81**
Wolfgang Keil

4.1   Introduction / 81
      4.1.1   Forensic Significance / 81
      4.1.2   Production, Sale, and Use of Condoms / 82
      4.1.3   Condom Production / 85
4.2   Examination for Condom Residue Traces / 88
4.3   Forensic Evaluation of the Substances and Examinations / 96
4.4   Case Studies / 105
References / 111

5. Latent Invisible Trace Evidence: Chemical Detection Strategies   **115**

Gabor Patonay, Brian Eckenrode, James John Krutak, Jozef Salon, and Lucjan Strekowski

5.1   Introduction / 115
5.2   Latent Bloodstain Detection / 117
5.3   Fingerprint Detection with Near-Infrared Dyes / 000
5.4   Pepper Spray Detection / 130
    5.4.1   Pepper Spray Detection Using Near-Infrared Fluorescent Dyes / 131
    5.4.2   Pepper Spray Detection Using Chemical Derivatization / 133
References / 138

6. Applications of Cathodoluminescence in Forensic Science   **141**

Christopher S. Palenik and JoAnn Buscaglia

6.1   Introduction / 141
6.2   Theory / 143
    6.2.1   Luminescence Terminology / 143
    6.2.2   Electron Source / 143
    6.2.3   Cathodoluminesence / 144
    6.2.4   Limitations / 147
6.3   Instrumentation / 147
    6.3.1   Electron Source / 147
    6.3.2   Microscope / 149
    6.3.3   Camera / 149
    6.3.4   Spectrometer / 150
    6.3.5   SEM-CL / 150
6.4   Techniques and Forensic Considerations / 151
    6.4.1   Instrumental Conditions / 151
    6.4.2   Sample Preparation and Preservation / 152
    6.4.3   Image Collection / 153
    6.4.4   Spectral Collection / 154
    6.4.5   Luminescence Fading / 155
    6.4.6   Sample Alteration / 155
6.5   Luminescent Minerals / 156
    6.5.1   Calcium Carbonate Group / 156
    6.5.2   Feldspar Group / 159
    6.5.3   Quartz / 160
    6.5.4   Accessory Minerals / 163
6.6   Forensic Applications / 164
    6.6.1   Screening and Comparison / 165

6.6.2 Identification / 165

6.6.3 Authentication / 166

6.6.4 Provenance / 166

6.7 Geological Samples: Soil and Sand / 167

6.8 Anthropogenic Materials / 168

6.8.1 Cement and Concrete / 168

6.8.2 Slag, Fly Ash, and Bottom Ash / 168

6.8.3 Glass / 169

6.8.4 Paint / 170

6.8.5 Duct Tape / 170

6.9 Conclusions and Outlook / 171

Acknowledgments / 171

References / 172

7. Forensic Application of DART™ (Direct Analysis in Real Time) Mass Spectrometry   **175**

James A. Laramée, Robert B. Cody, J. Michael Nilles, and H. Dupont Durst

7.1 Introduction / 175

7.2 Experimental / 176

7.3 Drug and Pharmaceutical Analysis / 177

7.3.1 Confiscated Samples / 178

7.3.2 Endogenous Drugs / 178

7.3.3 Drug Residues on Surfaces / 179

7.4 Samples from the Human Body / 180

7.4.1 Fingerprints / 180

7.4.2 Bodily Fluids / 181

7.5 Condom Lubricants / 184

7.6 Dyes / 184

7.6.1 Self-Defense Sprays / 184

7.6.2 Currency-Pack Dye / 185

7.7 Explosives / 185

7.8 Arson Accelerants / 188

7.9 Chemical Warfare Agents / 189

7.10 Elevated-Temperature DART for Material Identification / 190

7.11 Glues / 191

7.12 Plastics / 191

7.13 Fibers / 192

7.14 Identification of Inks / 193

7.15 Conclusion / 194

Acknowledgments / 194

References / 194

8. Forensic Analysis of Dyes in Fibers Via Mass Spectrometry   **197**

Linda A. Lewis and Michael E. Sigman

8.1   Introduction / 197

8.2   Conventional Fiber Color Comparison Methods Employed in Forensic Laboratories / 198

8.3   Shortcomings Associated with UV–Vis Based Comparative Analysis for Trace-Fiber Color Evaluations / 199

8.4   General Overview of Modern Dye Ionization Techniques for Mass Analysis / 203

8.5   Trace-Fiber Color Discrimination by Direct ESI-MS Analysis / 204

8.6   Examples of Negative Ion ESI-MS Analysis of Colored Nylon Windings / 207

8.7   Examples of Tandem Mass Spectrometry (MS/MS) Applications to Elucidate Structure / 208

8.8   LC-MS Analysis of Dyes Extracted from Trace Fibers / 210

8.9   Proposed Protocols to Compare Trace-Fiber Extracts / 214

      8.9.1   Direct Infusion MS/MS Protocol / 214

      8.9.2   Generalized LC-MS and LC-MS/MS Protocol / 216

8.10   Conclusions / 217

Acknowledgments / 217

References / 217

9. Characterization of Surface-Modified Fibers   **221**

Robert D. Blackledge and Kurt Gaenzle

9.1   Fibers as Associative Evidence / 221

9.2   Surface-Modified Fibers / 222

9.3   Preliminary Examinations / 222

      9.3.1   Infrared Spectra and Properties Measured by Polarized Light Microscopy / 223

      9.3.2   Infrared Mapping with an FTIR Microscope / 223

      9.3.3   Raman Mapping / 224

      9.3.4   AATCC Test Method 118-2002 / 224

      9.3.5   A Simple Example / 224

9.4   Distinguishing Tests / 225

      9.4.1   Scanning Electron Microscopy/Energy Dispersive Spectroscopy / 225

      9.4.2   Gas Chromatography/Mass Spectrometry / 231

      9.4.3   Pyrolysis Gas Chromatography/Mass Spectrometry / 234

Acknowledgments / 238

References / 238

10. Characterization of Smokeless Powders   **241**

Wayne Moorehead

10.1   Introduction / 241
10.2   Purpose of Analysis / 242
    10.2.1   Identification of Smokeless Powder / 242
    10.2.2   Determining Brand / 242
10.3   Brief History of Smokeless Powder / 243
10.4   Characterization Toward Smokeless Powder Identification / 245
10.5   Characterization Toward Brand Identification / 246
    10.5.1   Characterization by Morphology / 246
    10.5.2   Micromorphology / 247
    10.5.3   Other Characteristics / 257
10.6   Micrometry / 258
10.7   Mass / 259
10.8   FTIR Spectroscopy / 260
    10.8.1   Transmission Micro-FTIR / 260
    10.8.2   ATR-FTIR / 261
10.9   Chromatography with Mass Spectrometry / 262
    10.9.1   Gas Chromatography / 263
    10.9.2   Liquid Chromatography / 265
10.10   Conclusion / 266
References / 266

11. Glass Cuts   **269**

Helen R. Griffin

11.1   A Homicide / 270
11.2   A Robbery / 271
11.3   A Hit and Run / 274
11.4   Cutting Versus Tearing / 276
11.5   Slash Cuts Made by Glass / 279
    11.5.1   Associated Glass / 280
    11.5.2   Fabric Type / 280
    11.5.3   Blade Characteristics / 280
11.6   Conclusion / 287
Acknowledgments / 287
References / 289
Additional Sources / 289

12. Forensic Examination of Pressure Sensitive Tape   **291**

Jenny M. Smith

12.1   Introduction / 291
12.2   Product Variability / 292

12.3   Tape Construction / 293

    12.3.1   Tape Backings / 296
    12.3.2   Adhesive Formulations / 297
    12.3.3   Common Reinforcement Fabrics / 299

12.4   Duct Tape / 299
12.5   Electrical Tape / 307
12.6   Polypropylene Packaging Tape / 309

    12.6.1   Oriented Films / 310
    12.6.2   Polarized Light Microscopy Examinations of Packing
           Tapes / 310
    12.6.3   Is It MOPP or BOPP? / 311
    12.6.4   Thickness / 312
    12.6.5   Degree of Offset from the Machine Edge / 313

12.7   Strapping/Filament Tapes / 313
12.8   Masking Tape / 315
12.9   Initial Handling / 316

    12.9.1   Sharing Evidence with Other Sections / 317
    12.9.2   Untangling Tape and Recovering Trace Evidence / 318

12.10   Methods / 318

    12.10.1   Physical End Matching / 318
    12.10.2   Physical Characteristics / 319
    12.10.3   Separation of Backing, Reinforcement, and
           Adhesive / 319
    12.10.4   FTIR Analysis / 320
    12.10.5   Elemental Analysis / 320
    12.10.6   Polarized Light Microscopy / 325
    12.10.7   Pyrolysis GC/MS / 326
    12.10.8   Sourcing Tape Products to a Manufacturer / 326

12.11   Case Example / 327
Acknowledgments / 329
References / 329
Additional Sources / 331

**13. Discrimination of Forensic Analytical Chemical Data Using
Multivariate Statistics**   **333**
Stephen L. Morgan and Edward G. Bartick

13.1   Patterns in Data / 333
13.2   Experimental Design and Preprocessing / 336
13.3   Dimensionality Reduction by Principal Component Analysis for
       Visualizing Multivariate Data / 342
13.4   Visualizing Group Differences by Linear Discriminant
       Analysis / 348
13.5   Group Separation, Classification Accuracy, and Outlier
       Detection / 354

13.6 Selected Applications / 359
13.7 Conclusion / 366
Acknowledgments / 367
References / 367

14. The Color Determination of Optically Variable Flake Pigments   **375**
Michael R. Nofi

14.1 Introduction / 375
14.2 OVP: Form, Characteristics, and Function / 376
14.3 Color Measurement / 379
14.4 Color Blending / 382
14.5 Additive Color Theory / 383
14.6 Methods of Formulating OVP / 385
14.7 Blending of Pigments / 388
14.8 Microspectrophotometry / 389
14.9 Measurement Geometry / 390
14.10 Switching Objective Magnifications / 391
14.11 Determining Sample Size / 392
14.12 Measurement Uncertainty / 392
14.13 Sample Preparation and Measurement / 393
14.14 Spectral Profiling / 394
14.15 Statistical Methods of Evaluation / 395
14.16 Challenges for the Future / 395
14.17 Other forensic Methods / 396
Acknowledgments / 396
References / 396
Additional Sources / 397

15. Forensic Science Applications of Stable Isotope Ratio Analysis   **399**
James R. Ehleringer, Thure E. Cerling, and Jason B. West

15.1 What Are Stable Isotopes? / 400
15.2 What Are the Units for Expressing the Abundance of Stable Isotopes? / 401
15.3 What Is the Basis for Variations in Stable Isotope Abundances? / 401
15.4 What Instrumentation Is Needed for High-Precision Stable Isotope Measurements? / 402
15.5 How Can Stable Isotope Analyses Assist Forensics Cases? / 404
15.6 Stable Isotope Abundances in Forensic Evidence / 405
    15.6.1 Food Products, Food Authenticity, and Adulteration / 406
    15.6.2 Doping and Drugs of Abuse / 408

15.6.3 Sourcing of Humans, Animals, and Animal Products / 411

15.6.4 Humans: Bones, Hair, and Teeth / 413

15.6.5 Stable Isotope Abundances of Manufactured Items / 414

References / 416

Index **423**

Preface

The idea for this book had its genesis as a result of my organization of a symposium, "Forensic Analysis on the Cutting Edge," at the 2004 Pittcon held in Chicago. Three of my presenters, John Allison, Brian Eckenrode, and Wolfgang Keil, are also contributors. The symposium featured either types of trace evidence that were increasing in importance but had not previously received much attention, or new and better methods for trace-evidence characterization. In its various chapters this book follows that same theme.

Forensic Analysis on the Cutting Edge: New Methods for Trace Evidence Analysis is not intended to be a "coffee table book"! To be sure, mystery writers, attorneys, forensic science students, criminal investigators, or anyone wishing to go beyond being "CSI dilettantes" may find the various chapters both useful and fascinating. However, it is hoped that in forensic laboratories worldwide the book will become a well-thumbed reference on the shelves of both novice and experienced criminalists.

Each of the book's chapters is stand-alone. That is, no chapter assumes the reader has acquired information provided in a previous chapter. Since no chapter is intended to serve as a basic introduction to the principles, tools, and methods of trace evidence analysis, it is incumbent that some admonition be made regarding precautions against cross-contamination. The entire value of trace evidence lies in its ability to show an association between a suspect/victim/crime scene. Therefore, every effort must be made to prevent even the *possibility* of cross-contamination. This must begin with the crime scene and the collection and packaging of evidence. Suspects and victims must be transported in different vehicles and interviewed in different rooms. Search/examination areas in forensic laboratories must be scrupulously clean and should be located in areas with minimal foot traffic and not in areas where the opening and closing of doors or windows can create sudden breezes. Evidence items associated with suspects should if possible be examined in an area separate from those used for examination of evidence items from the victim. Labcoats and gloves worn by examiners must be changed when going between these areas. For a more detailed discussion readers are referred to *Scientific Working Group on Materials Analysis (SWIGMAT) Trace Evidence Recovery Guidelines*, http://www.fbi.gov/hq/lab/fsc/backissue/oct1999/trace.htm.

San Diego, California   ROBERT D. BLACKLEDGE
January 2007

Foreword

I am very pleased to have been invited by Robert Blackledge to write the Foreword for this book. I have long admired Bob's contributions to the field of forensic science, specifically criminalistics and trace evidence. We share much of the same philosophy about criminalistics. This might be termed a "classical philosophy of criminalistics." I will return to some of the concepts this term connotes further along in this Foreword. For the time being, I will point out that for many years I have been concerned that this philosophy is in danger of extinction.

This is an important book dealing with important topics. Experts in the field have contributed the chapters on a variety of trace-evidence topics and techniques. The book will be most useful to practicing forensic scientists (criminalists) engaged in trace-evidence analysis. A second group of users will be made up of students in university forensic science programs. It is to be hoped that a third group, already employed in forensic science laboratories, will be inspired to develop expertise with trace evidence and contribute to the field. Although these three groups will make up the bulk of the market for the book, others including attorneys, judges, and even mystery writers will be able to extract from it useful information and ideas. It should be understood that this is a book dealing with a variety of carefully selected trace-evidence techniques and problems. It is not a textbook on trace evidence. No such book exists. There are books that deal with case examples where trace evidence was critical to the case solution. Although case examples are used here, this is primarily a book about the recognition, analysis, and interpretation of types of trace evidence that may occur in nonroutine cases. The broad range of evidence types covered is only a very small portion of the possible trace-evidence types that can be expected to be encountered as trace evidence.

Trace evidence is at the core of criminalistcs. Surprisingly, this is not widely appreciated. If one accepts this fact, it is difficult to understand why there are so few trace-evidence analysts compared to the number of narrowly educated specialists employed in forensic science laboratories. To appreciate the potentials of trace evidence, it is important to consider the general question of the creation of physical evidence and the resulting record produced. Individual items of physical evidence can be viewed as the fundamental components of a natural record made of an event. Actions taking place during a human-initiated event create a record of varying degrees of detail through the interactions of energy and matter with the environment in which the event takes place. These actions cannot help but produce a record. This is a trace-evidence record. If the event of concern is a possible crime, decoding and interpreting this record becomes a problem for the criminalist.

There are two different conceptual views of trace evidence. In many quarters trace evidence is thought of as the small amounts of material that are transferred from one surface to another as a result of contact. This has been codified as the *Locard Exchange Principle.* One expression of this principle is that "every contact leaves a trace." Here the term "trace" seems to imply the exchange of a small amount of material. It also implies that a contact is necessary for the transfer to occur. However, this interpretation is by no means universal. The transfer may involve that of a pattern rather than that of material. Thus, an example would be indentation or striation-type toolmarks, where a pattern is created by the impression into, or the motion against, a softer material by a harder one, but where no material transfer is detected. The second view is broader. Thus, the word "trace" itself has a broader meaning and does not necessarily connote a small size and direct contact is not necessary. In its broadest concept a "trace" can be thought of as something left behind, a remnant, a vestige, or, more to the point, physical evidence of a prior interaction.

In this sense trace evidence is more than a collection of evidence types or techniques. More generally, it is an approach to problem solving in criminalistics. It is this approach that is most appropriate in all but the simplest of cases. Real cases demand more than unthinkingly applying "tests" on "items" of evidence. "Items" at the crime scene have to be recognized as having potential significance to the investigation before the most appropriate analytical approaches can be designed and decisions made as to what testing is warranted. This leads to better case solutions while simultaneously making the best use of laboratory resources.

It is unfortunate that the attention devoted to trace evidence has been diluted by a combination of factors, including rapid growth in the field, an increasing focus on technology at the expense of science, the desire for high throughput in forensic science laboratories, and the large scale hiring of new laboratory personnel educated in narrowly defined specialty areas. In addition, it is unfortunate that many of the recent, and unquestionably important positive, developments in the field such as methods standardization and laboratory

accreditation have had unforeseen or often unrecognized negative conse-
quences with respect to trace-evidence analysis. One truly regrettable result
is that there is a dearth of generalist scientists capable of recognizing trace
evidence and formulation testable scientific questions to maximize the information extracted from the available trace evidence in many cases. Evidence
that goes unrecognized cannot be analyzed and might as well have been nonexistent. Opportunities for the exculpation of the wrongly accused or inculpa-
tion of the guilty and solving a crime are lost.

In addition to providing valuable information for trace-evidence examiners,
it is to be hoped that this book will serve to illustrate the power and potential
of trace evidence and to stimulate interest in garnering more support for trace
evidence in the form of increased hiring of personnel educated with
respect to trace-evidence analysis problem solving approaches, and result in
increased support in the form of research funding. The work in trace evidence
is challenging and difficult, but it also provides an unmatched measure
of personal satisfaction. This book should serve to make the reader aware of
the possibilities.

Professor of Criminalistics   PETER R. DE FOREST
Director of the Master of Science in Forensic Science Program
Department of Sciences
John Jay College of Criminal Justice/CUNY
New York, New York

Contributors

JOHN ALLISON, Professor of Chemistry and Director of Forensic Chemistry, The College of New Jersey, Ewing, New Jersey.

EDWARD G. BARTICK, Retired, Research Chemist, FBI Academy, Counterterrorism and Forensic Science Research Unit, Quantico, Virginia. Present: Director Forensic Science Program, Department of Chemistry and Biochemistry, Suffolk University, Boston, Massachusetts.

ROBERT D. BLACKLEDGE, Retired, former Senior Chemist, Naval Criminal Investigative Service Regional Forensic Laboratory, San Diego, California.

JOANN BUSCAGLIA, Research Chemist, FBI Laboratory, Counterterrorism and Forensic Science Research Unit, FBI Academy, Quantico, Virginia.

THURE E. CERLING, IsoForensics, Inc., and Professor of Biology and Geology and Geophysics, University of Utah, Salt Lake City, Utah.

ROBERT B. CODY, Product Manager, JEOL USA, Inc., Peabody, Massachusetts.

H. DUPONT DURST, Research Chemist, Edgewood Chemical Biological Center, Edgewood, Maryland.

BRIAN ECKENRODE, Research Chemist, FBI Laboratory, Counterterrorism and Forensic Science Research Unit, FBI Academy, Quantico, Virginia.

JAMES R. EHLERINGER, IsoForensics, Inc., and Biology Department, University of Utah, Salt Lake City, Utah.

KURT GAENZLE, Senior Materials Engineer, Materials Engineering Laboratory, Naval Air Depot, North Island, San Diego, California.

HELEN R. GRIFFIN, Forensic Scientist III, Ventura County Sheriff's Department Forensic Sciences Laboratory, Ventura, California.

EDWIN L. JONES, Jr., Forensic Scientist III, Ventura County Sheriff's Department Forensic Sciences Laboratory, Ventura, California.

WOLFGANG KEIL, Professor of Legal Medicine, Department of Legal Medicine, Ludwig-Maximilians-University of Munich, Munich, Germany.

JAMES JOHN KRUTAK, Senior Scientist, FBI Engineering Research Facility, Operational Technology Division, Technical Operations Section, Quantico Virginia.

JAMES A. LARAMÉE, Principal Chemist, EAI Corporation, a subsidiary of SAIC, Inc., Abingdon, Maryland.

LINDA A. LEWIS, Research Staff Member, Oak Ridge National Laboratory, Oak Ridge, Tennessee.

WAYNE MOOREHEAD, Senior Forensic Scientist, Orange County Sheriff–Coroner, Forensic Science Services Division, Santa Ana, California.

STEPHEN L. MORGAN, Department of Chemistry and Biochemistry, University of South Carolina, Columbia, South Carolina.

J. MICHAEL NILLES, Senior Scientist, Geo-Centers, Inc., a Subsidiary of SAIC, Inc., Abingdon, Maryland.

MICHAEL R. NOFI, Senior Metrology Engineer, JDSU-Flex Products Group, Santa Rosa, California.

CHRISTOPHER S. PALENIK, Research Microscopist, Microtrace, Elgin, Illinois.

GABOR PATONAY, Department of Chemistry, Georgia State University, Atlanta, Georgia.

JOZEF SALON, Department of Chemistry, Georgia State University, Atlanta, Georgia.

GLENN D. SCHUBERT, Forensic Scientist, Illinois State Police, Forensic Science Command, Southern Illinois Forensic Science Centre, Carbondale, Illinois.

MICHAEL E. SIGMAN, Associate Professor of Chemistry and Assistant Director for Physical Evidence, National Center for Forensic Science, University of Central Florida, Orlando, Florida.

JENNY M. SMITH, Criminalist III, Missouri State Highway Patrol Crime Laboratory, Jefferson City, Missouri.

LUCJAN STREKOWSKI, Department of Chemistry, Georgia State University, Atlanta, Georgia.

JASON B. WEST, Research Assistant Professor, Biology Department, University of Utah, Salt Lake City, Utah.

Cover photo from:

LA ICPMS and IRMS Isotopic and Other Investigations in Relation to a Safe Burglary

Gerard J.Q. van der Peijl, PhD*, Netherlands Forensic Institute of the Netherlands, Ministry of Justice, PO Box 24044, The Hague, 2490 AA, Netherlands; and Shirly Montero, PhD, Wim Wiarda, Ing, and Peter de Joode, Ing, Netherlands Forensic Institute, PO Box 3110, Rijswijk, 2280 GC, Netherlands. Oral presentation B37, Criminalistics Section, American Academy of Forensic Sciences, 2005 Annual Meeting, New Orleans, USA. [Abstract available at: http://www.aafs.org/pdf/NewOrleansabstracts05.pdf]

Cover photo caption: (right) material in plastic bag with money at suspect's house and (left) safe wall filling as found at crime scene (sawdust & alum)

Isotopic ratio mass spectrometry (IRMS) results:

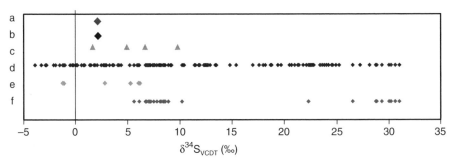

Alum 34S variation (a) safe, (b) suspect, (c) Dutch Alum samples, (d) alunite data from minerals – worldwide, (e) alunite Spain – Riaza and (f) alunite Spain – Rodalquilar

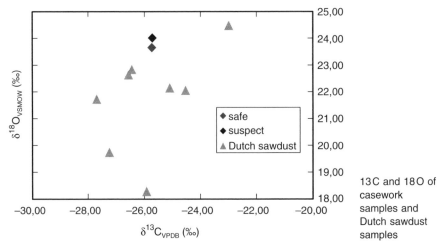

13C and 18O of casework samples and Dutch sawdust samples

Forensic Analysis on the Cutting Edge

1

All that Glitters Is *Gold!*

Robert D. Blackledge

Retired, former Senior Chemist, Naval Criminal Investigative Service Regional Forensic Laboratory, San Diego, California

Edwin L. Jones, Jr.

Forensic Scientist III, Ventura County Sheriff's Department Forensic Sciences Laboratory, Ventura, California

We've all heard the saying, "all that glitters is not gold." Although not made of gold, tiny particles of commercial glitter may be "as good as gold" in terms of their potential value as associative evidence in criminal cases.

1.1   WHAT IS GLITTER?

Glitter is entirely man-made. It may be tiny pieces of aluminum foil or plastic with a vapor-deposited aluminum layer, or it may consist of multiple layers of plastic with no metal layer at all. In the manufacturing process, before it is cut into individual tiny particles, it is in the form of rolled sheets of foil or plastic. Most often the sheets are cut to make particles that are hexagonal, square, or rectangular since these shapes can fully fill a two-dimensional surface with no waste material produced. Although glitter can be obtained in specialized shapes such as circles, stars, and crescent moons, these particles are usually much larger since it is intended that a viewer can see and recognize these shapes.

Forensic Analysis on the Cutting Edge: New Methods for Trace Evidence Analysis, Edited by Robert D. Blackledge.
Copyright © 2007 John Wiley & Sons, Inc.

Some cosmetics products are advertised as containing "shimmer." "Shimmer" is not glitter. Shimmer starts off as tiny pieces of mica. Although shimmer particles may fall into a certain size range, their shape is totally irregular and random. In order to increase their sparkle the mica pieces may be coated with titanium dioxide, and iron oxides or other pigments may be added to produce color. Although like glitter, shimmer has potential value as associative evidence, it will not be considered in this chapter.

Glitter has been around for many years, but until fairly recently its cosmetic use was primarily for special costume party events such as Halloween and Mardi Gras. But today glitter can be found in every possible variety of women's cosmetic products, and most are intended for everyday wear rather than just for special occasions. Glitter is even found in a number of different brands of sunscreen lotion. There are also many varieties of removable glitter tattoo transfers, and several brands of glitter gel pens in different colors. Glitter is also used extensively in arts and crafts projects. For this purpose it may be sold loose (no vehicle), in pencil sticks (also used in cosmetics), in glues, and in paints. As we shall see later in the chapter, there have been several cases where glitter that had its origination in children's arts and crafts projects became very important transfer evidence. Glitter is used commercially to add decoration to greeting cards, Christmas balls, and other ornaments, and it is used year round to make decorations on women's items of apparel such as T-shirts and sweatshirts. Glitter is also used to add eye appeal to many solid plastic items. In these cases the glitter is within the transparent plastic rather than just on the surface. Glitter is often used in fishing lures, and companies sell glitter-containing dough as well as gels into which lures may be dipped. One company offers a glitter-containing gel that has a fish-attracting maggot aroma! Yeech! We hope we never have a case involving that as trace evidence!

1.2 THE IDEAL CONTACT TRACE

In many ways, glitter may be the ideal contact trace. In her master's thesis, *Evidential Value of Glitter Particle Trace Evidence*, Klaya Aardahl [1] lists the properties of the ideal trace evidence: "(1) nearly invisible, (2) has a high probability of transfer and retention, (3) is highly individualistic, (4) can quickly be collected, separated, and concentrated, (5) the merest traces are easily characterized, and (6) is able to have computerized database capability." Let's see how well glitter meets these ideal properties. Like some other polymer evidence types such as paint and some fibers, they will survive most environmental insults like a hot car, a putrefying body, exposure to sunlight, and exposure to cleaning products.

1.2.1 Nearly Invisible

One glitter manufacturer (Meadowbrook Inventions, Bernardsville, New Jersey, USA) offers glitter in 13 different sizes ranging from 0.002 in. (50 μm) up to 0.250 in. (6250 μm), but 0.004 in. (100 μm) and 0.008 in. (200 μm) are more typical of glitter used in cosmetic products. Under typical viewing conditions we really don't "see" glitter particles, we just see the light reflected off them. It is like seeing a star at night. Although we see the light that long ago originated from the star (or if it is a planet in the solar system we see our sun's light reflected from it), the area of our field of vision that the star or planet occupies is far, far too small for us to actually see it. Unlike more obvious transfers like blood or lipstick smears, because of their small size a criminal is often not aware that some glitter particles have transferred to his/her skin, clothing, vehicle carpeting, and so on. Not being aware of their presence, a criminal takes no measures to remove them or get rid of the clothing items. In at least one case [2] the suspect was known to have unsuccessfully cleaned out his vehicle after committing an abduction/homicide.

1.2.2 High Probability of Transfer and Retention

Absent any unusual surface properties (e.g., cockleburs or resin-covered plant materials), small and lightweight particles are more likely to transfer and stick. For example, the larger gunshot residue particles have poor retention on the hands of a shooter. In a small trial reported by Aardahl [1, 3], glitter originating from selected cosmetic products was found to readily transfer and be retained. This is not surprising. For example, glitter made from several layers of polyethylene terephthalate (PET) might typically be 0.008 in. (200 μm) in diameter and 0.002 in. (50 μm) thick and have a specific gravity of 1.2. Indeed, small and light.

1.2.3 Highly Individualistic

Glitter is class type evidence. With class evidence the smaller the class into which it can be placed, the greater its value as associative evidence. For example, white cotton fibers are so common that they have no associative evidential value and are ignored by criminalists. Whereas the carpet fibers found on so many victims in the Atlanta child murders case could be traced back to a small mill that had only produced that carpet several years previously for a brief period of time and in limited quantity [4]. Therefore, those carpet fibers had great value as associative evidence. How individualistic are glitter particles? The more ways in which glitter from different sources can be shown to vary, the smaller the subclass into which an individual particle may be placed. The smaller the subclass, the more it has value as associative evidence. So let's count the various ways in which we may be able to categorize an individual glitter particle.

Shape. Is the particle hexagonal, square, or rectangular?

Size. What are its dimensions or its area? Many glitter manufacturers offer as many as 12 different size particles. (Square particles are listed as the length of a side and hexagonal particles are listed as the distance from one apex to its opposite apex.)

Thickness. What is its thickness? Depending on the number of plastic layers and the chemistry of each layer or if there is an aluminum layer, the thickness of glitter particles from different sources may vary. Looking at just one glitter manufacturer's website (Meadowbrook Inventions), for its various glitters the manufacturer lists the following thicknesses: 15 μm (Micronic Jewels), 18 μm (Metallic Jewels), 25 μm (Alpha Jewels HTMP, Cosmetic Jewels, Polyester Jewels), approximately 28–36 μm (various glitters in their Crystalina series), 50 μm (Alpha Jewels, Clear Polyester, Polyester Pearl Jewels), 65 μm (High Chroma Silver/Mirror Crystalina), 75 μm (Phosphorescent Poly Glitter), and 175 μm (Plastic Jewels).

Specific Gravity. Again at the Meadowbrook Inventions website, the manufacturer lists the following specific gravities: 1.2 (Clear Polyester, Crystalina 400), 1.25 (Polyester Pearl Jewels), 1.3 (High Chroma Silver/Mirror Crystalina), 1.36 (Crystalina 300, Crystalina 310, 311, Crystalina Colors), 1.4 (Alpha Jewels, Cosmetic Jewels, Micronic Jewels, Plastic Jewels, Polyester Jewels), 2.4 (Alpha Jewels HTMP, Metallic Jewels), and 2.5 (Phosphoresescent Poly Glitter).

Morphology. Although many glitter products have no distinctive surface appearance and appear the same on both sides, some have on one side an added pigment that is distinctive in both color and morphology (Figure 1.1).

Number and Thickness of Layers. As with paint chips, individual glitter particles may be cross-sectioned and the separate layers counted and measured (Figure 1.2).

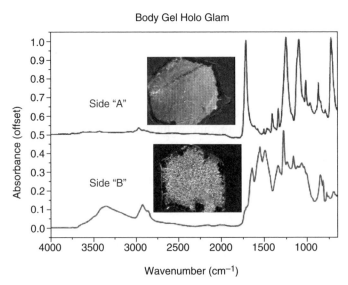

Figure 1.1 *(See color insert.) Glitter particle morphology. Two sides of the same particle (700× original magnification) with corresponding ATR FTIR spectra. (From Reference 5.)*

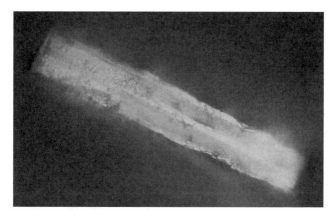

Figure 1.2 A glitter particle in cross section showing five distinct layers. (From Reference 6.)

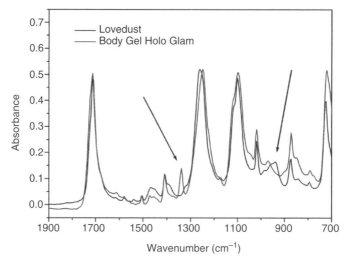

Figure 1.3 Overlay of ATR FTIR spectra from two different glitters shows reproducible differences. (From Reference 5.)

Chemistry of Each Layer. By using methods of infrared spectroscopy the chemical composition of each layer may be determined. The method has the sensitivity to detect even subtle differences in chemistry in particles from different sources (Figure 1.3).

Color. For some of their products manufacturers may offer as many as 48 different colors as well as glitters that fluoresce under ultraviolet light or have holographic properties. Art Institute Glitter, Inc., a company that sells glitter for arts and crafts, offers over 350 different colors in 11 different varieties at their website, www.artglitter.com. One variety even consists of tiny pieces of broken glass (Vintage Glass Glitter).

Glitter Manufacturers. If they examine enough properties and have adequate instrumentation, criminalists can usually distinguish paint samples originating from different manufacturers. Although it might not be easy, the same thing should be true of glitter. The more manufacturers there are of glitter, the more variety we would expect. It is not easy to pin down exactly how many manufacturers there are of glitter worldwide. Some companies only make the film and sell it to other companies to be cut into glitter. However, via Internet searches over a dozen manufacturers have been identified in countries including the United States, Germany, India, Pakistan, China, Taiwan, and South Korea. So, how individualistic is glitter? At its website, Meadowbrook Inventions claims over 20,000 varieties, and another large manufacturer, Glitterex Corporation, claims in excess of 10,000 (Table 1.1.).

Cutting Machine Characteristics. There is another way in which glitter might vary. In the manufacture of shoe outsoles, occasionally one of a company's

TABLE 1.1   Glitter Film and/or Particle Manufacturers

Engelhard Corp., Iselin, NJ, USA [makes film and sells it to glitter cutters]
    www.englehard.com
Meadowbrook Inventions Inc., Bernardsville, NJ, USA [world's largest glitter manufacturer, with >20,000 different products]
    www.meadowbrookinventions.com
Glitterex Corp., Cranford, NJ, USA [over 10,000 different glitter prodcts]
    www.glitterex.com
Spectratek Technologies, Los Angeles, CA, USA
    www.spectratek.net
Glitron Products, Pakistan
    www.glitron.com
A-Joo Industrial Co. Ltd., Korea
    www.ajootex.com
Advance Syntex Pvt. Ltd., India
    www.midasglitter.com
Spick Global Films, India [makes film and sells it to glitter cutters]
    www.spickglobal.com
Jincolor Co. Ltd., Taiwan
    www.jincolor.com
Metlon India Pvt. Ltd., India
    www.metlonindia.com
Ho Long Glitters Enterprises Co., Taiwan
    www.holongglitters.com.tw
WooSol Industry, Korea
    www.woosolind.com
Gaoyuan Glitter Materials Co., Ltd., China
    www.cngaoyuan.com
Masa Glitter, Germany
    www.masa-glitter.com
RJA Plastics GmbH, Germany
    www.rja-plastics.com

several molds for a specific size and right or left outsole will develop a slight defect. As long as this mold defect is barely noticeable, the company will continue to use that mold. Criminalists examining footwear impressions from crime scenes have to be very careful that they don't mistake this mold defect for accidental damage. If it were accidental damage that had occurred during the wearing of the shoe, then this mark could lead the examiner to conclude that no other shoe (than the shoe from the suspect) could have made that impression. But if the mark were due to a mold defect, then every outsole from that particular mold would have that mark. The mark would put the questioned impression into a smaller class, but it would be wrong to conclude that the questioned shoe and only the questioned shoe could have made that impression. In microscopically examining glitter particles from different sources, imperfections from a perfect hexagon or perfect square or rectangle have been noted in some particles in some samples. Although an area that requires more study (visit to glitter manufacturers and collecting samples from specific cutting machines), it is possible that glitter samples that are otherwise alike might be distinguished due to these imperfections. See the photomicrograph in Figure 1.4 for an example of a case where the same imperfection was

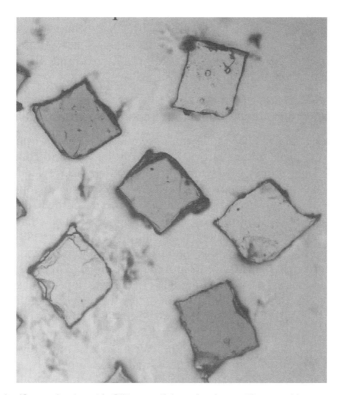

Figure 1.4 *(See color insert.) Glitter particles showing cutting machine anomalies (one rounded corner and protrusion on adjacent corner. (From Reference 6.)*

found in both glitter particles associated with the victim and those with the suspect, and therefore increased the likelihood that they had a common origin.

Vehicle. There is potentially at least one more way that glitter particles originating from different sources might be discriminated. When glitter is in cosmetic products it is generally in some vehicle (aerosol hair spray, roll-on, pump spray, lotion, eye liner, lipstick, etc.). Cosmetic products from two different companies might contain identical glitter particles if both companies had purchased the same type of glitter and from the same manufacturer. However, it would be unlikely that the vehicle used by both companies would have identical formulations. If traces of the vehicle still adhered to a recovered glitter particle and if these traces could be extracted and chemically identified, it might be possible to show that these traces did or did not match up with the ingredients in the glitter-containing cosmetic product in the victim's purse. For her master's thesis Chandra Weber [7] used gas chromatography with a mass selective detector and attenuated total reflectance FTIR microscopy in an attempt to recover and characterize vehicle traces from individual glitter particles. These attempts were largely unsuccessful. Either the vehicle was too volatile (hair spray), or it was rapidly absorbed by the skin (lotions). However, with more sensitive instrumentation methods it still might be possible to recover identifiable vehicle traces if the glitter had been in less volatile and less water-soluble cosmetics such as lipsticks and eyeliners. Additionally, glue and paint are frequently used as vehicles in glitter-bearing clothing decorations and in arts and crafts applications.

1.2.4   Quickly and Easily Collected, Separated, and Concentrated

Although a criminal may be totally unaware that glitter has transferred to him or his surroundings, glitter particles are easily located and collected when processing crime scenes or examining items of physical evidence. Smaller items may be searched directly using a stereobinocular microscope or a video microscope at low power and illumination with reflected light. Tape lifts can be made over larger items or areas and then the lifts can be examined under low magnification. However, if one is specifically searching for glitter particles it may be more efficient to search under low light conditions and use a flashlight or other bright light source to spot the reflections of individual particles. As a likely individual particle is found it is specifically tape lifted and the specific location where it was found recorded. For this purpose, the authors like to use Post-it® notes. The glue used on Post-it notes is strong enough to lift off the particle, but is weaker than that found on various brands of transparent tape. Because of this the examining criminalist can later more readily pick off the particle for subsequent examinations, and in the process be less likely to damage the particle or have the particle be contaminated with glue from the tape. Another advantage of the Post-it notes is that after collection they can be folded over and all the necessary identifying information written on the note prior to it being sealed in an evidence envelope, or the identifying information can be written on the same side as the adhering glitter particle and placed in a small ziplock bag where the evidence can be examined microscopically through the bag.

Location, Location, and Location! Sometimes the location where trace evidence is found is every bit as important as the evidence itself. For example, glitter particles recovered from the pubic hairs of a sexual assault suspect who allegedly forced his victim to perform fellatio might be far more incriminating that those collected from his exterior clothing.

1.2.5   Easily Characterized

Although several recovered particles all exhibiting the same properties would make for a better argument against accidental origination from the general environment, only one particle is necessary for characterization. Examples of results from various characterization methods will follow.

1.2.6   Computerized Database Capability

For associative evidence any data that will provide even a rough estimate of its rarity will assist a jury in evaluating its evidential value. For fibers numerous target studies have been published (see the references in Reference 3). In a target study a type of trace evidence having specific characteristics is preselected (e.g., orange nylon carpet fibers). Then locations (e.g., theater seats) or garments are randomly selected and sampled by making tape lifts. Then the tape lifts are searched microscopically for fibers exhibiting all the preselected characteristics. The results provide at least a rough idea of the frequency of occurrence of that fiber type in the environments selected. This kind of information may assist a jury in assessing the likelihood that the accused may have innocently picked up these traces from the general environment. A limited target glitter study [1, 3] was conducted in the San Diego, California area with four different types of glitter particles preselected. This study indicated a low probability for picking up from the general environment a specific type of glitter particle. Although target studies have value, they also have many limitations. Such studies implicitly assume a random distribution of the targeted items, whereas often they are concentrated in limited areas. In the cited target glitter study, from a total of 58 separate tape lifts 23 contained one or more glitter particles, but of a total of 169 particles recovered, 69 or 41% were recovered from one tape lift of one seat in a movie theater. While attending the 2003 Annual Meeting of the American Academy of Forensic Sciences, it was noticed that a seat in a meeting room was covered with hundreds of glitter particles while no indications of glitter were seen on adjacent seats or on seats in other rows. Upon approaching the hotel's information desk and explaining why we needed some pieces of transparent tape, we were told that the previous week there had been a convention for people who loved to dance to salsa music. Many of the female dancers had worn very decorative costumes. Had we sampled a different seat or from a different room or at a different hotel or at a different city or at a different time of year or even a different year (when the popularity of glitter in cosmetic items was either greater or lesser) our chances of recovering any glitter particles could have varied considerably. Some might argue that a compilation of the frequency with which a specific item is encountered in actual criminal cases would provide a better estimate of rarity. For many years as the FBI Laboratory has encountered fragments of broken glass in evidence items, it has entered their properties (density, refractive index, color (if any), and relative amounts of selected elements) into a database. With each new submission the laboratory can then determine how many times in the past it has encountered a glass sample with these same properties. This too has its limitations (e.g., a type of rope that had never previously been encountered by the FBI Laboratory might actually be fairly common in the village markets around Guadalajara, Mexico (based on transcripts from the trial for the kidnapping and murder of U.S. Drug Enforcement Agency Special Agent Enrique Camarena, July 1988, Los Angeles, CA)). If sufficient number of layers are present on a paint chip that originated from an automotive vehicle, in some cases it may be possible to identify the vehicle manufacturer (Ford, General Motors, Honda, etc.) and the years in which that particular formulation was in use. Manufacturing data may then tell one how many vehicles with that color and that formulation were produced, and thus provide a general assessment of rarity. Most of the properties enumerated for glitter particles (size, shape, thickness, specific gravity, metal layer? (yes/no), opaque/translucent?, different morphology on one side? (yes/no), number of layers, thickness of each layer, chemistry of each layer, and color) are capable of being entered in digital form into a computer-searchable database. Whether such a database was based on target glitter studies, data from previous glitter-containing cases, or data provided by manufacturers, it would at least provide some general estimate of rarity.

1.3 CHARACTERIZATION METHODS

In forensic trace-evidence comparisons the null hypothesis is always "Questioned sample, Q, did *not* originate from the same source as Known sample, K." The reason for this is that this statement can be proved, while the statement "Questioned sample, Q, originated from the same source as Known sample, K" cannot. (For the first statement, as soon as any test shows Q and K to be *different*, we have proved it to be correct. With the second statement, we can prove that it is incorrect (a test shows Q and K to be different), but no matter how many tests we run that give the same results for Q and K there will still always potentially be one more test (that we have not as yet run) that would show them to be different.) And since the object of our tests is to determine if the evidence shows a possible association between the suspect and the victim, if there were any rationale for a possible implicit bias on the part of the criminalist it would have to be in favor of the defendant (as soon as a test shows Q and K to be different there is no need to conduct additional tests on that evidence). Therefore, in formulating an analysis scheme criminalists tend to first do those tests that are nondestructive, quick, and easy. And since glitter particles are tiny and could easily be lost, those examinations requiring a minimum of particle manipulations would also have precedence. Last in order in any criminalist's examination scheme would be any tests that would actually destroy all or a part of the sample. So although a method such as pyrolysis gas chromatography/mass spectrometry might have high discrimination value, it would not be selected unless it could be shown that no combination of nondestructive tests were capable of discriminating between two samples that were in fact different.

1.3.1 Color

Potentially, color has great value for discriminating between glitter particles originating from different sources. For example, PolyFlake, just one of Glitterex Corporation's ten different glitter products, is offered in 44 different colors. One might think that you could just do a side-by-side microscopic comparison of questioned and known particles (or side-by-side comparison of questioned and known particles acquired and saved under identical lighting conditions on a video microscope system). In some cases this is true, but by no means all. Many glitter products contain no dyes or pigments at all. Their color as perceived by a human observer is based on the properties of the various plastic layers in the glitter flake. Some have color-shifting effects and their apparent color will vary according to the lighting and the angle of observation. Criminalists will also encounter this problem when comparing the colors of paint chips from automotive makes and models having optically variable pigments, and also when comparing tiny pieces of mica (shimmer or pearlescent particles). As we roughed out a draft of this chapter we came to the realization that the determination of color when performing trace-evidence comparisons was a rather complex subject and was deserving of its own chapter. See Chapter 14 for an in-depth discussion of this subject.

One problem with color determination as perceived by humans (rather than machines) is that it can be quite subjective. Color perception may be affected by lighting conditions. When samples Q and K are on different stages of a comparison microscope, whether using reflected or transmitted light, it is very difficult to get the lighting the same on both stages. When small particles that are colored alike are viewed/compared on different colored backgrounds, even those individuals having normal color vision may perceive them to be different. These problems can be somewhat alleviated by capturing the images of K and Q particles on the same stage of a video microscope with the same background and same lighting and magnification. However, for elimination of subjectivity and to enable color measurements to be entered into a searchable database, it would be desirable to have an objective, machine-determined measurement of color. In a kidnapping/sexual assault case involving glitter as associative evidence [8], a QD1 1000 Microspectrophotometer (Craic Technologies, Altadena, CA, USA) was used to make color comparisons.

1.3.2 Morphology

Although most glitter particles have a similar appearance when viewing either surface, some do not. Some may have an added pigment on one surface, and some may have an aluminized layer with plastic layers over. These characteristic morphologies may be viewed at higher magnifications (700× or higher), and the chemistries of the two surfaces may be determined by ATR infrared microscopy (Figure 1.1). One must ensure that what appears to be a surface morphology is not due to traces of vehicle on the surface of the particle or due to contamination. Additionally, some glitter products achieve holographic or color-shifting effects by virtue of (Mike Nofi, Flex Products, Inc.) "an embossed foil with a microstructure (grating), coated by an aluminum layer and finally protected by a colored or non-colored resin." Two examples, one with approximately 0.5 µm between grating lines and the other with approximately 1 µm between lines, are shown in Figure 1.5. The finding on either or both surfaces of a questioned glitter particle of a grating pattern and the mea-

Figure 1.5 *Surface of two different glitter particles illustrating an ~0.5 µm (left) and ~1.0 µm diffraction grating (right). Top is a stage micrometer with lines 10 µm apart. Meadowbrook Inventions, Alpha Jewels HTMP, P98000, silver, 0.008 in. × 0.008 in. (right), and Alpha Jewels, P9825HX, silver, 0.025 in. hexagonal (left). Conditions: Mounted in Permount™ under a cover glass, 60× dry objective with 0.80 NA and 6.7× photo eyepiece. (Photomicrographs by Edwin Jones, Ventura County (California) Sheriff's Department Forensic Sciences Laboratory.)*

surement of the distance between lines would place this glitter into a much smaller subclass of associative evidence.

1.3.3 Shape

Shape is readily determined by examination of glitter particles recovered on tape lifts using reflected light and a stereobinocular microscope or a video microscope at relatively low magnification. Hexagonal shaped particles were most frequently found in the commercial cosmetic products purchased and examined by the authors, as well as those products examined by Aardahl [1] and Kirkowski [5]. Hexagonal particles were also most frequently recovered from the general environment as reported in the target glitter study of Aardahl [1] and Aardahl et al. [3]. Square particles are next in popularity, followed by rectangular. It is rare to find a product that mixes different shapes, but in some cosmetic products containing glitter in a liquid vehicle we have seen mixtures consisting of many tiny glitter particles and much fewer large glitter particles in shapes such as stars or crescent moons. If such a mixture of particles recovered from evidence items associated with a suspect were found to match a product worn by the victim, their associative value would be great. Also, a mixture of particles having different shapes as well as other properties may be encountered if a victim was wearing two or more glitter-containing products. As with fibers, the more different kinds of glitter particles found on the suspect's clothing or in the environment associated with the suspect that match glitter in products worn by the victim, the stronger the association.

1.3.4 Size

Kirkowski [5] used a video microscope (Inf-500™ CCD Video Microscope (Moritex Corp., Tokyo, Japan)) with a PC that contained video capture software (VisionGauge™ (VISIONx Inc., Point-Claire, Canada)). Most images of individual particles were captured at either 280× or 700×. After a simple calibration both the area and the dimensions of an individual particle could be quickly acquired by simply clicking on the particle. Typically in forensic evidence comparisons, one has virtually an unlimited amount of the K sample and quite a limited amount of the Q sample. With a system similar to that used by Kirkowski, it would be a simple matter to acquire the average area, average dimensions, and standard deviations of each for the K sample. These properties for the Q particles could be determined and it could quickly be shown whether the Q particles were included or excluded by size as having a common origin with K.

1.3.5 Specific Gravity

The specific gravities of the various glitters (varying from 1.2 to 2.5) listed on the Meadowbrook Inventions website make this property potentially valuable for differentiating glitters. The sink float method as outlined in Kirk's [9] book on density and refractive index may be used. This technique involves mixing a liquid with a high specific gravity and a liquid with a low specific gravity together (the liquids must be miscible) along with a solid (glitter particle), which is being tested until the solid neither floats to the top nor sinks to the bottom. When the testing liquid reaches this point an additional solid test subject (glitter particle) can be added. If the additional solid test subject has a higher specific gravity than the first solid, it will sink while the first solid will remain suspended. If the two test subjects are suspended together, then this technique can be refined by using a rise or fall of temperature to alter the specific gravity of the suspending liquid, causing the test subjects to float to the top together and sink to the bottom together. The use of bromoform or tetrabromoethane as the heavy liquid and ethanol or bromobenzene as the lighter liquid should be performed in a chemical fume hood for safety. Water-based solutions using sodium polytungstate or potassium iodide do not require a chemical fume hood and may be more convenient. Any testing of this technique should be used on the K sample first to make sure that the color in the glitter is color fast and does not dissolve or change in the liquid.

1.3.6 Thickness

Thickness of glitter particles can be measured mechanically with a suitable micrometer gauge or optically with a calibrated microscope. A micrometer gauge, which is accurate to the nearest micron, can be used to measure glitter particles. A technique to perform this measurement would involve having the lone glitter particle lying flat on an adhesive tape or Post-it note and zeroing the gauge on the blank area adjacent to the particle, then measuring the particle while it remains on the adhesive. Measurement with a microscope requires a calibrated ocular micrometer. This technique requires the particle to be stood on its edge (with the aid of adhesive or clay) or cross-sectioned.

For those glitter particles that are translucent, the interference fringes present in their infrared spectra can provide an alternative method of thickness determination. The following procedure is taken from Smiths Detection Application Brief AB-060 [10].

Additionally, the optical thickness of glitter particles can be measured using the all-reflecting objective on the IlluminatIR. With a film thickness of 5 to 200 microns, the reflection from the top and bottom surfaces creates measurable interference. This interference pattern is a regular periodic oscillation in the transmittance (or absorbance values) and it is best observed in spectral regions where there are no strong absorption bands as shown in Figure 1.6 for nylon 6,6. The calculation of the thickness is performed as follows:

$$t'(\text{mm}) = N/2 \times [10/(Wh - Wl)]$$

$$t'(\text{mm}) = 2/2 \times (10/540) = 0.0185\,\text{mm}$$

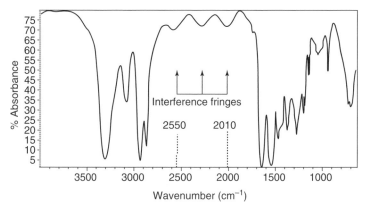

Figure 1.6 *IR spectrum of a nylon 6,6 fiber showing a pattern of interference fringes. (From Reference 10.)*

$$t'(\text{mm}) = optical\ path\ length = n \times t = 0.0185\,\text{mm}$$

$$t = 0.0185/n \text{ or } t = 0.0185/1.5 = 0.0123\,\text{mm}$$

The optical path (t') is calculated as shown above. Fringes are counted, starting at zero up to the maximum number (N). To calculate the true film thickness (t) requires knowing the refractive index (n) of the thin film. Since the composition of the plastic glitter particle can be determined from its spectrum, the refractive index can be found in any standard reference.

1.3.7 Cross Section

Cross sectioning of glitter can be achieved by hand with the aid of a stereo-binocular microscope or with the aid of embedding media and a microtome. To cross-section by hand simply place the glitter flat on a piece of adhesive tape or the adhesive of a Post-it note and cut at 90° with the aid of a stereo microscope. A fresh razor or scalpel blade and cardboard or other absorbing material (to protect the stage of your microscope from cut marks) may be used. If the objective of the cross sectioning is to obtain a clean layer of material for FTIR analysis, then the angle of the cut needs to be decreased to increase the area of sample available for testing. Encasing the glitter particle in a block of epoxy, Norland Optical Adhesive, or a suitable thermosetting plastic can give better control for separating the layers and producing a very thin section for FTIR analysis.

We have found it surprisingly difficult to obtain cross-sections with no smearing of layers or distortion of the particle. A solution may be to simply view the particle standing on end. Total thickness measurements may be made, the number of separate layers may be counted, and measurements of individual layer thickness may be made, all producing additional levels of

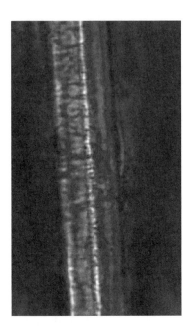

Figure 1.7 *(See color insert.) Glitter particle as viewed on edge at 400×. Transmitted light (left) and transmitted light with crossed polars (right). Meadowbrook Inventions, Inc., Crystalina 300 Series, #326, C3625HX, 0.025 in. hexagonal. (Photomicrographs by Michelle Siciliano (Intern) and Gene Lawrence (Criminalist Supervisor), San Diego County Sheriff's Crime Laboratory.)*

discrimination. When separate layers consist of chemically different polymers, viewing under crossed polars may heighten appearance differences between layers (Figure 1.7).

1.3.8 Infrared Spectra

Kirkowski [5] evaluated several methods of obtaining infrared spectra of individual glitter particles. Even glitter particles that had no opaque aluminum layer and that were translucent were optically too thick to obtain high quality spectra when viewed in transmitted light with an FTIR microscope. This was still true when he attempted to squash a particle in a diamond compression cell and obtain the particle's spectrum in transmission with the cell on the stage of the FTIR microscope. Poor quality spectra were also obtained when a particle was placed on the surface of a gold-coated mirror with the mirror on the stage of the FTIR microscope and the infrared spectrum of the particle's top surface obtained in the reflectance mode.

However, Kirkowski found that high quality infrared spectra of the two surfaces of a particle could be obtained by examining an individual particle using an FTIR microscope having an ATR (attenuated total reflectance) objective. No sample preparation was required other than to ensure that the

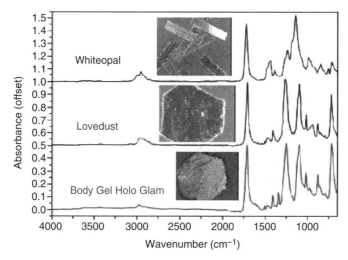

Figure 1.8 *(See color insert.) ATR FTIR spectra of individual glitter particles in three different commercial cosmetic products. (From Reference 5.)*

particle was clean and that no traces of the vehicle (if the particle had originated from a cosmetic product) or adhesive (if the particle had been recovered using a tape lift) were present. Figure 1.8 shows the FTIR spectra obtained (in absorbance) from individual glitter particles present in three different commercial cosmetic products. The spectra in this figure were obtained with the IlluminatIR™ Infrared Microspectrometer (Smiths Detection (formerly SensIR Technologies), Danbury, CT, USA) and the ContactIR objective. The IlluminatIR system consists of a high quality optical microscope (Olympus BX51) and the Smiths Detection FTIR accessory. In addition to a standard 10× optical objective, two infrared objectives are available on the turret. The All Reflecting IR objective is useful for samples such as thin films on metals and obtains the spectra in reflected light. The ContactIR objective employs a diamond to make direct sample contact and obtain spectra using ATR. Because of the design of the ContactIR objective and an included video camera and monitor, the operator can simultaneously view the sample and see when the diamond has made contact and also see that good contact has been made as the sample's infrared spectrum appears on the monitor. Of course, only the infrared spectrum of the outer layer of the side of the glitter particle contacting the diamond is obtained. However, it is a simple matter to turn the particle over and then obtain the spectrum of the other side. To obtain the spectra of any interior layers the particle must be cross-sectioned. As with multilayered paint chips, in order to increase the width of an interior layer it may be necessary to cross-section a particle at an angle rather than at 90° to the particle's surface. In practice, the glitter particles tend to stick to the diamond objective. After each spectrum is obtained one needs to unscrew the objective, view it at low magnification under a stereobinocular microscope, and pick off the particle. A cotton swab moistened with methanol is then used to clean off the diamond. Any methanol traces will quickly evaporate and the diamond objective can be returned to the turret and the system is ready for the next sample.

Figure 1.3 shows an overlay of the infrared spectra obtained for glitter particles from two different commercial products. Even though their spectra are quite similar, there are reproducible differences. So although polyethylene terephthalate (PET) is the most common polymer used in glitter, oftentimes a layer may be a polyester copolymer and FTIR may be able to discriminate between samples. PET is a biaxially oriented polyester film. Mylar® (Dupont) is one of the better known tradenames.

The IlluminatIR system also includes various libraries and search capabilities. Figure 1.9 shows that a search of the middle spectrum in Figure 1.8 (Lovedust) against an ATR library of various polymers (ATR-V01 with 243 entries) gives PET as the closest match with a hit quality of 91.1 out of a possible best of 99.9. Figure 1.10 shows that a search of the top spectrum in Figure 1.8 (Whiteopal) against the same library gives a different polymer, poly(methyl methacrylate), as the closest match with a hit quality of 96.4.

FTIR microscope mapping is a relatively new technique that may be suitable for the examination/comparison of glitter particles in cross section. Let's assume that two or more layers in a glitter particle consist of different poly-

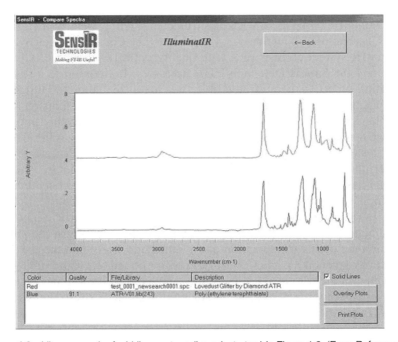

Figure 1.9 *Library search of middle spectrum (Lovedust—top) in Figure 1.8. (From Reference 5.)*

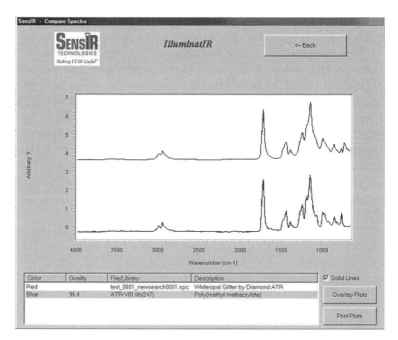

Figure 1.10 Library search of top spectrum (Whiteopal—top) in Figure 1.8. (From Reference 5.)

mers (e.g., PET and poly(methyl methacrylate)). Wavelength bands in the infrared are selected that are specific to one polymer but not found in the other. The entire cross section is scanned for just these two selected wavelengths. For an *x,y* printout one color is selected for one wavelength (e.g., blue for the wavelength specific to PET) and another color (e.g., red for the wavelength specific to poly(methyl methacrylate)) for the other. The printout would then produce a map of the cross section with all PET layers in blue and all poly(methyl methacrylate) layers in red. However, preliminary attempts at FTIR mapping of glitter particles have been thwarted due to difficulties in obtaining high quality cross sections and because the layers are frequently quite thin.

1.3.9 Raman Microspectroscopy

Raman microspectroscopy would appear to be ideally suited for the characterization of the different polymer layers in a glitter particle. Whereas an FTIR microscope with ATR objective can do an excellent job of characterizing the two surfaces of a glitter particle, interior layers can only be characterized subsequent to preparation of a cross section, and even then may present problems if individual layers are quite thin. With a Raman microspectrophotometer the depth of focus can be adjusted so that polymer layers lying below the surface may be located and identified. Following are partial results when four different glitter products were examined with a JASCO NRS-3100 Raman system fitted with 532 nm and 785 nm lasers and a motorized (mapping) x-y-z stage [11]. Experimental conditions: 100 mW 532 green or 500 mW 785 nm deep red solid-state lasers; 1800 gr/mm (532 nm) or 600 gr/mm (785 nm) holographic grating; dichroic beamsplitter used with 785 nm; 100× UMPLFL objective lens; 100 nm precision automated stage. Crystalina 321 and 421 (both products purchased from Meadowbrook Inventions) were examined and compared. Confocal depth mapping was used. A glitter particle was placed flat on the stage and probed by changing the sample position relative to the laser spot in 1 μm increments, from the surface toward the center. This produced an effective sampling volume of approximately 1 μm diameter and 2 μm thickness (z). The depth resolution was therefore about 2 μm. Figure 1.11 shows the Crystalina 321 particle.

As shown in Figures 1.12 and 1.13, the Crystalina 421 sample has a clearly defined acrylic surface layer of about 3–4 μm on a polyester core. There was

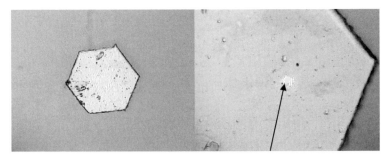

Figure 1.11 *Crystalina 321 sample image captured with 5× and 20× objective lenses. Position of laser spot is visible in 20× image at right. (From Reference 11.)*

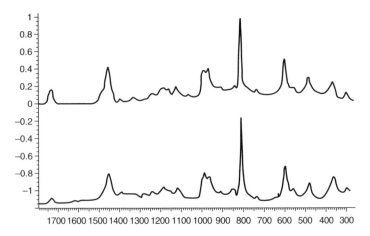

Figure 1.12 *Crystalina 421 glitter surface Raman spectrum with 5 μm depth spectrum subtracted (bottom), compared with database poly(methyl methacrylate) (PMMA) spectrum (top). (From Reference 11.)*

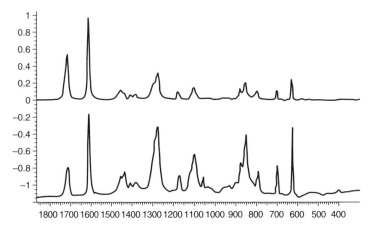

Figure 1.13 *Crystalina 421 glitter surface depth map 5 µm Raman spectrum with surface spectrum subtracted (bottom), compared to database polyester (PET) spectrum (top). (From Reference 11.)*

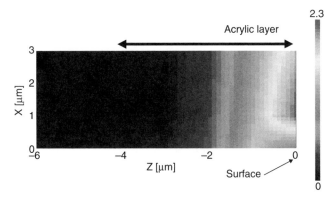

Figure 1.14 *(See color insert.) Crystalina 421 peak (2 vs. 1) ratio x-z confocal map. (From Reference 11.)*

some inevitable interference by scattering from the adjacent layers because the effective sampling volume z resolution is about 2 µm, so for positive ID the adjacent layer spectrum was subtracted before running the database searches.

For producing an x-z confocal map (Figure 1.14) raw peak area data were used and a peak found in PMMA but not in PET (peak 2) was ratioed against a peak found only in PET (peak 1).

Although Raman confocal mapping of the data from the Crystalina 421 glitter showed a clearly defined surface PMMA layer over a PET core, with the Crystalina 321 glitter the polymers appear to be mixed (copolymer or

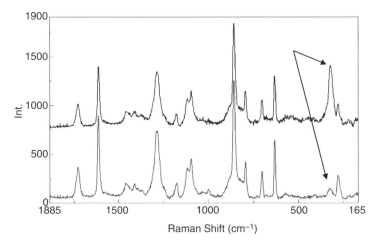

Figure 1.15 *Overlaid Raman spectra of glitter from two different commercial products demonstrate that they are similar polyester resins but may differ due to polyester density (marked). (From Reference 11.)*

blended). The surface was polyester rich, with a shallow layer slightly richer in the acrylic phase and then becoming more polyester rich at 8–10 μm depth. (Full data will be reported elsewhere.)

The other two glitter samples were from different commercial cosmetic products. Each had red hexagonal-shaped glitter particles of the same size, but one contained a single layer of aluminum while the other had two. When laid flat on the stage they only gave weak surface spectra as well as a high background (due to the metallized layers). This was not improved after switching the laser wavelength from 532 nm to 785 nm. Therefore, these samples were examined edge on rather than face on. Both samples were shown to have only PET layers in addition to the aluminum. However, an overlay of their Raman spectra (Figure 1.15) showed consistent differences between them.

1.3.10 Scanning Electron Microscopy/Energy Dispersive Spectroscopy

The scanning electron microscope (SEM) with energy dispersive spectroscopy (EDS) can be used to examine a glitter particle at very high magnification and to analyze its elemental content. The high magnification could be helpful while examining the cross sections. The aluminum layers can be identified with EDS. One of the authors was able to detect the aluminum layer through the 1 μm thick color layer on a red glitter particle that only had one layer of aluminum. Flipping this glitter particle to its other side showed that EDS was unable to detect the aluminum through the 25 μm of Mylar film and color layer. Even though this glitter had a top side and a bottom side, low power examination with the stereo microscope did not show this. Figure 1.16 shows SEM

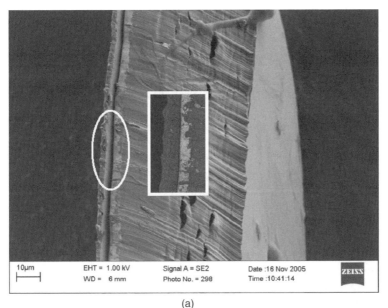

| 10µm | EHT = 1.00 kV | Signal A = SE2 | Date :16 Nov 2005 | ZEISS |
| | WD = 6 mm | Photo No. = 298 | Time :10:41:14 | |

(a)

| 20µm | EHT = 1.00 kV | Signal A = SE2 | Date :16 Nov 2005 | ZEISS |
| | WD = 6 mm | Photo No. = 295 | Time :10:33:08 | |

(b)

Figure 1.16 SEM backscatter images of the same two glitter products examined in Figure 1.15. Both have a hexagonal shape and similar dimensions; both are red, and they are about 220–250 µm in diameter. However, the glitter in part (a) has just one thin aluminum layer while the glitter in part (b) has two. The rectangular insets show the backscatter images that correspond to the areas within the ovals. Because of its higher atomic number compared to the carbon, oxygen, and hydrogen in the polymer layer, the single aluminum layer appears brighter against the background in the backscatter image. Although partially obscured by the smearing of the polymer when the glitter particle was cross-sectioned with a scalpel, two separate thin aluminum layers with a polymer layer in between can be seen in the rectangular inset (backscatter image) for the second glitter product. (From Reference 12.)

[12] images in cross section of the same two very similar (but different) glitter particles whose Raman spectra of their polyester resins are shown in Figure 1.15.

1.4   GLITTER AS EVIDENCE IN CRIMINAL CASES

The first paper we could find on glitter as trace evidence was authored by Michael Grieve [13] and published in the *Journal of the Forensic Science Society*. This was toward the end of the Cold War, and the U.S. Army still had a crime laboratory in Frankfurt, West Germany. Germany has a festive season called *Fasching*, and as with Mardi Gras in New Orleans the people dress up in costume. The women especially wear costumes and makeup that contain glitter, colored feathers, beads, and so on. In the cases reported by Grieve the victims were German women dressed up for Fasching and they claimed to have been sexually assaulted by male members of the U.S. Armed Forces. Glitter particles were some of the trace evidence found on the clothing of suspects. Grieve not only compared these recovered particles with standards obtained from the victims, he also acquired glitter from other sources. He found that glitter from different sources exhibited quite a bit of variety and therefore glitter could be valuable associative evidence.

The next case (*Alexander v. State*) [14] is interesting from several standpoints. First, rather than originating from the victim, the source of the glitter was the suspect's environment. Second, the glitter was from arts and crafts use rather than cosmetics. Third, rather than direct transfer (victim to suspect or suspect to victim), the case involves secondary or even tertiary transfer. In 1987 a high school girl in Fairbanks, Alaska left her home one evening and drove her car back to school so that she could retrieve a book she had left in her locker. She never returned home. Her car was found in the school parking lot with the engine still running. Her body was found late the next day lying on the side of a road on the outskirts of Fairbanks. Glitter was found on her clothing. A suspect was developed and glitter was found in his car. Investigation developed information that the suspect was married and had small children, but was currently estranged from his wife. His wife surrendered several bottles of glitter that had been in the family residence and been used by the children in arts and crafts work. At trial, FBI Special Agent Robert Webb testified as an expert on polymers. He testified "that glitter particles found on the decedent's body and clothing, and the medical examiner's sheets, were *identical* or *consistent* with the glitter samples surrendered by the suspect's wife."

The FBI was also involved in another homicide case involving glitter for arts and crafts. In September 1994 during a house burglary a young woman and her five-year-old daughter were murdered. The little girl had been using glitter in an arts and crafts project and in the bedroom where the bodies were found glitter was strewn on the bed and carpeting. An unusual aspect of this case is that the suspect was originally arrested, but the court later dismissed the complaint. Then in 1999 (five years later) he was rearrested when glitter was found in his car and the FBI Laboratory determined it to be *consistent* with glitter found at the scene of the murders. For eight months after the murders the defendant used the car (including using it for transporting his children). The car then sat in his yard inoperable for some time until it was towed to a junkyard. It remained there until the police towed it to their facility for inspection. In a pretrial motion to suppress the glitter as evidence the defense claimed that the defendant's "children played with glitter during that time and that they occasionally played in the vehicle while it sat inoperable in his yard, and that no glitter was found in the vehicle when it was first searched right after the murders in 1994." At the Annual Meeting of the American Academy of Forensic Sciences held in Chicago in February 2003, Dr. Maureen Bradley of the FBI Laboratory presented a poster based on this case, "Glitter: The Analysis and Significance of an Atypical Trace Evidence Examination" [15]. The poster disclosed "a total of ten pieces of glitter of four different colors were recovered from the driver's side carpet of the suspect's car." (Remember what we said earlier about "location, location, and location!" If the glitter had been transferred when his children played in his car, how likely would it be that it would *only* be deposited on the driver's side carpet?) The abstract for the poster goes on to state: "Each piece of glitter was comparable in size, shape, and color to corresponding glitter from the crime scene. All were approximately $1\,mm^2$ with an aluminum substrate and each had a distinctive notch on one side. The coating on each piece of glitter was analyzed using Attenuated Total Reflectance (ATR) infrared spectroscopy and scanning electron microscopy with energy dispersive X-ray analysis (SEM/EDXA)." In addition, the FBI obtained 11 different commercially available glitters and compared them with the glitter in the case. All but one of these 11 samples could be distinguished from the case glitter based on physical attributes and/or ATR. SEM/EDXA was then able to distinguish three of the four colors in the case glitters from this sample.

Glitter that was sprinkled onto the victim's head during a 4th of July party was valuable associative evidence in a case in Moorpark, California, where a serial rapist's crimes eventually escalated to abduction and murder [2]. The morning after the abduction, the suspect was observed cleaning out his vehicle for several hours. Weeks latter the vehicle was tape lifted with Helmac lint rollers and 20 glitter particles were recovered (15 of them from the front seat). One month after the abduction, the victim's body was recovered in an advanced state of decomposition. Ten (10) glitter particles were picked out of the victim's hair with a stereo microscope and fine forceps. Despite the one-month exposure to the outdoor elements and the decomposition products exuding from the body, the glitter particles compared favorably with the particles recovered from the suspect's truck and the empty glitter bottle used to sprinkle the glitter onto the victim. Investigation into the source of the glitter revealed that 10,000 bottles of that glitter were produced exclusively for a national retail store.

Jones tried without success (approximately two months after the crime) to obtain a sample from three of the local retail stores including the store where the original bottle was purchased.

The U.S. Army Criminal Investigation Laboratory (USACIL) in Atlanta, Georgia had a sexual assault case where glitter was found on the underwear of the male suspect as well as on two used condoms left at the crime scene [16].

In Hannibal, Missouri in 2001 a man was accused of the abduction, sexual assault, and murder of a thirteen-year old girl. Glitter particles found on a pair of jeans seized from the back of the suspect's car had the same physical and chemical characteristics as those found on the victim's bedspread. This included an anomaly in the shape of the particles. See the photomicrographs in Figure 1.17 showing glitter particles recovered from the suspect's blue jeans; glitter particles recovered from the victim's bedspread; and glitter particles recovered from the victim's blue jeans that were found on the floor of her bedroom. Although the particles are generally square, one of the corners is rounded and there is a protrusion on an adjacent corner. This anomaly suggests that all of the particles may have been cut on the same machine. Figure 1.18 shows the infrared spectra of individual glitter particles recovered from the three sources. In addition, glitter particles from all three sources were examined in cross section. Particles were also measured in terms of size and thickness (25–30 µm). In addition to visual microscopic comparisons, colors of individual particles were also measured with a microspectrophotometer in the visible region [6].

An individual was found guilty of kidnapping and sexual assault in a case in Illinois [8]. This case was interesting in that it involved transfer of particles of glitter from a design on the victim's shirt to the clothing of the suspect. A twelve-year-old girl had been attacked in a wooded area in southern Illinois. She was beaten and her throat was cut, but she bit the attacker, kicked him in the groin, and was able to escape before any sexual act was accomplished. She ran home and her mother rushed her to the hospital and en route contacted law enforcement. Sheriff's department personnel searched the area but were unable to locate a suspect. An Illinois State Police Crime Scene Investigator was called in to do a sketch of the suspect, and this was distributed to the local media. The day after the sketch appeared in the local newspaper, a citizen came forward and said he saw someone matching that description in the city park, not too far from the location of the attack. The sheriff's department searched the woods near the park and discovered a small shack. An occupant matching the description of the suspect came out of the shack with a knife in his hand, but dropped it when he realized he was staring down the barrel of a 45 caliber handgun. Upon arrest and interrogation the suspect provided some details of the incident, including the location of the jacket he had been wearing during the attack. The jacket, his clothing, and the victim's clothing were turned over to the Illinois State Police Crime Laboratory at Carbondale for analysis. The suspect's clothing was examined first. Some blood-like stains were observed on his leather jacket. His jacket and other clothing were taped

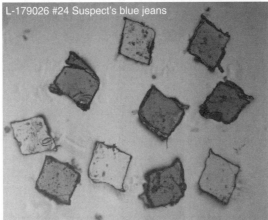

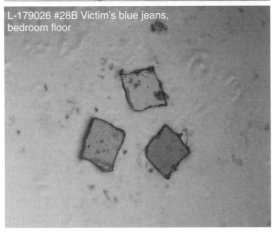

Figure 1.17 *(See color insert.) Glitter particles recovered from victim's bedspread (*top*), suspect's blue jeans (*middle*), and victim's blue jeans (*bottom*). (From Reference 6.)*

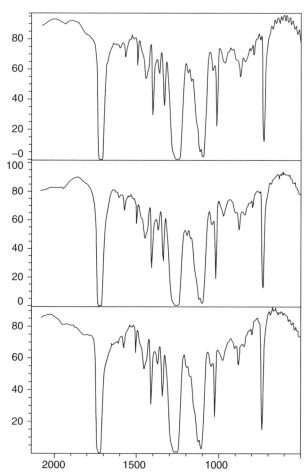

Figure 1.18 *Infrared spectrum of glitter particle recovered from victim's bedspread (*top*); suspect's jeans (*middle*), and victim's blue jeans (*bottom*). (From Reference 6.)*

with clear book tape to remove trace debris. Microscopic examination showed that the collected trace debris included numerous small silver-appearing particles. Could these be like the glitter from the victim's shirt? Higher magnification showed the particles to be small hexagon-shaped glitter particles. A total of five silver-appearing, hexagon-shaped glitter particles were recovered from the suspect's clothing—three from his jeans, one from his sweatshirt, and one from his jacket. In a different room, the victim's clothing was examined. The jacket and shirt she had been wearing at the time of the assault had both been cut several inches down the front center, apparently by the suspect. The word "WARNING" was written in silver-colored glitter on the top front center of the shirt and the cut extended right through this area (Figure 1.19). Glitter particles from the suspect and glitter from the victim's shirt were examined and compared on a comparison microscope under transmitted and reflected light (Figure 1.20). All of the glitter particles recovered from the suspect

Figure 1.19 *Victim's T-shirt with cut through area where the design was created with silver-appearing glitter. (From Reference 8.)*

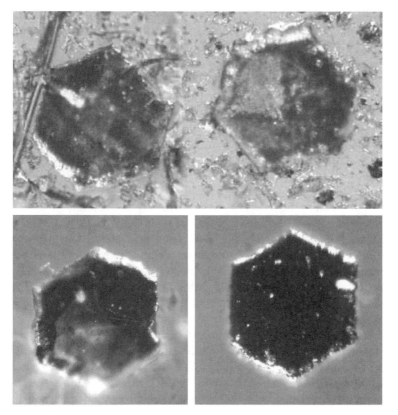

Figure 1.20 *Tape lift photomicrograph in reflected light (top) of glitter particle recovered from the suspect's jacket (left) and the victim's shirt (right). Photomicrographs after the glitter particles were picked off the tape and placed on microscope slides (bottom). (From Reference 8.)*

appeared similar to those from the shirt when examined under high magnification. These silver-appearing glitter particles were hexagon shaped and measured about 250 μm across. Most had a silver-appearing (actually aluminum) layer on both sides with a clear polymer layer in the middle. The two metal outer layers were removed from one particle from the victim's shirt and then the middle polymer layer was placed in a diamond compression cell and the infrared spectrum of this layer was obtained with an Olympus Continuum FTIR microscope. The same procedure was followed with one glitter particle recovered from the suspect's clothing. With each the polymer was identified as polyethylene terephthalate (PET). Next, the analyst did a color comparison using a CRAIC microspectrophotometer. The metallic layers were thin enough to do the analysis in the transmission mode. Although not showing the same level of sharpness and repeatability as infrared spectra, the analysis did show that the glitter recovered from the suspect's clothing was within the same range of color as the glitter from the victim's shirt. (A reader might well say, "I could tell that by just looking at them under a microscope," but increasingly today the courts, laboratory accreditation bodies like ASCLD-LAB (American Society of Crime Laboratory Directors–Lab), guidelines provided by groups like SWGMAT (Scientific Working Group on Materials; i.e. trace evidence), and defense attorneys clamor for testing methods that are not subjective, do not vary from laboratory to laboratory or between analysts, and can provide data in digital form so that database entries may be made, libraries searched, frequencies of occurrence determined, and error rates calculated.)

At the suspect's trial, the analyst testified that he did not find any differences between the glitter recovered from the suspect's clothing and that from the victim's shirt, and that the victim's shirt could be the source of the glitter that was recovered from the suspect's clothing. A DNA analyst testified that blood recovered from the suspect's jacket matched the DNA profile of the victim. The defendant was found guilty of Predatory Criminal Sexual Assault, Aggravated Kidnapping, and Aggravated Battery of a Child.

Glitter was one of the many types of associative evidence in a bizarre case in Albuquerque, New Mexico [17]. This case has already been the subject of several television programs. Although the victim's body has yet to be found, her husband and his girlfriend were convicted of homicide. One of the items of evidence was a bloody tarpaulin (DNA showed the blood to be from the victim). Glitter particles were found on the tarp and were a match for those used by the girlfriend in arts and crafts work.

One last case is worth mentioning because it demonstrates the value of glitter as associative evidence in a totally different type of case [18]. Oftentimes in hit-and-run vehicle accidents or other types of automotive accident investigations, it is not too difficult to identify the vehicle(s) involved. The really difficult question is: Who was driving? This problem is even exacerbated today by the prevalence of cell phones. All a drunk driver has to do is abandon his/her vehicle at the accident scene, whip out a cell phone, and report to the police that his/her car was stolen! Or sometimes the driver will survive a crash but the passenger is killed, and when the police arrive the survivor denies being the driver. In a case in 2004 in Florida a mother and her daughter were killed when a pickup truck driven by an intoxicated woman slammed into the rear of their car. When police arrived no driver was present, but she was later found nearby hiding in the brush. She denied being the driver, but she was wearing cosmetic glitter and at the moment of impact some of it was transferred to the driver's side airbag!

REFERENCES

1. Aardahl K (2003). Evidential value of glitter particle trace evidence. Master's Thesis, National University, San Diego, CA, USA.
2. Jones EL Jr (2004). Trace evidence and bloodstain interpretation from the Sanchez/ Barroso case. Presented at the Fall Seminar of the California Association of Criminalists, Oct. 2004, Ventura, CA, USA.
3. Aardahl K, Kirkowski S, Blackledge RD (2005). A target glitter study. *Science & Justice* **45**: 7–12.
4. Deadman HA (1994). Fiber evidence and the Wayne Williams trial. *FBI Law Enforcement Bulletin* **53**(3 & 5), March (part 1): 12–20; May (part 2): 10–19.
5. Kirkowski S (2003). The forensic characterization of cosmetic glitter particles. Master's Thesis, National University, San Diego, CA, USA.
6. Smith J (2005). Personal communication, Criminalist III, Missouri State Highway Patrol Crime Lab, Jefferson City, MO, USA. See also http://caselaw.lp.findlaw.com/scripts/getcase.pl?court=mo&vol=/supreme/062004/&invol=40608_104.
7. Weber C (2004). Glitter as trace evidence. Master's Thesis, National University, San Diego, CA, USA.
8. Schubert GD (2004). Personal communication, Forensic Scientist, Illinois State Police, Southern Illinois Forensic Science Centre, Carbondale, IL, USA.
9. Kirk PL (1951). *Density and Refractive Index: Their Application in Criminal Investigation*, Charles C. Thomas, Springfield, IL, USA.
10. Smiths Detection (2005). Trace analysis of glitter particles, Application Brief AB-060. www.smithsdetection.com.
11. Larsen R (2006). Personal communication, Scientific Applications Manager, JASCO, Inc., Easton, MD, USA.
12. Gaenzle K (2006). Personal communication, Senior Metrology Engineer, Materials Analysis Laboratory, Naval Air Station, North Island, San Diego, CA, USA.
13. Grieve MC (1987). Glitter particles—an unusual source of trace evidence? *Journal of the Forensic Science Society* **27**:405–412.
14. *Alexander v. State* (8/21/92) ap-1242, appellate brief at: www.touchngo.com/ap/html/ap-1242.htm.
15. Bradley MJ, Lowe PC, Ward DC (2003). Glitter: the analysis and significance of an atypical trace evidence examination. Poster presentation (Paper B24), at the Annual Meeting of the American Academy of Forensic Sciences, Chicago, IL, USA, Feb. 2003. See also www.lycolaw.org/cases/opinions/kramer010904a.PDF.

16. Taylor CE (2004). Personal communication, Forensic Chemist, United States Army Criminal Investigation Laboratory, Forest Park, GA, USA.

17. Horner M (2000). Splitting hairs. www.markhorner.com/Hoss/splitting_hairs.html.

18. *The Standard*, Arrest made in the death of mother and daughter. Baker County, MacClenny, FL, USA, 28 May 2003. www.bcstandard.com/News/2003/0528/Front_Page/001.html.

2

Forensic Analysis of Automotive Airbag Contact—Not Just a Bag of Hot Air

Glenn D. Schubert

Forensic Scientist, Illinois State Police, Forensic Science Command, Southern Illinois Forensic Science Centre, Carbondale, Illinois

Automotive airbags are designed to serve as supplements to seatbelts in reducing injuries and death to drivers and passengers of vehicles involved in frontal collisions. The frontal airbags are not designed to deploy in side impacts, rear impacts, or rollovers. The purpose of the front airbags is to slow the occupants' speed to zero without doing damage to their bodies. When used in conjunction with seatbelts, airbags reduce drivers' deaths by 20% and reduce the chance of head injury by 75% (www.autoliv.com).

If the question arises as to who was driving a vehicle that was stolen and crashed, involved in a hit-and-run accident, or involved in a reckless homicide, forensic science might provide the answer. Since the airbags must deploy in a fraction of a second, it happens with considerable force. In keeping with Edmond Locard's Principle of Exchange, when an airbag deploys and the occupant strikes it, there is likely going to be an exchange of material between

Forensic Analysis on the Cutting Edge: New Methods for Trace Evidence Analysis, Edited by Robert D. Blackledge.
Copyright © 2007 John Wiley & Sons, Inc.

them. This can be in the form of hairs, fibers, or body fluids from the occupants onto the airbags, and fibers and particles from the airbags onto the occupants' clothing. Also, the means by which the airbags deploy may leave telltale signs on the clothing of the occupants. This physical evidence might provide information as to the position of the occupants in the front seat when the airbags deploy. These findings have only recently been reported [1, 2].

2.1   HISTORY OF AIRBAGS

Allen Breed invented the first automobile crash sensor and airbag safety system in 1968; however, it took several years to perfect the system (www.inventor.about.com). General Motors was the first manufacturer to offer airbags as an option for 1975 models (www.edmunds.com). In 1986, Mercedes-Benz was the first company to install airbags in all of its models. In the late 1980s, airbags were routinely installed on the driver's side of many other manufacturers' automobiles. In the early 1990s, they were routinely installed on the passenger side of many automobiles. Starting with 1998 models, federal regulations have required driver and passenger airbags on all new passenger cars (Automotive Occupational Restraint Council). This was also the first year that depowered airbags, also known as second-generation airbags, were required. The requirements for airbags were extended to sport utility vehicles, vans, and light trucks for 1999 models.

2.2   HOW DO AIRBAGS WORK?

Most vehicles are equipped with one to three sensors, which detect a sudden forward deceleration, not necessarily a collision, by flipping a mechanical switch when there is a mass shift that closes an electrical contact. This sends an electrical current to the detonator in the inflator mechanism, which causes the airbag to inflate and deploy with gas either from a storage canister or from a chemical reaction. Airbags inflate in less than 1/20th of a second, at a speed of 200 miles per hour (www.howstuffworks.com). One to three vent holes are usually located in the back of the airbag, so it begins to deflate immediately, making it a better cushion to absorb the impact of the occupant. Before 1998, driver side and passenger side airbag units were "full powered" to meet the National Highway Transportation Safety Administration (NHTSA) crash test criteria. The depowered airbags that were introduced in 1998 inflate with 20% less velocity.

The first commercial airbags were inflated with compressed gas. The initial obstacle with the compressed gas systems was finding room for the canisters, and the question of the gas remaining contained under pressure for an extended period was also an issue. Most driver and passenger side airbags now use some type of solid propellant pyrotechnic inflator. Sodium azide was the most commonly used solid propellant through the 1990s. It is still in limited use but is being phased out because of its toxicity and reactivity (personal communication, airbag manufacturer, name withheld by request). Sodium hydroxide is one by-product of some sodium azide systems. Some people who have experienced an airbag deployment have reported burning in their nasal passages. There have even been reports of chemical and thermal burns to occupants' arms, faces, and bodies [3–5]. High efficiency filters are now used to filter out the solid caustic products, but this means additional size and weight. Other types of injuries and even deaths due to airbag deployments have also been reported [6–8]. The use of internal tethers in some driver side airbags has helped reduce injuries. Alternative types of noncaustic solid propellants now being used include guanidine nitrate (also known as GuNi) and 5-aminotetrazole. The pyrotechnic inflators produce a hot gas that inflates the airbags. Some companies also manufacture a hybrid inflation system, which uses a stored gas, usually argon, which is released after a solid propellant heats the gas and ruptures the container. These are used mainly in passenger side and side impact applications. Blown down systems are the simplest types of inflator in that they just rupture a bottle and let the gas flow. These are found in rollover curtain systems. Helium is the gas of choice for these because it does not cool as much as most gases, so it can keep the bag pumped up longer.

2.3 TYPES OF FORENSIC EVIDENCE TO LOOK FOR

Occupants who are wearing their seatbelts contact the airbag mainly with their face, chest, and arms. Occupants who are unrestrained can also contact the airbag with their abdomen and legs. Although it is not packed in a "clean room," the surface of the airbag should be free of most debris prior to deployment (personal communication, airbag manufacturer, name withheld by request). Therefore, the outer surface of the deployed airbag should be examined for the possible transfer of hairs, fibers, and biological fluids from the occupant. Recovery of damaged hairs or fibers could be significant. Cosmetics may also be recovered on the surface of the airbag (see Chapter 1 on glitter). If the airbag is going to be examined for the presence of body fluids from the occupants, precautions should be used, in accordance with your laboratory guidelines, to avoid contamination of these items. You should coordinate these examinations with your DNA analysts to optimize the value of the evidence. Fiber transfers from the airbag to the occupant's clothing, mainly loose ones from manufacturing, may also be possible. If seatbelts were worn, they should also be examined for possible fiber transfers.

Most airbags are made of long, round, colorless or lightly dyed nylon fibers, 25–30 µm in diameter, in zero-twist yarns. The stitching threads are usually made of nylon fibers also, frequently dyed a contrasting color such as red or green. The fabric is generally in a plain weave pattern, but the front and back panels of a driver side airbag can differ in construction, weave size, and fiber crimp. Passenger side airbags can have a raised, textured pattern in their weave. The driver side airbags are round and 23–28 inches in diameter, depending on the make and model of the vehicle for which they were designed. Passenger side airbags are rectangular in shape and $2\frac{1}{2}$ to 3 times larger than the driver side.

Most driver side airbags manufactured through the late 1990s have a black neoprene lining on the inner surface of the front or both panels, to prevent the hot gas from leaking through the fabric weave. These types of airbags are packed with cornstarch as a lubricant to prevent the neoprene from sticking to itself during deployment, which could be several years after it was manufactured. Talc is also used as the lubricant, but not as often as cornstarch. Passenger side airbags do not contain a neoprene lining, so they are not packed with starch. Some starch may be on the outer surface of the driver side airbag or it can become airborne through the vent holes, in the "cloud of smoke," when the deployment occurs. This means that some starch particles might be recovered from the driver's clothing, while little or none should appear on the passenger's clothing. These items should be screened for trace debris by either taping with a clear tape, such as 3M™ book tape, or by scraping. The tape can be examined directly under a polarized light microscope for starch particles, as well as for hairs and fibers. These starch particles will appear as single colorless polyhedral or subspherical granules, 5–15 μm in diameter, with a small dark air bubble in the center [9]. Under crossed polars, the granules will appear white to gray with well-marked black crosses (Figure 2.1). If you are looking at a taping through a polarized light microscope with crossed polars, the starch granules may have alternating blue and orange quadrants, due to the birefringence of the tape polymer, depending on the orientation of the analyzer. You

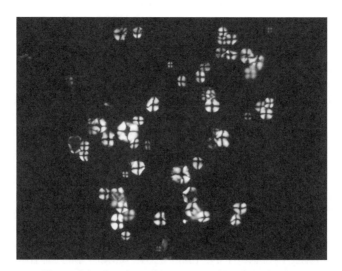

Figure 2.1 Starch particle at 400× with crossed polars.

can rotate the microscope stage to get the best contrast or birefringence. Starch from food contamination will usually appear as agglomerates, rather than single granules. Talc will appear mostly as colorless, translucent, fibrous and platy particles or aggregates [9]. The talc particles can be 1–100 µm (or more) long, depending on how the talc was processed. Talc is a biaxial (−) crystal, with a birefringence of ~0.050. Under crossed polars, the particles will have moderate birefringence colors and with a 530 nm waveplate, more birefringence colors will be present. Talc particles can also be identified by Fourier transform infrared (FTIR) analysis. Make sure to wear powder-free gloves when examining clothing items for the presence of starch or talc.

More recently, driver side airbags have been produced with smaller neoprene lined areas, such as just the tethers, so a smaller amount of starch is used. The newest driver side airbags are now being produced without any neoprene lining, so no starch or talc is used. These airbags use silicone as the sealant on the inside surface of the panels. The timeline of the sealant change varies by make and model, but neoprene and cornstarch (or talc) have not been used for the last several years. So, when an investigator submits clothing to you and asks you to look for "airbag powder," it is important to also have the driver side airbag to determine if it has a neoprene lining and utilizes cornstarch or talc.

Systems that use sodium azide in a pyrotechnic inflator produce hot N_2 gas to inflate the airbags. The gas discharged from the driver side airbag inflator can be as hot as 600 °C and as hot as 700 °C from the passenger side inflator. Systems that use guanidine nitrate in the inflator produce a gas mixture of H_2O, CO_2, and N_2. The gas discharged from inflators in these systems can be even hotter than the sodium azide systems. The gas starts to cool immediately but is still hot when it reaches the inner surface of the airbag. Since the airbags are designed to be inflated for only a short time, they are not completely airtight. The hot gas escapes through the vent holes in the back, as designed, but it can also be forced through the small holes from the stitching seams in the front of the airbag, which tend to be a weak point. If clothing from the occupants comes into contact with the airbag when it is at or near maximum inflation, the hot gas leaking through the seams can cause singe patterns on the front of the clothing, characteristic of the airbag seam patterns. The singe patterns appear as a series of small black dots or smears, 2–3 mm apart, in a single line or multiple parallel lines (Figure 2.2), depending on the number of airbag seam rows. Singe patterns have been observed on clothing made of natural fibers like cotton, silk, or wool, but have not yet been observed by this author on clothing made of man-made fibers. The singeing can occur on the clothing covering the chest and arms of a restrained front seat occupant. Singeing can also occur on clothing covering the abdominal area of an unrestrained front seat occupant. The singe pattern may be difficult to see on dark clothing, though it may produce a pattern of dots, discolored by the heat [10] (see Case Sample 9). Wearing a seatbelt will likely reduce the opportunity for clothing to become singed.

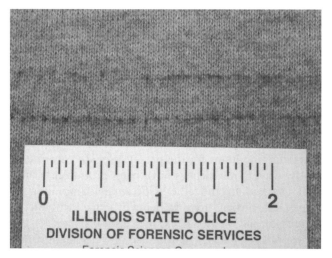

Figure 2.2 Example of a singe pattern on clothing produced by contact with a deployed airbag.

Since the driver side airbag is round with round seams, it will usually produce arc shaped singe patterns. If there is more than one seam on the outer edge of the driver side airbag, it will likely produce parallel or concentric singe patterns about $\frac{1}{4}$ inch apart. Some driver side airbags utilize tethers on the inside to prevent them from ballooning out too far. These airbags will have a seam on the front center of the airbag 6–8 inches in diameter, in single or multiple rows, to attach the tethers. A singe pattern produced from hot gas leaking through these seams will appear in a tighter arc. Again, there might be more than one row of seams in this area, so the singe pattern produced should have more than one row. The vent holes in the back can vary in number (0–4), location, and diameter ($\frac{3}{8}$ to 1 inch). They may cause a solid, round singe pattern on the cuffs of a driver's long sleeve shirt, or a burn on the inside of the wrist if a short sleeve shirt is worn, if the driver is still holding the steering wheel when the airbag deploys. In a personal communication with a driver who experienced an airbag deployment, he stated that he received some minor burns on his upper thighs. As he was wearing shorts, these were likely produced by exhaust from the vent holes. Because there are differences in the design and seam patterns of driver side airbags, there can be different class characteristics in the singe patterns produced, thereby possibly including or eliminating certain airbags as the source.

The passenger side airbag is $2\frac{1}{2}$ to 3 times larger than the driver side airbag and rectangular in shape, with straight seams in single or multiple rows, and some may have rounded corners. Hot gas leaking through the seams of a passenger side airbag will produce mostly straight singe patterns on the occupant's clothing, in single or multiple parallel rows, depending on the number of seams. Again, since there are differences in the design and seam patterns of passenger side airbags, there can be different class characteristics in the singe patterns produced, thereby possibly including or eliminating certain airbags as the source. Because there are differences in the design and construction between driver side and passenger side airbags in every vehicle, it is possible to differentiate the singe patterns produced by them, in most cases, and make a likely determination of the front seat occupants' positions when the airbags deployed. It is important to have both front airbags when making such a determination.

2.4 AIRBAG CASE REPORTS AND EXAMPLES

Case Sample 1 A 24-year-old male reported to his local law enforcement agency that his 1996 Dodge Ram pick-up truck had been stolen. The truck was located about one mile from the owner's house, crashed into a tree. Both airbags were deployed. When the owner was questioned, the investigator saw some unusual impression patterns on the front of the light gray sweatshirt that he was wearing. He also noticed the smell of alcohol on the owner's breath. The owner denied any involvement in the crashed truck. The investigator collected the sweatshirt from the owner, the driver side airbag, and the steering wheel to submit to the Illinois State Police Crime Laboratory in Rockford. The evidence was then transferred to a fabric impression specialist at the Illinois State Police Crime Laboratory in Carbondale.

The impression patterns on the sweatshirt, which was composed of 100% cotton, were examined, and it was determined that they were singed into the fabric. The surface of the sweatshirt and the airbag were then tape lifted with clear book tape, in separate examination rooms, to remove any trace materials. The arc shaped patterns were in a series of small dark dots or smears, 2–3 mm apart. The manufacturer of the airbag was contacted and the maker stated that the hot gas that inflates the airbag can leak through the seams and produce singe patterns on an occupant's clothing (personal communication, airbag manufacturer, name withheld by request), as had been seen previously. With this information, the singe patterns on the sweatshirt were compared to the seams of the airbag. Some of the airbag seam threads appeared to be charred and slightly melted. Two parallel singe lines observed on the abdominal area (Figure 2.3) and left sleeve of the sweatshirt corresponded to the two outer seams of the airbag (Figure 2.4). This also indicated that the driver was not likely wearing the lap belt. Singe patterns observed on the upper left quadrant of the sweatshirt (Figure 2.5) corresponded to the smaller single central seam on the airbag (Figure 2.6), which attached the inner tethers. This also indicated that the driver was not likely wearing the shoulder restraint. The $\frac{3}{8}$ inch oblong singe pattern observed on the inside left sleeve of the sweatshirt

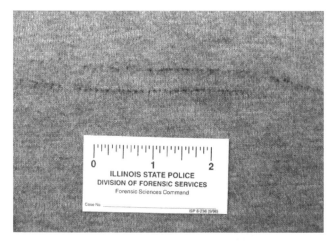

Figure 2.3 *Double row singe pattern on the lower abdominal area of the sweatshirt from the suspected driver of a crashed 1996 Dodge Ram pick-up truck.*

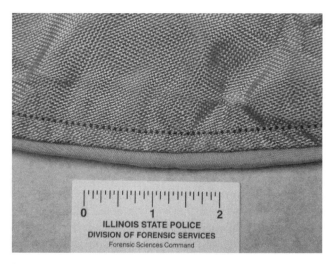

Figure 2.4 *Double outer seam area of the deployed driver side airbag from a 1996 Dodge Ram pick-up truck. Some of the top seam threads are charred and slightly melted.*

(Figure 2.7) corresponded to the single vent hole located on the back of the airbag (Figure 2.8). This indicated that the driver was still holding the steering wheel when the airbags deployed.

The taping from the front outside of the airbag was examined under high magnification with a polarized light microscope. Although numerous white cotton fibers were observed on this taping, no comparisons to the sweatshirt fibers were conducted since white cotton is very common. Numerous corn-starch particles were also observed in this taping. This airbag had a neoprene

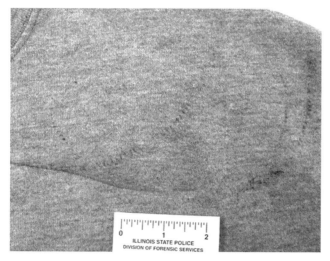

Figure 2.5 *Single row singe patterns on the front upper left quadrant of the sweatshirt from the suspected driver of a crashed 1996 Dodge Ram pick-up truck.*

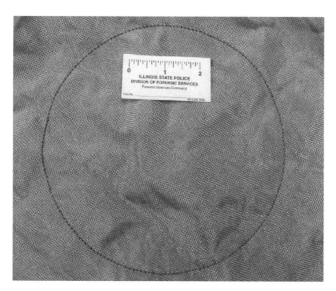

Figure 2.6 *Single central seam area of the deployed driver side airbag from a 1996 Dodge Ram pick-up truck.*

lining on the inside of the front panel and used cornstarch as a lubricant (personal communication, airbag manufacturer, name withheld by request). The taping from the front of the sweatshirt was also examined with a polarized light microscope. Numerous cornstarch particles were observed, which were microscopically consistent with the starch from the airbag (Figure 2.9). Three round colorless nylon fibers were also observed on this taping. These were

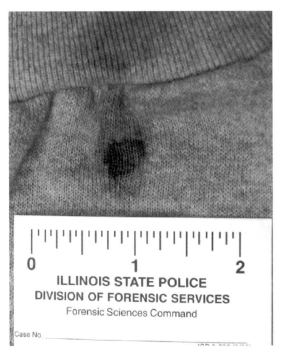

Figure 2.7 Singe pattern on the lower left sleeve of the sweatshirt from the suspected driver of a crashed 1996 Dodge Ram pick-up truck.

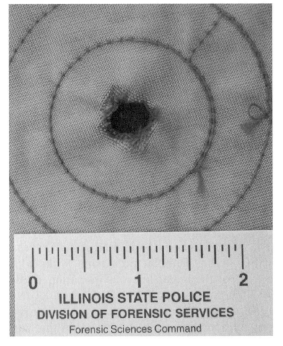

Figure 2.8 Single vent hole on the back of the deployed driver side airbag from a 1996 Dodge Ram pick-up truck.

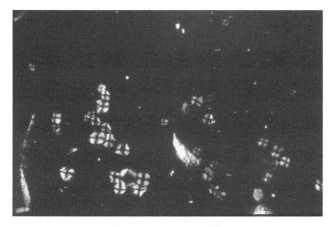

Figure 2.9 *Comparison of a debris taping with cornstarch particles collected from the front of the deployed driver side airbag from a 1996 Dodge Ram pick-up truck (left) to the debris taping with cornstarch particles collected from the front of the sweatshirt from the suspected driver of the crashed truck (right), viewed with crossed polars at 400×.*

found to be consistent with the fibers that composed the back panel of the airbag, but were eliminated as coming from the front panel of the airbag because of a different crimp. At trial, the analyst testified that the owner was in the driver's seat of this truck, or a vehicle that used a similar airbag, when the airbags deployed. The owner was found guilty of several offenses.

Case Sample 2   In July 2002, two southern Illinois musicians were returning home late at night from a performance, in their 1972 Chevrolet pick-up truck. As they entered an intersection, a 1996 Kia Sephia, with one black male and one white female occupant, ran through a stop sign and struck the truck on the passenger side. The truck overturned and caught fire, killing the two occupants. After striking the truck, the Sephia hit a tree on the passenger side. When emergency personnel arrived, they found the female occupant of the Sephia outside the car and the male occupant, who had to be extricated, in the passenger seat. Some witnesses said that the male was in the driver's seat when they arrived and he crawled over to the passenger seat, although he denied being the driver. Both airbags in the Sephia had deployed. A toxicology screening was done on the male occupant. His blood alcohol level was 0.056 g/dL (the legal limit in Illinois is 0.08 g/dL) three hours after the accident. Cocaine and THC were also detected in his blood. His shirt, which was blood soaked from a large head wound, was disposed of at the emergency room. Since criminal charges could potentially be filed, the investigators decided that they would submit evidence to the Illinois State Police Crime Laboratory in Carbondale to help them determine who was driving the Sephia. The evidence consisted mainly of blood swabs and debris tapings from the inside of the car. The female occupant's shirt and both airbags, which contained bloodstains, were submitted at a later date.

After a visual examination, the shirt and the airbags were taped with clear book tape, in separate examination rooms, to remove any trace debris. Upon a more detailed examination, a diagonal singe pattern of small dots in a straight single line about 2 inches long was observed on the upper left chest area of the female's shirt (Figure 2.10). Under UV light (366 nm), the cotton fabric of this shirt fluoresced brightly while the singed areas were a contrasting black, so this enhanced row of dots and smears now appeared to be about $3\frac{3}{4}$ inches long (Figure 2.11). The driver side airbag had a double outer seam in

Figure 2.10   *Straight single row singe pattern on the upper left chest area of the shirt worn by the suspected passenger of a crashed 1996 Kia Sephia.*

Figure 2.11   *Same area as in Figure 2.10 viewed under 366 nm UV light.*

a large arc shape and a double central seam in a tighter arc pattern. This was not consistent with the pattern on the shirt, so the driver side airbag was eliminated as a source of the singe pattern. The passenger side airbag seams were all in straight lines with many single and some double rows. Further comparison showed that the passenger airbag could not be eliminated as the source of the singe pattern (Figure 2.12).

A damaged Caucasian head hair recovered from the front of the passenger airbag was found to be microscopically consistent with the head hair standard from the female occupant and dissimilar to the male occupant. A small clump of Caucasian head hairs, with their proximal ends broken off, was recovered from the front passenger seat. These hairs were also found to be microscopically consistent with the head hair standard from the female occupant. A Negroid hair fragment, unsuitable for comparison, was recovered from the driver's seat.

No cornstarch particles were observed on the taping from the female passenger's shirt. A small area of neoprene was used on the driver side airbag from this car, so it had been packed with some cornstarch. Cornstarch particles were observed in the taping from the outside of the driver side airbag, but not in the taping from the outside of the passenger side airbag. Blood on the driver side airbag and driver side door matched the DNA profile of the male occupant. At the trial of the suspect, a forensic microscopy analyst testified that, because of the evidence he observed, his opinion was that the female was sitting in the passenger seat when the airbags deployed. The testimony of the DNA analyst supported the theory that the male was driving the car when the

Figure 2.12 *Straight single seam area on the deployed passenger side airbag from a 1996 Kia Sephia.*

accident occurred. A forensic toxicologist testified about the toxicology results but was not allowed to extrapolate the blood alcohol level back to the time of the accident. The defense council hired an engineer who testified that before the airbags deployed on the third impact, which was sideways into a tree, the female occupant, who weighed about 200 lb, flew from the driver's seat to in front of the male passenger, who is 6 ft 1 in. and weighs 220 lb. The jury was out for eighteen hours, but found the male occupant guilty of Reckless Homicide, and Aggravated Driving Under the Influence of Drugs and Alcohol, in June 2004.

Case Sample 3   In a rural southern Illinois community, two males looking for drugs and money broke into the trailer of a couple. Both occupants of the trailer were beaten, shot, and left for dead. The female died from her wounds, but the male survived and was able to call for help after the assailants left. The assailants sped away from the scene in a 1994 Chrysler New Yorker. A short distance from the crime scene the driver lost control and crashed the car into a tree. Both airbags deployed. The assailants were apparently not seriously injured and fled from the accident scene on foot. Through a thorough and diligent investigation by the Illinois State Police, some suspects were soon developed. The investigators were able to collect clothing that the suspects might have been wearing during the commission of the crime, although they believe the driver of the car was not wearing a shirt. They submitted this clothing along with evidence collected from the crime scene and from the crashed car, including the deployed airbags, to the Illinois State Police Forensic Science Laboratory in Carbondale.

Bloodstains were observed on both airbags and some of the clothing, and portions were collected for analysis. Trace debris was also collected from both airbags and some of the clothing, and examined. The driver side airbag had a black neoprene lining. Among the material collected and observed from this airbag were small, flat crystal-like particles. They were identified as talc particles. The manufacturer of this airbag verified that talc was used as the lubricant in this model (personal communication, airbag manufacturer, name withheld by request). Among the particles collected from the passenger side airbag were several black cotton fibers. Among the clothing items from the suspects that the investigators submitted were two black T-shirts and one pair of black sweatpants, all at least partially composed of cotton. These items were examined for the presence of any singe patterns and none were observed. Since the singe patterns would not be very visible on a black shirt, the analyst tried using different light sources, such as UV light and a Lumalite, and still no singe patterns were observed. Debris was collected from these clothing items by taping, and no talc particles were observed. The analyst did a microscopic comparison of the black cotton fibers recovered from the passenger side airbag with the three black clothing items. The black cotton fibers from these three items were indistinguishable microscopically, so the microscopic comparison was not going to be very useful. The analyst decided to do a comparison of the fiber dyes on a CRAIC 1000 microspectrophotometer. With this instrument, the analyst was able to distinguish the black cotton from these three clothing items, since the spectra from each one was different, indicating that different dyes were used. The color of the black cotton fibers recovered from the passenger side airbag was then compared to the three clothing items and found to be consistent with one of the black T-shirts, which happened to be the shirt that the investigators believe the suspect in the passenger seat was wearing, and dissimilar to the other two items. Bloodstains recovered from this black T-shirt and the driver side airbag matched the DNA profile of the male victim. The two suspects and one accomplice were all found guilty or plead guilty to homicide charges.

Case Sample 4 A 23-year-old female was driving her 2001 Saturn LS in city traffic. She failed to sufficiently slow down her car at a red light and struck the vehicle in front of her, causing the airbags to deploy. She was wearing her seatbelt and was not injured, but the car did sustain moderate damage. Although this was not a criminal case, this author was able to obtain the driver's shirt for examination. Attempts to obtain the airbags from the vehicle were unsuccessful.

The driver's shirt was light gray in color and made of 100% cotton. Visual examination did not show any type of singe marks on the front of the shirt. Examination with a UV light (366 nm) also did not reveal any type of singe patterns. The front of the shirt was tape lifted for debris and examined with a polarized light microscope. Most of the debris consisted of colorless cotton fibers. No starch or talc particles were observed.

The manufacturer of the airbags for this car was contacted (name withheld by request) to obtain some information. The manufacturer stated that since this was a depowered airbag, the driver was wearing her seatbelt, and this was not a high-speed impact, it would be less likely to see any type of singe pattern on the driver's shirt. The manufacturer also said that this airbag does not have a neoprene lining, so no lubricant particles are used. This is a good example to show that the fabric singeing does not occur in every airbag deployment, and that the often quoted phrase by Dr. Carl Sagan of "absence of evidence is not evidence of absence" may hold true in some cases.

Case Sample 5 Four occupants were traveling in a 1995 Chevrolet Camaro that left the road and struck a utility pole, causing both airbags to deploy. None of the occupants were wearing their seatbelts, and they were all ejected or partially ejected from the car, with one fatality occurring. The owner of the car had a blood alcohol level that was over the legal limit for Illinois, but he claimed that he was not driving at the time of the accident. Both of the airbags from the vehicle and the clothing from the fatal occupant were submitted to the Illinois State Police Forensic Science Laboratory in Fairview Heights for examination. The clothing from the other occupants was not obtained. The investigator wanted to know if the fatal occupant could have been the driver.

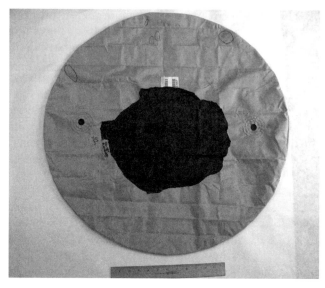

Figure 2.13   *Back side of the driver side airbag from a 1995 Chevrolet Camaro, showing the neoprene lining.*

The shirt that was submitted was a red long sleeve button-up shirt made of 100% cotton. It contained some fabric damage and bloodstains. The exterior of the shirt was tape lifted to remove any adhering debris. No starch particles or nylon fibers were observed in the tape lifts. The driver side airbag was constructed of gray fabric and had just a single outer seam and no center seam. It contained a full neoprene lining (Figure 2.13) and used cornstarch as a lubricant, which was observed in the tape lift from the airbag exterior. Also observed in the tape lift from the driver side airbag were some red cotton fibers. When analyzed with a microspectrophotometer, these fibers were determined to be dissimilar in color from the red shirt. One Caucasian head hair was recovered from the passenger side airbag. No head hair standards were received from the occupants of the car for comparison, and the deceased victim had already been buried. The analyst reported: "No evidence of transfer between the victim and the driver's side airbag was observed."

Case Sample 6   Information was received from one airbag manufacturer (name withheld by request) during this research concerning an automobile accident that had been investigated. In 1992 a women driving a 1991 Lincoln Towncar rear-ended another vehicle and the airbags deployed. The driver was not wearing her seatbelt but, thanks to the airbag, did not sustain any injuries. Contact with the deployed airbag produced a double parallel row of pin-point singe marks, approximately 160 mm long, across the top of her shorts (Figure 2.14). The personnel at this airbag company called this "The Case of the Stained Shorts" (although we in the forensic field have a different defini-

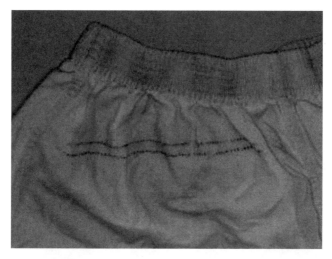

Figure 2.14 *Singe pattern on the shorts of the driver of a 1991 Lincoln Towncar.*

Figure 2.15 *Possible singe patterns on the shirt of the driver of a 1996 Pontiac Sunfire after an airbag deployment.*

tion of stained shorts). This was the earliest example of the singe patterns produced by airbags that this author could document.

Case Sample 7   In September 2003 an accident reconstruction class was conducted in southern Illinois. This author was unable to attend the class but did receive some videos and photographs of the events. The finale of the class was a head-on crash of a 1996 and a 1997 Pontiac Sunfire. After the crash, one of the drivers proudly showed off the abrasions on his forearms that were produced by the airbag deployment. In a high resolution photograph of this driver, what appears to be a single row of singe mark dots is visible on the lower chest and left sleeve of his light gray T-shirt (Figure 2.15). Although

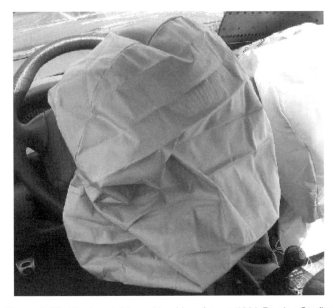

Figure 2.16   *Deployed driver side airbag from a 1996 Pontiac Sunfire.*

attempts were made to obtain this shirt, it was never made available. Photographs of the driver side airbag from the Sunfire that he was driving show just a single outer seam and no inner seam (Figure 2.16). This would seem to indicate that the marks on his shirt were made by the deployed airbag, but could not be verified.

Case Sample 8   A shirt that was worn by a crash test dummy during some airbag deployment testing was obtained from an airbag manufacturer (name withheld by request). This shirt was light blue in color, made of 100% cotton, and had some vertical ribbing in the construction. An arc shaped singe pattern, approximately 8 inches long, was observed across the lower front region of this shirt (Figure 2.17). This appeared to have been made by an airbag with a single outer seam, although this could not be confirmed, and had a slightly smeared pattern. The ribbed texture of the fabric did affect the appearance of the singe pattern and made it more difficult to see individual dots and smears. Some other discolored areas, apparently from previous testing, were also observed on this shirt.

Case Sample 9   In a case from Michigan, the hot gas escaping through the airbag seam caused a different effect on the shirt from a driver [10]. A 911 call came in for a property damage accident (PDA) where two females in a vehicle were struck by another vehicle at an intersection. The victim's vehicle was turning left when the suspect vehicle attempted to pass on the left and struck

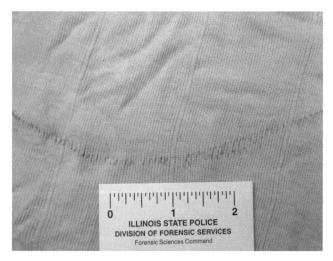

Figure 2.17   *Singe pattern observed on the lower front region of the shirt from a crash test dummy.*

the victim vehicle. Neither of the victims saw who was driving the suspect vehicle. The responding officer arrived on scene and found that the driver at fault for the accident had fled the scene leaving his 1992 Cadillac Deville behind. The airbags on the Deville had deployed. The officer ran the Cadillac's license plate and it came back registered to a male owner. The officer proceeded to the man's house but he was not home. As the officer was speaking with the owner's next-door neighbor, another 911 phone call was received in regard to a stolen vehicle. The person making the call was the car owner. He claimed that his vehicle had been stolen from the local bowling alley while he and his fellow bowling team members were celebrating a tournament win. He stated that he had left his keys in the ignition with the vehicle unlocked because of an ignition problem (the officer later checked the ignition and it was in perfect working condition). The officer spoke with some witnesses at the bowling alley to see if they could corroborate the owner's story. One of the witnesses stated that the owner had contacted her right after the accident. He stated he left the accident scene because he was drunk and didn't want to be arrested. The Cadillac's owner was soon located and placed into custody. His coat and T-shirt where collected as evidence. No injuries were observed on the suspect. The responding officer also removed the driver side airbag from the Cadillac Deville at a local body shop. The suspect was charged with filing a false police report (stolen vehicle) and leaving the scene of a PDA.

The evidence (airbag, T-shirt, and black leather jacket) was submitted to the Michigan State Police Crime Laboratory in Northville for airbag singe pattern analysis and fiber analysis. Examination of the airbag did not reveal any fiber transfers from the suspect's black T-shirt, which was 100% cotton. The leather jacket did not appear to have any damage consistent with making

Figure 2.18 *(See color insert.) Discolored dot pattern on the bottom front of the shirt from the suspected driver of a crashed 1992 Cadillac Deville.*

contact with the airbag. One questionable black smear was noted on the air bag, but it could not be conclusively associated or eliminated from being similar to the leather coat.

Examination of the T-shirt did not reveal any starch particles, obvious singe patterns, or airbag fibers. However, two areas of evenly spaced red dots were observed on the front. One of these red dotted patterns consisted of fourteen dots near the bottom left seam on the front of the T-shirt (Figure 2.18). Another smaller pattern of five red dots was located on the front right stomach area of the T-shirt. Stereoscopic examination of these areas did not reveal burnt fibers or any apparent singeing. The color in these areas appeared to be washed out or discolored. The red dotted pattern was not visible when the shirt was turned inside out. A comparison was made between the spacing of the holes in the airbag outer seam and the red dotted patterns. The spacing lined up perfectly between the dotted patterns and the seam holes (Figure 2.19).

A cutting of known fibers was then taken from the submitted black T-shirt for experimentation purposes to determine the effect of heat on these black cotton fibers. A portion of this cloth was exposed to an open flame and allowed to briefly catch fire. A stereoscopic examination of the cloth revealed a progression of color change in the fibers. The portion of cloth that was actually on fire charred and turned black. It then progressed from black to brown, orange, red, pale black (almost purple), and to the original black color of the T-shirt. Except for the blackened, charred areas, the discolored fibers didn't have the appearance of being singed or altered in their appearance. The only difference detected was in color.

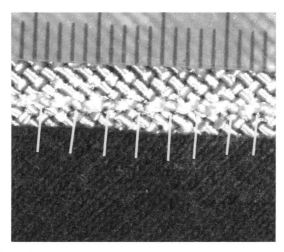

Figure 2.19 *(See color insert.) Comparison of the discolored dot pattern on the shirt with the seam of the deployed driver side airbag from a crashed 1992 Cadillac Deville.*

Another portion of the cloth was exposed to a hot clothing iron for a period of time. This also caused the fibers exposed to the heat to significantly discolor after a period of time. The color ranged from orange to red, pale black, to the original black color of the T-shirt. Once again the fibers that changed color did not have the appearance of being melted or charred.

Bundles of the known black cotton fibers from the T-shirt were then dry mounted on microscope slides with cover slips. Using a microscope equipped with a hot stage, these fibers were exposed to controlled heating. The temperature was started at 100 °C and gradually increased. The color change appeared to start at approximately 270 °C. At this temperature for approximately one minute, the fibers ranged in color from pale black to pale black with a reddish tint. The fibers turned a red color (most similar to those seen in the airbag singe pattern) at approximately 275 °C. At 283 °C the fibers turned to a pale orange-red color, and at 290 °C they turned orange. Fiber samples from each of the above temperatures were then mounted in Cargille 1.53 E high dispersion oil and compared to discolored fibers from the T-shirt pattern. The fibers exposed to 275 °C were the most consistent in color to the fibers from the singe pattern on the T-shirt (Figure 2.20).

Based on the results of this experimentation and examination of the patterns of red dots on the T-shirt, it was determined that hot gas leaking through the airbag seam likely caused the black dye to discolor. It was concluded in the lab report that the pattern of red dots observed on the submitted T-shirt likely resulted from it coming into contact with the submitted airbag or another airbag with similar seam spacing when it deployed. As a result of this examination and witness statements, the suspect plead guilty to reckless driving and leaving the scene of a property damage accident (PDA).

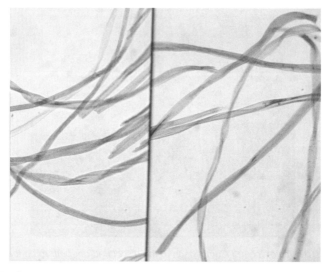

Figure 2.20 Comparison of fibers from the black T-shirt exposed to 275 °C on a hotplate on the left with fibers from the discolored pattern on the black T-shirt on the right at 200×.

2.5 CHANGES THAT ARE OCCURRING

As mentioned previously, changes have occurred in the design of automotive airbags over the last few years. The airbag manufacturers have dealt with injuries and legal issues from airbags for several years, and most of these changes have occurred with occupant safety in mind. The manufacturers have apparently known for some time that clothing from the front seat occupants, or even the occupants themselves, can be singed by the hot gases that fill the airbags leaking through the seams. Some manufacturers are now adding extra material at the seams to reduce this leaking. With the new two-staged or newer multistaged systems, the airbags deploy with less force at a low speed impact. The newest advanced airbag systems have sensors that can detect the size of the occupant, how close they are sitting, the severity of the impact, and whether the occupants are wearing their seatbelts, to determine the extent to inflate the airbag (www.edmunds.com). The Occupant Restraint Control Module in the vehicle controls this.

This author just purchased a 2005 Jeep Grand Cherokee. The passenger seat in this vehicle is equipped with an Occupant Classification System, which determines if the seat is empty or occupied by someone who is classified in the "small child" category (below 40 lb) and will not deploy the airbag (Jeep owner's manual). This type of system could help prevent the claim of the "phantom driver" that occurs in some cases.

The result of all these changes is likely to affect the degree to which the singe patterns on the clothing may be observed. The absence of a singe pattern does not exclude that person from being a driver or passenger. The one advantage for forensic examinations is that the solid propellant materials now being used generally produce a gas hotter than that from the previous sodium azide-based propellants (personal communication, airbag manufacturer, name withheld by request).

2.6 FINAL DISCUSSION

Since airbags have been required by federal law to be installed since 1998 on all new cars (1999 for SUVs, vans, and light trucks), their deployment can provide another method to help determine who was in the driver or passenger seat of a crashed vehicle when it is in question. Not only can the airbags be examined for hairs, fibers, and body fluid transfers from the occupants to help determine their seating positions in the vehicle, but also the means by which the airbags deploy can provide a way to help make that determination by the singe patterns that can be produced. Forensic scientists as well as investigators and accident reconstructionists should be aware of the potential value of this evidence, and what to look for. Both driver and passenger upper body clothing, if applicable, and both airbags should be collected and sealed in separate paper bags and submitted to the forensic laboratory for examination. Examination should be conducted in separate rooms or at different times as recommended in the SWGMAT (Scientific Working Group on Materials Analysis) examination guidelines [11], to prevent any type of cross contamination.

Note: While the airbag manufacturers that I contacted during the course of this research were very helpful when it dealt with information for the criminal cases, they preferred to have their names withheld because of legal issues. Some manufacturer information can be obtained from the Automotive Occupant Restraint Council (www.aorc.org). If you need additional contact information for a manufacturer, please contact this author. Additional airbag safety information can be obtained from the Automotive Coalition for Traffic Safety (www.actsinc.org).

REFERENCES

1. Schubert GD (2004). Oral presentations, SWGMAT, March 2004, Inter/Micro, July 2004, MAFS, September 2004.
2. Schubert, GD (2005) Forensic value of pattern and particle transfers from deployed automotive airbag contact. *Journal of Forensic Sciences* **50**(6):1411–1416.
3. Hallock GG (1997). Mechanism of burn injury secondary to airbag deployment. *Annals of Plastic Surgery* **39**(2):111–113.
4. Vitello W, Kim M, Johnson RM, Miller S (2000). Full-thickness burn to the hand from an automotive airbag. *Journal of Burn Care Rehabilitation* **21**(3):288–289.
5. Jernigan MV, Rath AL, Duma SM (2004). Analysis of burn injuries in frontal automotive crashes. *Journal of Burn Care Rehabilitation* **25**(4):357–362.

6. Jumbelic M (1995). Fatal injuries in a minor traffic accident. *Journal of Forensic Sciences* **40**(3):492–494.

7. Dumas SM, Kress TA, Porta DJ, Woods CD, Snider JD, Fuller PM, Simmons RJ (1996). Airbag induced eye injuries: a report of 25 cases. *The Journal of Trauma* **41**(1):114–119.

8. Copper JT, Balding LE, Jordan FB (1998). Airbag mediated death of a two-year-old child wearing a shoulder/lap belt. *Journal of Forensic Sciences* **40**(3):1077–1081.

9. McCrone WC, Delly JG (1973). *The Particle Atlas.* Ann Arbor Science Publisher, Ann Arbor, MI.

10. Nutter G (2005). Forensic Scientist, Michigan State Police, Northville Laboratory, Northville, MI, USA, personal communication.

11. Scientific Working Group on Materials Analysis (1999) Trace evidence recovery guidelines. *Forensic Science Communication* **1**(3).

3

Ink Analysis Using UV Laser Desorption Mass Spectrometry

John Allison

Professor of Chemistry and Director of Forensic Chemistry, The College of New Jersey, Ewing, New Jersey

3.1 INTRODUCTION

Laser desorption mass spectrometry (LDMS), a technique developed decades ago, has recently been rediscovered and reinvented as a sensitive and powerful tool for biomolecule analysis, matrix-assisted laser desorption (MALDI) time-of-flight (ToF) mass spectrometry. This chapter discusses the work that has been done with a modern LDMS instrument, using a nitrogen laser that emits UV photons at 337 nm, in the identification of colorants, such as dyes found in pen inks and pigments found in artists' paints.

Forensic labs may have a number of spectroscopic tools at their disposal for characterizing materials on surfaces. In reflectance IR, light impinges on a surface and the infrared light that comes off the surface is used to generate an IR spectrum. In reflectance UV–visible spectroscopy, again, light impinges

Forensic Analysis on the Cutting Edge: New Methods for Trace Evidence Analysis, Edited by Robert D. Blackledge.
Copyright © 2007 John Wiley & Sons, Inc.

on a surface, and the light that comes off that surface is used to construct a UV–visible spectrum. This is a very similar experiment, in which light impinges on a surface. Instead of characterizing the light that comes off the surface, singly charged gas phase ions that are formed are characterized—again representative of the absorbing molecules on that surface.

One may ask if the analysis of pen inks continues to be important to the forensic analytical community [1]. We spend many more hours at a computer than using a pen, in this technological age. Nonetheless, authentic signatures are still unique identifiers, and even computer-generated documents contain dyes and pigments from the printer. Ink analysis may be used in cases where wills are contested, when checks have been modified, or when pen-written entries in a patient's chart are questioned in a medical malpractice lawsuit. (Many doctors still use pens to write on patients' charts at the end of their hospital beds.) Pen ink analysis gained renewed importance in 2001 when the anthrax letters appeared; these were all addressed by hand. As will be shown, some pen inks change over time when on paper (a process that can be monitored by LDMS), and there has been considerable interest in using such information to determine the age of the document. It has been cited as a very important goal: "The ability to determine when a document was written would rate as one of the major breakthroughs in forensic science, having a significant impact on the detection of all kinds of fraud. This would result in a corresponding financial benefit to both State and Federal administrations" [2].

With an ability to analyze dyes in pen inks, one also has the capability for characterizing the clothing dyes and food coloring, including dyes found in pills such as those containing illicit drugs.

A few practical examples are presented here, to help the reader appreciate how experiments are designed, and how these types of mass spectral data are interpreted.

3.2 THE INSTRUMENTATION

Mass spectrometry (often incorrectly referred to as mass spectroscopy, even though light is not detected) is an instrumental method in which analyte molecules are converted into gas phase ions. The masses (m) of these ions, actually the mass-to-charge ratio (m/z), are determined by the mass analyzer (again, a commonly used term that is incorrect, since no mass spectrometer measures molecular mass directly). The distribution of gas phase ionic products from a particular analyte (the m/z values of the ions formed and their relative abundances) is summarized in graphical form in what is known as a mass spectrum.

In the history of mass spectrometry, electron impact ionization (EI) has been used to generate ions for subsequent MS analysis. This continues to be the most commonly used ionization method in the gas chromatography/mass spectrometry experiment. In EIMS, gas phase analyte molecules, such as those exiting a GC column, are bombarded with high energy (70 electron volt) electrons and ions representative of the analyte are formed. EI requires that the analyte molecules be available in the gas phase, and thus its use is limited to compounds that have detectable vapor pressures.

While EIMS had been used to solve structural and quantitative problems for years, many compounds could not be analyzed by MS because they were either nonvolatile or thermally labile. For example, one can heat up acetylsalicylic acid (aspirin) in the ion source of a mass spectrometer and create a vapor pressure sufficient to obtain a mass spectrum. In contrast, the experiment is more difficult to do with sucrose, another small molecule that exists as a solid at room temperature. The reason is not that sucrose is nonvolatile, but that it is thermally labile. When heated, sucrose will decompose faster, and at a lower temperature, than it will vaporize.

A number of "tricks" have been developed to allow for the mass spectral analysis of compounds that were thought to be inappropriate for MS analysis due to the volatility/thermal stability issues. That is, a wider range of "solids" can be analyzed by MS using one of a set of ionization methods known as desorption/ionization (D/I) methods. As the name implies, a process occurs in which a sample is changed from a condensed state to the gaseous state (desorption) and, during the process, some portion of the vaporized molecules is ionized as well. The D/I label is not meant to imply an order of events, rather, in some way, condensed phase analyte molecules are converted into gas phase ions. Many of the D/I methods add energy to a sample such that the temperature of a small volume is rapidly heated to a temperature (typically several hundred degrees) where desorption rates compete favorably with degradation rates. (At lower temperatures, degradation dominates.) One example is fast atom bombardment (FAB) [3]. In this experiment, analyte molecules are typically dissolved in a viscous, polar solvent such as glycerol. A microliter of this low vapor pressure liquid is introduced into the FAB ion source where 6–10 keV (fast!) Ar or Xe atoms bombard the solution. At the point of fast atom impact, energy is deposited (momentum is transferred) that is sufficient to desorb and ionize glycerol ("matrix") molecules, some in an ionic state. Also, analyte molecules dissolved in the liquid matrix are desorbed or ionized. FAB MS has been used successfully to analyze azo dyes [4]. Other methods such as field desorption were developed but were not implemented as extensively as was FAB. In contrast, electrospray ionization is a D/I method that has become very popular. While it will not be discussed here, we note that it has also been used to analyze colorants [5].

Another method evaluated was laser desorption as the basis for a mass spectrometric ion source [6]. In this experiment a laser, often an infrared laser, delivers a large flux of photons, in a focused beam, onto a solid analyte target. If light absorption occurs, a large number of ions will result. In this experiment, photons from a pulsed laser rapidly deposit energy into the analyte crystals. In some cases, desorption/ionization occurs. In other cases, usually when the analyte molecules has a MW of 3000 or more, degradation will dominate over D/I. Nonetheless, LDMS allowed for mass spectra of molecules, with molecular weights (MWs) up to 3000, to be obtained; this was a substantial improvement over EIMS, with a MW limit of approximately 600.

While LDMS was an interesting new method, it only "works well" (generates ions that can be structurally related to the analyte) for certain kinds of molecules. It had not been widely used, in part because it represented a serious financial commitment. Mass spectrometers are always expensive. To have to add the additional expense of a laser, and probably to have to hire a spectroscopist to operate the laser, made this option cost-prohibitive for many years.

D/I methods stimulated developments in MS by making demands on the mass spectrometer. Some methods, notably LDMS, used pulsed lasers. That is, the laser may only "fire" a few times a second and may only generate light (and thus ions) for a few nanoseconds per pulse. This could not be used as the basis for an ionization source with "CW" mass spectrometers that require a continuous flow of ions from the ion source. Methods that worked well with pulsed ionization sources were time-of-flight MS and Fourier-transform (FT) MS. This is an interesting contrast—involving one of the simplest and least expensive mass spectrometers (ToF MS) and one of the most expensive (FTMS).

When ToF MS is used with an EI ion source, the source would be pulsed. In a short time period, a collection of energetic electrons would be made in the ion source, resulting in a set of analyte ions, which would then be extracted and analyzed, based on the time required for them to travel approximately 1 meter in a low pressure pipe (the "flight tube"). Thus, LD seemed to be most appropriately used with ToF MS. However, ToF MS suffered from low resolution and from a speed that introduced complications. Complete mass spectra are generated with each pulse of the laser in LD ToF MS, on the microsecond time scale. In the 1950s when ToF MS was developed, spectra could not be captured, digitized, and collected because of the speed at which information was generated [7]. However, developments in high speed acquisition methods, such as the integrating transient recorder [8], allowed for all of the spectra generated by a ToF MS to be collected and utilized, even when the experiment was a GC/MS experiment that lasted for an hour. The remaining problem for ToF MS was that it was inherently a low resolution method. This was a problem for EI ToF MS and made LD ToF MS, which generated higher m/z ions, difficult to seriously consider as a viable method.

Mass spectrometers could usually be used to answer structural questions for chemists, but biochemists were frequently turned away from MS facilities until relatively recently. Their interests were often in characterization of biomolecules (peptides, proteins, oligosaccharides, oligonucleotides) with MWs that were too large for MS analysis. Also, they often isolated materials from biological sources at the subnanomole level, below detection limits for MS. All of this changed as scientists focused on the issue of MW limits in MS.

Experiments were pursued in which analyte molecules were deposited on a surface in the presence of a large excess of matrix molecules. The resulting "dirty crystals" of matrix molecules, with analyte molecules present as a minor component, gave exciting results in LDMS. Instead of depositing energy directly into analyte molecules, the matrix molecules absorb the energy. If any damage such as degradation occurs, it will be for these absorbers. Energy deposition into a small portion of a solid crystal/surface results in D/I. Embedded analyte molecules become desorbed and ionized as thousands of matrix molecules around them are vaporizing as well. In such experiments, matrix molecules are compounds such as sinapinic acid—a small aromatic compound with an absorption maximum close to 337 nm. This is the wavelength at which UV photons are formed by nitrogen lasers. Matrix molecules are solids at room temperatures but have low heats of vaporization. Through such experiments, LDMS reemerged on the MS scene as matrix-assisted laser desorption time-of-flight mass spectrometry, MALDI ToF MS [9]. MALDI was an unusual D/I method because it could generate ions for compounds with MWs above 100,000, and for analyte amounts three orders of magnitude below any other ionization method! This sudden ability to use MS for the analysis of biomolecules was the impetus for several groups to reinvent both ToF MS hardware [10] and data systems to allow for high quality spectra to be collected. Also, there was the change in focus from using IR lasers to UV lasers. While IR lasers can "heat" a larger number of compounds, IR photons are of relatively low energy, so many photons are required to induce a D/I event. In contrast, while all molecules do not absorb 337 nm UV light, these high energy photons can be efficiently generated and utilized. UV lasers continue to become smaller and more reliable, making it easier to incorporate them into a MS package. (*Note*: N_2 lasers are not, technically, real lasers. They are superradiant light sources.)

Figure 3.1 shows a simple commercial (Voyager DE ToF MS; PE Biosystems, Framingham, MA, USA) MALDI ToF MS. This instrument was used to collect all of the spectra that are discussed in this chapter. Light from the UV laser can be attenuated to a specific power before it enters the vacuum chamber (the ion source housing). The radiation then impinges on a small area of the sample plate target. Ions formed, when the sample is irradiated, are accelerated away from the target, toward the flight tube, using electric fields created between the plate and wire grids. One experimental variable is the electrical potential on the target, which can be a positive voltage (to extract positive ions) or a negative voltage (to extract negative ions). All ions leave the ion source with the same kinetic energy ($KE = \frac{1}{2}mv^2$), determined by the electric potential (V_{accel}) through which they pass ($KE = eV_{accel}$). Since they all have the same KE, ions of different masses have different velocities. Since time = distance/velocity, ions of different mass will travel the 1 meter flight tube distance to the detector in different times. The m/z values can be computed from these accurately measured times. Electrical signals representing arrival of the ions are generated by an electron multiplier detector. A typical sample

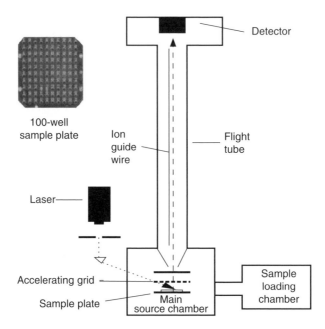

Figure 3.1 *A linear laser desorption time-of-flight mass spectrometer, as currently sold for performing matrix-assisted laser desorption/ionization mass spectrometry. Light from a UV laser is directed through a window into the ion source housing, where it is focused on a sample plate (see inset). Ions formed are accelerated through a high potential and travel a distance of approximately 1 meter (the flight tube length) to the detector.*

plate target for this instrument is shown as an inset in Figure 3.1. Shown is a gold-plated 100 well plate. Thus, up to 100 samples can be introduced, and the plate moved around such that the laser impinges on any position on that plate. The plate dimensions are 2.5 in. × 2.5 in.

While the MALDI ToF MS experiment has been very attractive for biomolecule analysis, the instrument shown in Figure 3.1 is technically not a MALDI instrument but a LDMS instrument. It is sold for performing the MALDI MS experiment, but can do both. Almost all users embed their biomolecule analytes in matrix and take advantage of what MALDI MS has to offer; however, matrix does not need to be used. Colorants—dyes and pigments—are smaller molecules than most of the compounds classified as biomolecules and are generally amenable to analysis by LDMS directly ("matrix-free" MALDI!). As will be shown, there are advantages to doing so. We have greatly benefited by the activity of chemists and instrumentation specialists due to their interest in biomolecule analysis. Their work resulted in an efficient and high performance LDMS system that never would have been developed without being driven by the interests and funding of biochemistry research.

3.3   THE ANALYTE TARGET MOLECULES

LDMS can be used for the selective analysis of colorants. The implication of the word "selective" is that they can be detected in the presence of other compounds. "Colorants" is a general term that can include dyes and pigments. These may be found in inks, on questioned documents, works of art, and so on. Human beings have an interesting relationship with color, in that we add color to almost everything we touch. We don't just build cars, we color them inside and out. We color the walls of our houses, our hair, our skin, our fingernails, even our food.

The term "dye" usually refers to a colorant that is soluble in the solvent system (referred to as the "vehicle") used to deliver it to its desired destination. Many pen inks contain soluble dyes. In contrast, pigments are usually insoluble in the solvent system and are introduced as a suspension. Pigments may be organic molecules, organometallic molecules, or inorganic compounds. Pigments may be found in a wide range of systems from pen inks to artists' paints.

Since much of this chapter will discuss the analysis of colorants in pen inks, a definition is warranted. Ballpoint pens have evolved into complex chemical and mechanical devices. In a pen, the ink has many components, only one of which is the dye (the colorant that you *see* on the page in the final product). There is also a vehicle that carries the dye out of the pen—both aqueous and nonaqueous systems have been used. Many chemical reactions can take place in such a system, and most of these are now addressed through additives. Ink components can degrade. Ink components can react with metal components of the pen tip. In response to these concerns, buffers and anticorrosion agents are now added to some inks. Biocides are added to prevent the growth of organisms inside the ink cartridge. Compounds similar to "sunblock" may be added to provide stability to the ink, so that it will resist photobleaching over time. Surfactants and humectants may be present to determine how the ink flows onto the paper and how quickly it dries. All of these components are deposited on the surface of paper on which you write. Solvents may evaporate or diffuse into the paper, but many of the components remain as part of the written line. They are not colorants and are present in small amounts, so they are essentially invisible to the eye. It is important to realize that, while the focus is frequently on what we see on a page, the actual dye or pigment, the ink itself is much more than just color.

Analyzing ink on paper, such as a signature on a check, is a challenge from an analytical perspective. A typical ballpoint pen cartridge contains about 0.6 gram of ink [11]. A typical width of the line it draws is 0.36 mm. If one assumes that the ink is 20% dye by weight and that the dye is an organic molecule with a molecular weight in the 300–400 g/mole range, the ink cartridge would contain only about 0.3 millimole of dye. A pen can write a line approximately 3000 meters long. These numbers suggest that, in a penstroke, the molar surface coverage per unit area is 0.3 mmol/m^2. In the experiments discussed here, a laser is focused on a sample, and the laser spot is approximately $0.03\,cm^2$. Thus, when the laser fires in an LDMS experiment, the total amount of sample in that beam is approximately 10 nanomoles. In the spectra of pen inks shown in the following sections, the spectra were obtained for single penstrokes on paper, and multiple spectra of high quality can be obtained from irradiation of a single spot. Thus, the sensitivity is well below the nanomole level. This is not surprising since, in the development of LDMS, detection limits as low as 10^{-19} gram were reported for some molecules [6].

3.4   LDMS FOR THE ANALYSIS OF DYES IN PEN INKS

LDMS experiments have the potential for identifying, at a molecular level, the dye present in a particular pen ink (for comparison with other samples or with information on which dyes are used by which manufacturers) and to obtain some information on the age of a document. There has been considerable interest in determining the age of a written document. Was Grandma Smith's last will and testament really signed 30 years ago, or last week? Some chemists working in this field consider the challenge as impossible—there are so many variables such as the pen, the paper, and storage conditions (light, humidity, temperature). However, many attempts have been made and results are encouraging. When LDMS is used as the analytical method, one needs to understand how to interpret the data. In the examples presented here, information will be provided on the general appearance of LDMS spectra, how these spectra are interpreted, and what methods have been developed for making identifications and structure determinations.

Example 1   Currently, ballpoint pens are probably the most widely used of writing instruments. At least 80% of all questioned document examinations requiring ink analysis were written with ballpoint pen ink [12], so this is an appropriate field of study. In this first example, the laser is focused onto a small portion of a single pen stroke from a Bic Stic® ballpoint pen on paper. For these experiments, a sample introduction plate (as shown in Figure 3.1) has been milled out such that there is a shallow depression cut into the top. It is deep enough that one can introduce a small piece of paper (containing a short segment of a pen stroke, 0.5 cm long or less) into the instrument. Such a sample is introduced by carefully taping it down, around its edges, or by using double-stick tape. If paper is taped down onto the metal surface, it is best to tape around 90% but leave a very small section untaped. The reason is that the plate is introduced into the mass spectrometer's vacuum system. Under the paper is air. The air can either slowly leak out—leaving the operator to wonder if there is an air leak in the system—or a path can be provided for that space to be easily evacuated.

In these experiments, both positive and negative ion spectra can and should be obtained. In this case, the negative ion mass spectrum was relatively uninformative—containing only two low mass peaks at m/z 35 and 37 (in a 3:1 ratio), representing Cl⁻ ions. The positive ion spectrum is more informative and is shown in Figure 3.2a [3]. There is a dominant mass spectrometric peak at m/z 372, which represents the blue dye molecules. If the atoms that make up the ion exist in more than one isotopic form in nature, "isotopic peaks" at higher m/z values will accompany the base (largest) peak. From the peak at m/z 373, it appears that the compound associated with this peak is an organic molecule containing C, H, and possibly O and/or N atoms. The naturally occurring isotopes of C (notably ^{13}C) are largely responsible for the isotopic peak at m/z 373, that accompanies the main peak. Readers unfamiliar with isotopic distributions in mass spectrometry should refer to a text such as Smith [13].

Figure 3.3 shows the structure of the dye that is represented by the spectrum shown in Figure 3.2a. Several important points should be made here on the LDMS of dyes. This dye is a cationic dye, meaning that the organic molecule shown, with the formula $C_{25}H_{30}N_3^+$, is positively charged (Cat⁺), accompanied by a small counteranion (Cl⁻). It is purchased and used as a salt, [Cat⁺][Cl⁻]. It desorbs as the cation with essentially no fragmentation—yielding an unusual mass spectrum. This is a desorption/ionization experiment, probably more

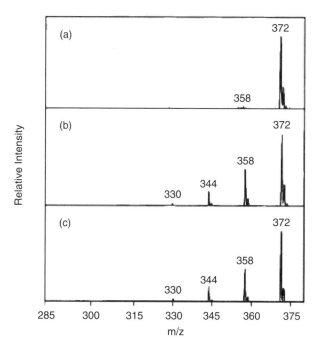

Figure 3.2 A portion of the positive ion laser desorption mass spectrum of ballpoint pen ink on paper: (a) ink from a new document, (b) ink naturally aged for 38 months in an ink library, and (c) ink from a document that has been irradiated for 6.25 hours with UV light.

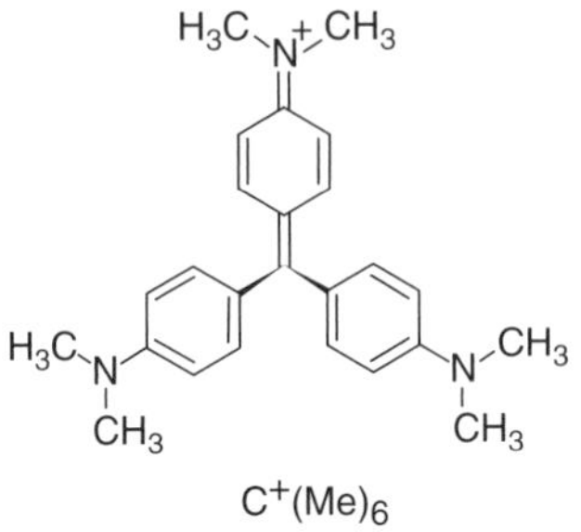

m/z	Structure
372	$C^+(Me)_6$
358	$C^+(Me)_5H_1$
344	$C^+(Me)_4H_2$
330	$C^+(Me)_3H_3$
316	$C^+(Me)_2H_4$
302	$C^+(Me)_1H_5$
288	C^+H_6

Figure 3.3 *Structures for a family of cationic blue dyes.*

precisely just a desorption experiment, in which "preionized" species (in contrast to a covalent, neutral compound) in the condensed phase are desorbed into the gas phase.

There are two dyes, methyl violet and crystal violet, that can be purchased. Each has a specific structure, and they are very similar. The structure shown in Figure 3.3 with six methyl groups on the six N atoms is called crystal violet and that with five methyl groups and one H atom is called methyl violet. Since similar compounds are available with varying numbers of methyl groups, the cation is designated here as $C^+(X)_y$. Thus, the crystal violet ion is $C^+(CH_3)_6$ and the cationic portion of methyl violet is designated as $C^+(CH_3)_5H$. No matter which is purchased, the same spectrum, Figure 3.2a, results. Chemists may find it unusual that, while specific names and structures exist, dyes and pigments can be impure. In some cases, one appears to be buying a color rather than a certain molecular structure.

(For readers who are unfamiliar with the field of dyes and pigments, which also includes stains and indicators, be aware that a single compound can take several names. Crystal violet is also known as Basic Violet 3 and Genetian Violet. Colorants are also identified by a color index number. In this case, crystal violet is C.I. 42555 [14].)

Crystal violet, with six methyl groups on the three nitrogen atoms of the dye, yields the peak at m/z 372. In Figure 3.2a there is also a small peak at m/z 358. This would correspond to the cation of methyl violet. It is 14 atomic mass units (amu) lower than the peak for crystal violet because one methyl group, weighing 15 amu, is replaced with a hydrogen atom, weighing 1 amu, resulting in a net difference in mass of −14. The peak at m/z 358 in Figure 3.2a shows that the dye used to make the ink is a mixture. That is, the peaks in the spectra represent intact chemical species (and their isotopic variants), in this case cationic gas phase species from cationic condensed phase species, not fragment ions. Another important feature to note is that, while ink and the paper it is on contain a substantial number of components, the spectrum is dominated by peaks representing the dye. The dye, in this case essentially the only component that absorbs light at 337 nm, is selectively excited. Only these molecules absorb UV light and quickly convert the energy into vibrational energy. The laser selectively "heats" the dye molecules to a temperature (at a rate of hundreds of degrees per nanosecond) at which they desorb faster than they decompose. Thus, the LDMS method is a good method for studying dyes and pigments—since most colorants absorb in the visible and into the UV. Alternatively, it is not useful for identifying other components.

Is the information in Figure 3.2a sufficient to identify the colorant? Certainly, there are many possible formulas that would lead to a molecular weight of 372. However, by knowing that this is from a blue ink, by considering the isotopic pattern, and by observing that positive ions are formed but not negative ions (except for Cl⁻ ions), we know that the absorbing portion of the dye is cationic. Such considerations and the available literature on the construction and manufacture of ballpoint pens can limit possibilities. Crystal/methyl violet is the most common dye found in blue ballpoint pen inks. Other data such as UV–visible spectra or TLC retention data can be helpful as well. Mass spectra are rarely the only information available on an analyte: visual descriptions and the source of the material (ballpoint pen, felt-tip pen, fountain pen) can be key parts to the puzzle, as the following examples will show.

Figure 3.2b shows the positive ion LDMS spectrum of a blue ink that had been on paper, aging naturally for 38 months, in the dark under controlled humidity and temperature conditions. Again, the peak at m/z 372 shows that the ink contains methyl violet. In this case, there are three dominant peaks at lower m/z values (358, 344, 330), each separated from the other by 14 mass units. These are degradation products, formed naturally over time in a process called oxidative demethylation. Figure 3.3 lists the products—the crystal violet cation naturally ages by demethylating, in which all six methyl groups can slowly be replaced by H atoms. Thus, in Figure 3.2b, the single dye molecules have become a four-component mixture of dyes, each blue, with a very similar UV–visible absorption spectrum. (Readers know that ballpoint pen inks can change over time, and the process can be accelerated by certain environmental stimuli such as light. When cars are taken to get an oil change, many companies put a small tag on the car's windshield with a reminder of the next oil change. If a ballpoint pen is used, the writing frequently fades before the 3000 miles are up! This is a relevant observation. Experience tells us that inks can change over time and the change can be accelerated using light.) If crystal violet degrades over time, can this be used to determine when a document was created—more specifically, when the ink was first introduced to the substrate? Grim et al. [3, 15, 16] have studied this question in detail from the perspective of dye analysis. In aging studies, it is certainly useful to have an accelerated aging approach in addition to a library of naturally aged samples. If one needs a 15-year-old sample and it is not available, an "accelerated aging" approach is a method used to generate an equivalent sample in a much shorter time period. Both UV and visible (incandescent) light can be used to accelerate crystal violet aging on paper. The same oxidative demethylation products are formed in both the natural and artificial processes, which is a requirement for a viable accelerated aging method. In controlled experiments, ink that initially resembles that shown in Figure 3.2a can be converted into that shown in Figure 3.2a after subjecting it to UV illumination for 6.25 hours [15]. Thus, for a particular UV lamp and lamp-to-sample distance, new ink can be aged to resemble 38-month-old ink in a little over 6 hours. (Of course, in another laboratory, a different lamp would require a separate correlation to be made.) Thus, if the analytical technique used focuses on the dye(s), there may be predictable changes that occur over time. Other changes occur as time passes. Solvent slowly evaporates from paper, possibly over a time period as long as several years. Solvent content can thus also be measured to estimate age. Solvent evaporation is not accelerated by light but can be accelerated by using heat [17].

There are, possibly, two or more "camps" concerning the question of age determination. One group suggests that it can be done, by appropriate selection of an analytical method to determine a specific component. Others suggest that there are just too many variables—the ink, the pen, the paper, storage conditions (gases, temperature, light)—to ever get an accurate determination.

An interesting aspect of this work is that one can quickly create a document that looks "old" with respect to the dye by using light to accelerate dye degradation, and old with respect to solvent loss, by applying heat. With a source of light and heat, it would be fairly easy to accelerate aging of a ballpoint pen ink, from the standpoint of many of the components, such that it would resemble a document written in 1900—years before ballpoint pens were invented!

An aside should also be made on the use of heat to accelerate aging. The concept, used in hundreds of fields [18] as the basis for accelerated aging experiments, is often based on van't Hoff's rule. This is an approximation, frequently found in physical or organic chemistry texts, which states that the rate of a chemical reaction approximately doubles with a 10 degree increase in temperature. To the extent that "aging" involves chemical processes, one might expect to make all processes occur faster with increased temperature. The rule, which holds in some cases, seems to have been extended to assume that rates approximately double with *each* 10 degree increase in temperature. One could easily reach unreasonable conclusions with such an approximation. For example, the oxidative demethylation of crystal violet has been discussed. With what does the dye react? How does it lose a methyl group and what is the source of the H atom that replaces it? Assume for this discussion that the other reactant is H_2O. If one heats a document sample from room temperature ($25\,^\circ C$) to $105\,^\circ C$, one expectation is that the eight 10 degree steps will increase the reaction rate by 2^8, that is, by a factor of 256. However, at temperatures above $100\,^\circ C$, a document will quickly "dry out"; all water would be eliminated, so the rate of change of the dye would drop to zero. Simply put, there is no one general approach that will make reactions proceed faster in a controlled, predictable way, without first being able to define the processes at a molecular level.

Example 2   This example is for a black ballpoint pen ink dye that is detected in negative ion mode. The spectrum is shown in Figure 3.4 [19]. The dominant peak in the negative ion mass spectrum is at m/z 352. The dye represented by the peak is an anionic dye, usually available as a sodium salt, $[Na^+][An^-]$. The dye is called metanil yellow, a yellow dye that contains a sulfonate group, —SO_3^-, sold as the Na^+ salt. The structure is shown as the inset in Figure 3.4. In the positive ion spectrum of the pen ink of Example 1, there was a single dominant peak at m/z 372, which represents methyl violet. Thin layer chromatography (TLC) of the ink shows that there are two dye components—a blue dye and a yellow dye. This is one way to make black ink. In this analysis, the TLC and the positive ion LD mass spectrum confirmed the presence of methyl violet. The TLC identified a second component, a yellow compound. The LDMS spectrum gave an m/z value for the intact component of an anionic dye. Again, a review of patents and the literature on black ballpoint pens quickly suggest possibilities. The identification can be made from the mass spectrum and other experiments, such as optical spectroscopy to determine the wavelength of maximum absorbance.

Examples 3 and 4   A review of the literature showed that the cationic dyes rhodamine B and rhodamine 6G, shown in Figure 3.5 [19], are commonly used in red ballpoint pen inks. The compounds both have the formula $C_{28}H_{31}ClN_2O_3$ and would form intact $C_{28}H_{31}N_2O_3^+$ cations at m/z 443. This is in fact the case. The spectrum for one of these red dyes is shown in Figure 3.6a. How can these two isomass species be distinguished? The m/z values are the same and the dyes are both red. As shown in Figure 3.5, the structures are different, although

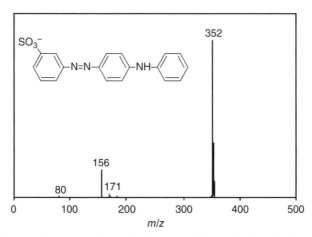

Figure 3.4   *The negative ion spectrum of a black ballpoint pen ink. The structure for the anionic dye metanil yellow is shown.*

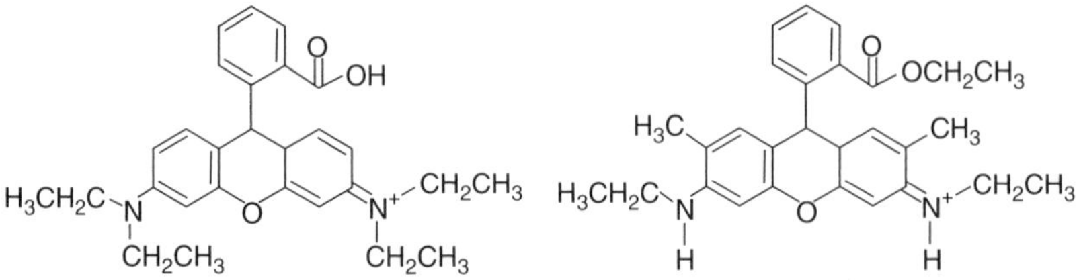

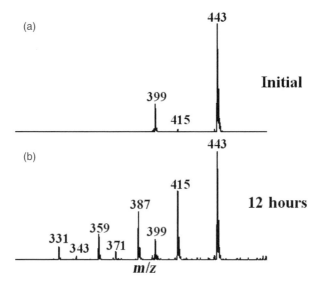

Figure 3.5   *The structures for two cationic red dyes.*

Figure 3.6   *(a) Positive ion spectrum of a red pen ink. The peak at m/z 443 could represent either rhodamine 6G or rhodamine B. (b) The same sample after exposure to incandescent light for 12 hours. Additional peaks represent photodegradation products.*

not dramatically so. Apart from the multiring core, the molecules contain eight additional C atoms. In rhodamine B, these are represented by the four ethyl (C_2H_5) groups on the two N atoms. In contrast, the eight additional C atoms are present in rhodamine 6G as two ethyl groups (one on each N), two methyl groups on the ring system, and the acid (—COOH) group of rhodamine B is an ethyl ester in rhodamine 6G. The structures differ. How can we take advantage of this?

The answer was already established in our work with crystal violet. The blue cationic dye, with methyl groups on N atoms, degraded by oxidative demethylation, and the process could be accelerated using light. Suppose we did the same with these red cationic dyes, with ethyl groups on N atoms. If —C_2H_5 groups can be lost and replaced by H atoms, the mass difference will be $1 - 29 = -28$ amu. Rhodamine B, with four ethyl groups on N atoms could lose 28 mass units up to 4 times; only two for rhodamine 6G. This prediction was borne out by working with pure standard dyes [19]. Figure 3.6b shows the negative ion LDMS spectrum of the red ballpoint pen ink dye of Figure 3.6a—with a peak at m/z 443. After exposing the red ink on paper to incandescent light for 12 hours, a set of degradation products are formed—each is desorbed in the LDMS experiment to yield a mass spectral peak. The dominant peaks at m/z 415, 387, 359, and 331, four peaks separated by 28 amu, confirmed that this pen ink was made with rhodamine B, and this is what is represented by the single peak in Figure 3.6a. If it had been rhodamine 6G, only m/z 415 and 387 would have been observed following photoaccelerated oxidative deethylation. Thus, we have shown that photodegradation of cationic dyes, a useful tool for accelerated aging studies, can also be powerful for *structure* determination as well.

Example 5   With a rich variety of "designer pens" on the market, with new colors, liquid inks, color-changing dyes, and the addition of glitter, it's not surprising that the solvent systems and types of colorants encountered are changing. Examples 5 and 6 focus on colorants that are not soluble in the vehicle, but are instead suspended in it like powdered colorant in paint. Such colorants are usually referred to as pigments. While historically pigments are most often inorganic compounds, there are a number of insoluble organic compounds that are considered pigments as well. A blue gel ink pen by Zebra (J-Roller®) was used to create a penstroke on paper for subsequent LDMS analysis. In contrast to the other examples, the analyte dye formed both positive ions at m/z 575 and negative ions at m/z 575 (with accompanying isotopic peaks). An exemplary spectrum, the positive ion spectrum, is shown in Figure 3.7. The

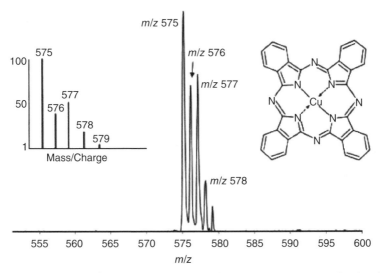

Figure 3.7   *A portion of the positive ion spectrum of a blue pen ink that contains the pigment copper phthalocyanine. Also shown is the structure of this molecule and the theoretical isotopic distribution.*

formation of positive and negative ions of similar m/z values can indicate that the analyte molecules carry no fixed charge (neither anionic nor cationic). The isotope pattern is different from those seen previously. The first peak (m/z 575) is largest. The peaks at m/z 576, 577, 578, and so on alternate in relative size. Such a pattern cannot be formed if the molecule contains just C, H, N, and O. While experienced mass spectrometrists may initially interpret the pattern as indicating the presence of a Cl atom, this is not the case. Chlorine atoms exist in nature as either ^{35}Cl or ^{37}Cl, in a 3:1 ratio. There is another element in the periodic table with a similar pattern—copper. The two isotopes, ^{63}Cu and ^{65}Cu, exist in nature in roughly a 3:1 ratio. The blue compound represented by this mass spectral data is copper phthalocyanine; the structure is shown as the inset in Figure 3.7, as is a theoretical isotopic distribution for the molecular ion of the molecule. Copper phthalocyanine is widely used because of its large molar absorptivity and its light fastness. It can be found at low levels in many white products such as paper and brighteners. It is one of the pigments used to print U.S. paper currency [20] and can be found in automobile paints. It also has an exceedingly rich chemistry [21]. Electron withdrawing or donating groups can be added around the perimeter of the molecule (up to 16 groups—replacing H atoms) to shift the color to either the blue or red, resulting in a class of compounds that are widely used commercially. Repp and Allison [22] encountered copper phthalocyanine as a pigment used for printing size and manufacturer's information on metal pipe, such as that used to construct pipe bombs. It is also commonly encountered as a pigment used in artists' paints [23].

Example 6   This example shows how inorganic pigments may be identified using LDMS. In such cases, it is always useful to have spectra of standards to compare with those of unknowns; however, one can frequently identify the compound from first principles, by interpreting the spectrum. In this case, the ink was a red ink on a questioned document—a page of the Qur'an purported to be from the 1600s [24]. The goal was to determine if the colorants (black, red, and gold were on the page) were compounds used during that time period in that part of the world. When the laser was focused on the red ink of the document, rich spectra were obtained. The negative ion spectrum, which was most informative, is shown in Figure 3.8a. Clearly, it is not a modern red pen ink with a rhodamine dye in it (see Example 2). At low m/z values, there is a set of peaks separated by 32 amu, with the first peak of the series at m/z 32. A total of 32 amu could correspond to the mass of two oxygen atoms or one sulfur atom (or, less likely, 32 hydrogen atoms!). The presence of an [A+2] isotope (^{34}S) and its abundance compared to ^{32}S suggested that these represent sulfur cluster anions—from S^- to S_6^-. Thus, the compound is some sort of inorganic sulfide. It is not elemental sulfur ($S_8(s)$), which may well yield similar ions, because the most common form of sulfur is yellow. In addition, at higher m/z values there is another set of peaks, also separated by 32 amu. These are shown in expanded form in Figure 3.8b. The isotopic distribution is much more complex and closely resembles that for mercury (shown as the inset in

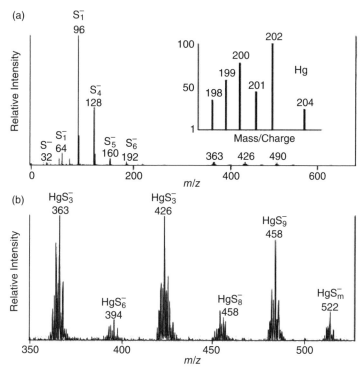

Figure 3.8 *The negative ion LDMS spectrum of the red ink/dye region of a page of the Qur'an, from the 1600s: (a) full spectrum and (b) expanded view of the higher m/z portion of the spectrum.*

Figure 3.8a). The most abundant isotope of Hg is ^{202}Hg. If these ions contain Hg, then m/z 365 (and associated isotopomers) would correspond exactly to HgS_5^-. Thus, ions representing HgS_{5-10}^- were formed. While there has not been sufficient LDMS studies of inorganic compounds to know exactly what ions might be expected, or even whether positive or negative ion formation would be preferred, it is clear that the compound is a mercury sulfide. Seventeenth century artists used a variety of red pigments including red lead and vermilion (HgS), also known as cinnabar [25]. In this case, the mass spectrum alone clearly indicated that the analyte was HgS. Additional information such as the fact that the compound was red, and historical information on colorants, quickly confirmed the assignment, as did the spectrum obtained from pure (modern) HgS.

The spectrum of HgS, a rich spectrum with signals at many different m/z values, is very different from that for a compound such as copper phthalocyanine (see Example 5). Well-defined molecules of copper phthalocyanine exist in the blue dye. However, there are no "HgS molecules" in HgS. The formula HgS does not represent discrete molecular species, but a stoichiometry. HgS(s) is a matrix of Hg and S atoms, bound together such that they exist in a 1:1

ratio. Thus, in laser desorption, there is no one species that would be expected to be desorbed and ionized from an inorganic solid. Instead, fragments from the crystal structure are formed, containing one to a dozen atoms depending on which lattice bonds are cleaved. While there are a number of ionic peaks to consider in interpreting the spectrum, this is different from a typical mass spectral interpretation (using EI ionization), where a single chemical species is ionized and fragmentation follows. In this case, each species is emitted from the laser-excited surface. By considering the m/z values, isotopic patterns, and differences in mass between peaks, one can obtain information on the atoms that make up the material. In this case, the HgS was undoubtedly mixed with some binder/vehicle, making essentially a paint, which dried to leave the red pen strokes. Even though the HgS particles are in a transparent matrix, ions representative of the colorant still result upon laser irradiation.

3.5 RELATED APPLICATIONS

It may be instructive for the reader to appreciate the kinds of items that have been introduced into the LDMS instrument, for which informative colorant spectra have been obtained. While each cannot be fully discussed, at least a partial list can be shared. We have had experience with analyses involving paper currency using LDMS [20]. Initially, we were interested in dyes used in security devices—in which a robber may be tear gassed and sprayed with a red security dye during a robbery attempt. These red inks can easily be detected on paper currency, even after the bills have been laundered with dye removers. It was intriguing to see that different spectra resulted when the laser impinged on the black and green ink used to print U.S. currency [20]. In the black ink, copper phthalocyanine appears to be present and Cl_{16}–copper phthalocyanine appears in the green ink. Spectra of colorants have also been obtained from postage stamps, cotton material, and single cloth fibers [20, 26].

A discussion with conservators at the Detroit Institute of Art made us aware of the artist and photographer Lewis Hine, whose black-and-white photographs are well known and sell for $2000–3000 each. A few years ago, a number of prints had "sufaced" [27] and sold well. They were unusual in that they were signed on the back, which Hine normally did not do. Is a "Hine" just a print from the negative of the photograph that the photographer took, or is a Hine photograph something that is also printed by the artist's hands? Clearly, it is the latter, and questions arose on whether these were recently made prints from Hine's negatives. The question of whether there was a good method for determining the age of a black-and-white photograph was raised. There does not appear to be one. LDMS can be used to generate spectra from black-and-white photographs, forming peaks representing silver ions and silver cluster ions [26]. Perhaps a method can be developed using LDMS. In the end, art conservators, people as clever as forensic scientists, used a much simpler approach. The conservators illuminated the prints with UV light and there was a fluorescence observed. This is due to whiteners in the paper. Whiteners were introduced into paper, including photographic paper, in the 1950s, years after Hine's death in 1940. Thus, the questioned photographs could not have been made by his own hand.

3.6 LDMS ANALYSES THAT "DON'T WORK"

LDMS has been used to characterize a large number of colorants. While many samples "work well," that is, yield informative spectra, some colorants do not undergo the desorption/ionization process when subjected to pulsed UV irradiation. A number of pen inks give either no ions or very low m/z species upon laser irradiation in the LDMS experiment. These frequently were gel pens. There are a number of dyes used in such pens that are different, compared to those discussed previously, in that they are polyanionic. That is, the dye is an organic molecule containing multiple negative charges. If a molecule contained two $—SO_3^-$ groups, it may be commercially available as the disodium salt, $(Na^+)_2(An^{2-})$. What will be generated when such a species is irradiated with laser energy? In all of the cases to date, the ions formed were singly charged positive or negative ions. There is a reason for this. Insufficient energy is deposited to generate gas phase species with more than a single charge. There are dyes that may be encountered that contain several charges. There are even more complex systems where a dye may contain both cationic and anionic functional groups. If an intact dye molecule cannot desorb as a multiply charged species, it has two choices: fragment ions with fewer charges may be formed, or no ions may be formed.

We had experienced a very similar situation in analyzing oligonucleotides using laser-based MS methods, specifically MALDI MS, because each phosphate group along the sugar phosphate backbone can carry a charge. Even a short oligonucleotide can carry many negative charges. The solution was to develop an additive to the MALDI matrix that would donate H^+ ions to the negatively charged phosphate groups, so that the oligonucleotide would have a mechanism available through which it could desorb with a single charge [28]. The details are beyond the scope of this chapter, but it was determined that the MALDI matrix known as HABA (2-(4-hydroxyphenylazo) benzoic acid) is useful for the analysis of multiply charged dyes, when used with the proton-donating additive diammonium hydrogen citrate [29].

One would think that two very different experiments, LDMS and MALDI MS, are involved here. In the LDMS experiment, molecules are desorbed directly from a penstroke on paper. In contrast, MALDI MS involves depositing a matrix/analyte solution onto a metal plate, letting the solvent evaporate, and irradiating the resulting crystals. The two can be combined. Starting with a penstroke on paper, a few microliters of matrix solution can be pipetted directly onto the paper. As the solvent evaporates, a small amount of dye is solubilized and cocrystallizes with the matrix on the paper. The resulting target can be analyzed and ions representative of the multiply charged analyte will be formed.

Two interesting features of this experiment are noteworthy. The first is that, without a matrix, a highly charged dye may fragment, yielding pieces that provide information on what kind of dye it is, but not which one. For example, multiply charged sulfonate dyes frequently form a negative ion with m/z 80, most likely SO_3^-. While this doesn't identify the dye, it identifies the type of dye. The MALDI MS spectrum then provides MW and isotopic information. The second interesting aspect is that we have come full circle. LDMS led to MALDI MS, and developments of MALDI MS allowed access to commercial, vastly improved LDMS instruments. Now we close the circle by finding that, for multiply charged dyes, LDMS is insufficient, but they can be identified successfully by MALDI MS.

3.7 CONCLUSIONS

While LDMS is a powerful tool that can provide chemical information on small amounts of sample, it is unlikely that instruments selling for more than $100,000 will find wide use in forensic laboratories other than in the best, biggest, and most well funded. However, the ultimate goal of research in forensic applications of LDMS need not be the development of a set of LDMS-based analyses. One approach is to appreciate LDMS data for the molecular/chemical information it can provide, then use the basic knowledge to develop other types of methods. For example, we have seen that some organic dyes degrade over time to form a set of molecules. The distribution of these may allow for age determination for a questioned document. Now that we know what the compounds are, one could develop chromatographic methods for their identification. Unique chemical features could be the basis of an IR or UV–visible spectroscopic method that would use the fundamental chemical information to estimate age. The most important aspect is to bring forensic science to a molecular level. That is the ultimate level of identification, and from it, scientists are free to create other methods with the basic information that mass spectrometry can provide. For example, Grim, Siegel, and Allison [16] considered the situation where a pen, already several years old, is used to sign a document. Does the ink age in the pen, or does that process only begin when the ink leaves the pen? Results were published using LDMS to show that, in some older pens, "aged ink" may be found, but not usually. Similar studies on the inhomogeneity and aging of inks inside pen cartridges were performed using a less expensive HPLC instrument a few years later [30]. Use of HPLC for studying aging in pen inks is becoming a popular approach (e.g., see Reference 31).

The work presented here focuses on the research pursued by a number of talented graduate and undergraduate students in the Allison laboratories at both Michigan State University and The College of New Jersey. Some of the results have been summarized (and additional information is available on pen inks) in an evolving pen and dye database that is available on the Internet, at http://www.chemistry.msu.edu/peninks/Pens_main.htm.

ACKNOWLEDGMENTS

The author would like to acknowledge the many scientists who contributed to this research, including Dr. Jay Siegel, Dr. Donna Mohr, Jamie Dunn, Leah Balko, Terry Sherlock, Dr. George Sarkisian, Beverly Chamberlin, Jason Schumaker, and Matthew Repp. Also acknowledged is partial support of this work from Michigan State University, The College of New Jersey, and the National Institute of Justice (Award 2002-RB-CX-K002).

REFERENCES

1. Cantu AA (1991). Analytical methods for detecting fraudulent documents. *Analytical Chemistry* **63**:847–854.
2. Tebbett IR (1991). Chromatographic analysis of inks for forensic science applications. *Forensic Science Review* **71**:72–82.
3. Grim DM, Siegel J, Allison J (2001). Evaluation of desorption/ionization mass spectrometric methods in the forensic applications of the analysis of inks on paper. *Journal of Forensic Sciences* **46**:1411–1420.
4. Monaghan JJ, Barber M, Bordoli RS, Sedgewick RD, Tyler AN (1982). Fast atom bombardment mass spectra of involatile sulphonated azo dyestuffs. *Organic Mass Spectrometry* **17**:569–574.
5. Ng LK, Lafontaine P, Brazeau L (2002). Ballpoint pen inks: characterization by positive and negative ion electrospray ionization mass spectrometry for the forensic examination of writing inks. *Journal of Forensic Sciences* **47**:1238–1247.
6. Conzemius RJ, Capellan JM (1980). A review of the applications to solids of the laser ion source in mass spectrometry. *International Journal of Mass Spectrometry and Ion Physics* **34**:197–271.
7. Holland JF, Enke CG, Allison J, Stults JT, Pinkston JD, Newcome B, Watson JT (1983). Mass spectrometry on the chromatographic time scale: realistic expectations. *Analytical Chemistry* **55**:997A–1004A.
8. Holland JF, Newcome B, Tecklenburg RE, Davenport M, Allison J, Watson JT, Enke CG (1991). Design, construction and evaluation of an integrating transient recorder for data acquisition in capillary gas chromatography/time-of-flight mass spectrometry. *Review of Scientific Instruments* **62**:69–76.
9. Karas M, Bachmann D, Hillenkamp F (1985). Influence of the wavelength in high irradiance ultraviolet-laser desorption mass-spectrometry of organic molecules. *Analytical Chemistry* **57**:2935–2939.
10. Juhasz P, Roskey MT, Smirnov I, Haff LA, Vestal ML, Martin SA (1996). Applications of delayed extraction matrix-assisted laser desorption ionization time-

of-flight mass spectrometry to oligonucleotide analysis. *Analytical Chemistry* **68**:941–946.

11. Kunjappu JT (1999). Molecular thinking in ink chemistry. *Ink World* **Dec**:50–53.

12. Andrasko J (2001). HPLC analysis of ballpoint pen inks stored at different light conditions. *Journal of Forensic Sciences* **46**:21–30.

13. Smith RM (2004). *Understanding Mass Spectra: A Basic Approach*, 2nd ed. John Wiley & Sons, Hoboken, NJ.

14. Green FJ (1991). *The Sigma–Aldrich Handbook of Stains, Dyes and Indicators*. Aldrich Chemical Company, Milwaukee, USA.

15. Grim DM, Siegel J, Allison J (2002). Evaluation of laser desorption mass spectrometry and UV accelerated aging of dyes on paper as tools for the evaluation of a questioned document. *Journal of Forensic Sciences* **47**:1265–1273.

16. Grim DM, Siegel J, Allison J (2002). Does ink age inside of a pen cartridge? *Journal of Forensic Sciences* **47**:1294–1297.

17. Browning BL (1977). *Analysis of Paper*. Marcel Dekker, New York.

18. Lambert BJ, Tang FW (2000). Rationale for practical medical device accelerated aging programs. *The Association for the Advancement of Medical Instrumentation Technical Information Report TIR 17* **57**:349–353.

19. Dunn JD, Siegel J, Allison J (2003). Photodegradation and laser desorption mass spectrometry for the characterization of dyes used in red pen inks. *Journal of Forensic Sciences* **48**:652–657.

20. Balko L, Allison J (2003). The direct detection and identification of staining dyes from security inks in the presence of other colorants, on currency and fabrics, by laser desorption mass spectrometry. *Journal of Forensic Sciences* **48**:1172–1178.

21. Herbst W, Hunger K (1997). *Industrial Organic Pigments: Production, Properties, Applications*, 2nd ed. VCH Publishing, Berlin, Germany.

22. Repp M, Allison J (2005). Identification of dyes and pigments found in inks used to label iron pipes: implications in the analysis of pipe bombs. *The Canadian Journal of Police and Security Services* **3**:77–83.

23. Grim DM, Allison J (2003). Identification of colorants as used in watercolor and oil paintings in UV laser desorption mass spectrometry. *International Journal of Mass Spectrometry* **222**:85–99.

24. Grim DM, Allison J (2004). Laser desorption mass spectrometry as a tool for the analysis of colorants: identification of pigments used in illuminated manuscripts. *Archaeometry* **46**:283–299.

25. Ball P (2001). *Bright Earth*. Farrar, Strauss and Giroux, New York.

26. Balko LG (2002). Laser desorption/ionization mass spectrometric analysis of samples of forensic interest: chemical characterization of colorants and dyes. MS Thesis, Department of Chemistry, Michigan State University, East Lansing, MI, USA.

27. Falkenstein M (2000). The Hine question. *Art News* May.

28. Distler AM, Allison J (2002). Additives for the stabilization of double-stranded DNA in UV-MALDI MS. *Journal of the American Society for Mass Spectrometry* **13**:1129–1137.

29. Dunn JD (2004). The detection of ink dyes by laser desorption mass spectrometry coupled with thin-layer chromatography and the use of photochemistry for dye characterization. MS Thesis, School of Criminal Justice, Michigan State University, East Lansing, MI, USA.

30. Andrasko J, Kunicki M (2005). Inhomogeneity and aging of ballpoint pen inks inside of pen cartridges. *Journal of Forensic Sciences* **50**:1–6.

31. Hofer RJ (2004). Dating of ballpoint pen ink. *Journal of Forensic Sciences* **49**:1–5.

4

Condom Trace Evidence in Sexual Assaults: Recovery and Characterization

Wolfgang Keil

Professor of Legal Medicine, Department of Legal Medicine,
Ludwig-Maximilians-University of Munich, Munich, Germany

4.1 INTRODUCTION

4.1.1 Forensic Significance

Condoms possess coatings on their surfaces that leave traces of contact upon unpacking and using them. The residues from these superficial coatings are especially easy to find in body orifices, on the skin, and, under certain conditions, on pieces of clothing. Using forensic analytical methods, these traces can be proved through swabs taken from the skin, vagina, rectum, mouth, or samples of clothing.

The analysis of condom-associated substances can be of enormous importance in judging sexual crimes, particularly in cases where no sperm is found

Forensic Analysis on the Cutting Edge: New Methods for Trace Evidence Analysis, Edited by Robert D. Blackledge.
Copyright © 2007 John Wiley & Sons, Inc.

on vaginal swabs or on those taken from other body orifices. The examination of condom residues has several goals:

1. It may provide a reference or proof that a condom either was used or can be completely excluded.
2. In a positive examination result, as many typical profiles of condom residues as possible should be evidenced. In specialist literature [1], the concept of "fingerprinting" is also used in relation to condom coatings. On the one hand, this makes possible the assignment of examination results of a link to a specific sort of condom. On the other hand, "condom fingerprints" taken from various swabs (e.g., from the penis and vagina) can be compared for a direct match. In court, this sort of analytical finding is treated as the best form of evidence.
3. It may help substantiate either the victim's or the suspect's account of the incident.
4. It may provide an evidential link if similar traces are found on both suspect and victim evidence items.
5. It may substantiate the SART (sexual assault response team) findings of the examination of the victim and help show that penetration occurred. In many jurisdictions the law provides for several levels of sexual assault. Evidence of penetration will raise the charges to a higher level.

Berkefeld [2] was the first person to become aware of the significance of condom-associated substances seen from a forensic point of view. Ten years prior, Blackledge and Cabiness [3] had proved the value of the analysis of lubrication residues in cases of rape and sodomy. At that time, residues of commercial lubricants such as Vaseline® rather than condom residues were examined.

4.1.2 Production, Sale, and Use of Condoms

Condoms have been in use for 12,000 years but they have only been industrially manufactured since the nineteenth century. Condoms quickly developed into a one-time use, throwaway article and soon became an item for the masses [4].

Since the beginning of the 1980s, the number of manufactured condoms has increased worldwide. The cause of this was prevention of HIV infections. Health awareness measures in many countries today have the goal of reducing the number of new HIV infections through the use of condoms.

In Germany, the total number of manufactured condoms per year has almost doubled from 1978 to today. According to data from the Association of the German Condom Industry, in 2002 more than 200 million condoms were produced and sold. The preference of clients for certain types of condoms is

TABLE 4.1   Market Share of the Ten Most Sold Condom Types in Germany at the End of 2003[a]

Manufacturer	Product Type or Brand Name	Market Share (%)
MAPA	Billy Boy	22.0
MAPA	Fromms	14.9
SSL International	SSL Durex Gefühlsecht	10.2
Ritex	Ritex RR1	7.7
Ritex	Ritex ideal	6.6
Condomi AG	Condomi ES 2	4.3
Condomi AG	Condomi Com	4.0
MAPA	Blausiegel	3.6
MAPA	Big Ben	2.5
SSL International	SSL Durex Performa	2.1
Total		77.9

[a]According to a statement made by the Association of the German Condom Industry.

quite varied. It is worth noting that often a basic condom type is offered in rather different models. The models differ in form, color, and surface texture. Moreover, the same basic condom type is not uncommonly produced with or without a flavoring element and with a normal or extreme lubricant coating.

One can see from Table 4.1 which sorts of condoms were most widely sold in Germany at the end of 2003. Dominant are condoms made by German companies such as MAPA, Ritex, and Condomi AG. All other brands had a market share of just 0.1–1.6%.

In the year 2005, roughly 200 different types of condoms were offered on the German market. As a result of the trade in condoms in specialty shops and the possibility of ordering condoms from overseas via the Internet, the variety of available condom types is hardly reckonable. This also applies to international standards. In Australia as early as 2001, at least 71 different condom types were registered officially [5]. In Italy in 1995, 47 different basic condom types were produced that were sold in 135 different variations, for example, with or without flavor elements [6].

Health awareness measures running parallel to the increasing production numbers have indeed led to an actual increase in condom use. Meanwhile, in many countries condoms have attained social acceptance. In the year 1999 it was claimed that 6–9 billion condoms were used worldwide [4]. Figure 4.1 reflects condom usage in Germany since 1998. Based on these data it can be estimated that in Germany alone 360 condoms are used per minute.

Based on the increasingly extensive use of condoms it can be deducted that the number of sexual crimes where condom use plays a role will likewise increase. As shown in Figure 4.2, this hypothesis is confirmed through data collected in Germany [8].

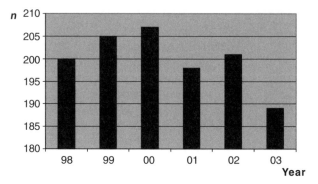

Figure 4.1 *Condom usage in Germany [7] (http://www.kondomberater.com/kondomwelt/zahlen.php).*

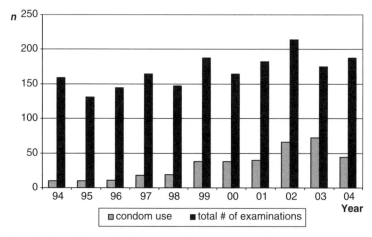

Figure 4.2 *Number of forensically examined sexual crimes where condom usage played a role. Cases are taken from the Institute of Forensic Medicine at the University of Munich and a population of 6.1 million.*

It can be assumed that sexual assailants consciously use condoms to a greater degree so that their DNA profiles cannot be ascertained. Blackledge and Vincenti [9] conveyed that already in 1993 in Las Vegas, and according to claims made by victims, a condom was used in one-quarter of the suspected sexual offences. Australian experts have also recently reported increased condom usage in sexual offences [5].

4.1.3 Condom Production

A prerequisite for examining condom residues is being familiar with the process of condom production. From a forensic, analytical standpoint the only meaningful steps in the process are those where substances are applied whose residues can be ascertained after condom use. For this reason and from here on we shall only look at these steps while other important manufacturing segments, such as the control and packaging of condoms, will not be mentioned.

Essentially, the production of the actual hood, or rough condom, is to be differentiated from the further refined processing of the same rough condom. Within the parameters of refined processing, the surface of the rough condom is coated in various ways.

The essential steps in the production of latex condoms are as follows: the *production of the rough condom through vulcanization*, then the *silicone treatment of the rough condom* followed by the *powdering* of the condom surface, and finally the *lubricant coating* of the condom's surface. The silicone treatment of the rough condom consists of an aqueous silicone emulsion. The silicones present in this slurry are much lower in molecular weight than those used if a silicone oil is added as a lubricant coating. It is these low molecular weight silicones used in the washing slurry that may be detected by GC/MS. The silicone oils used as a lubricant are not sufficiently volatile to pass through a GC column. Additionally, with a molecular weight range from approximately 3000 to well beyond 15,000 daltons with a mean of between 6000 and 8000 daltons [9], they are well beyond the mass range of GC/MS instruments.

When rough condoms are simply powdered, they are called dry condoms. These are currently sold in small amounts. When one speaks of wet condoms, the reference is to the application of a wet coating following the powdering of a condom. Due to the risk of dehydration, these sorts of condoms must be vacuum packed. According to estimates of the German condom industry, wet condoms represent about 98% of all condoms produced today [10].

Since the beginning of industrial production, condoms have been manufactured for decades out of natural rubber. Even today more than 90% of all condoms are made of latex. Since latex can lead to allergic reactions, latex-free condoms made of polyurethane and polyethylene have been available for some years now. Extensive experience with these condom materials is not documented [11]. Latex-free condoms compose a comparatively small market share.

Production of Rough Condoms Through Vulcanization

Latex Proteins   The rough condom is produced from natural rubber through vulcanization. Latex is a complex mixture of various components. Its protein content is between 2% and 3%. In extracts taken from ready-to-use condoms, protein concentrations of just 0.05–0.1% were established [12]. Certain of these proteins are responsible for a latex allergy.

Vulcanization Acceleration   In usual practice dithiocarbamates are applied to accelerate the process of vulcanization. The dithiocarbamates are often not completely used up in the vulcanization process.

Nitrosamines   Vulcanization accelerators can set nitrosamines free. They have been identified in extremely small quantities in numerous types of condoms [13].

Silicone Treatment of Rough Condoms   The rough condom is treated with silicone oil that penetrates the rubber. The silicone oil functions as a softener for the rubber.

Powdering

Cornstarch   Through powdering (namely, the coating of the condom surface with particles), the condom rolled up for packaging is less likely to stick together and the task of unrolling it is simplified. On many types of condoms cornstarch is used.

Polyethylene   Powder derived from polyethylene is also used as a coating. Various types of condoms are coated with a mixture of cornstarch and polyethylene particles. As such, the proportion of starch is five times greater than that of polyethylene [14].

Lycopodium Spores   These spores also possess the quality of not sticking together and present an excellent separating agent. Lycopodium powder is partially combined with cornstarch. In the beginning of the 1990s, Berkefeld [2] confirmed that 80% of the condoms he examined on the German market were treated with lycopodium spores. Today, the spores are only rarely applied.

Talc   Talcum powder also works against stickiness on latex surfaces. It is seldom used to powder condom surfaces, however.

Antioxidants and Preservatives   Antioxidants are primarily added to retard the degradation of latex. Although used in small quantities, they may be extracted from evidence items and identified by GC/MS. An antioxidant example is Wingstay-L®, a phenol derivative that is a butylated reaction product of *p*-cresol and dicyclopentadiene. Another is butylated hydroxytoluene (BHT). Kathon CG is an example of a preservative that kills bacteria.

Silica   Silica constitutes a filler material (or extender) for natural rubber. It has been reported from the United States that silica has been used as a component in condom coatings [9].

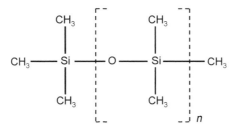

Figure 4.3 *Structural formula of polydimethylsiloxane (PDMS). Note that the repeat unit (area within dashed brackets) has a mass of 74 daltons. Therefore, any mass spectrum of PDMS should show a series of peaks each spaced 74 daltons apart.*

Lubricant Coating

Lubricants   As in powdering, a lubricant coating also simplifies the task of unrolling a condom. Lubricated condoms are much more slippery than dry condoms. Moreover, lubricants have a beneficial effect with regard to the shelf life of packaged condoms. Substances derived from a silicone base are most widely applied as lubricants. Polydimethylsiloxane is used most often (Figure 4.3).

PDMS is colorless, odorless, and water insoluble and does not react with other organic substances. In rarer cases, water-soluble substances such as polyethylene glycol are used as lubricants [5]. In Anglo-American usage, condoms treated with water-soluble substances are known as "wet condoms," whereas those coated with silicone combinations are referred to as "dry condoms" [9]. Approximately 150–300 milligrams of lubricant is applied to each condom before packaging [10].

Spermicides   Roughly 10% of available lubricated condoms contain a spermicide in the lubricant coating. These types of condoms are advertised as "doubly secure" with regard to their role as a contraceptive. Nonoxynol-9 is used exclusively as a spermicide. The proportion of nonoxynol-9 represents roughly 2–5% of the total lubricant coating [10].

Flavorings and Aromas   Numerous companies offer condoms that contain an aroma in the lubricant coating. The scent's are derived from the food industry. The scent's composition is usually unknown to condom manufacturers. There are condoms with the most diverse flavorings, for example, strawberry, banana, vanilla, and chocolate [15]. The flavorings are composed primarily of mixtures of numerous chemical components that, in turn, consist of esters.

Surface Anesthetics   Currently, benzocaine or lidocaine is being added to the lubricant coating of certain types of condoms. This leads to the male's delayed ejaculation [16].

4.2 EXAMINATION FOR CONDOM RESIDUE TRACES

Just a portion of the substances used in the production of condoms has been examined from a forensic point of view. The methods and results of analysis are compiled in the following discussion.

Rough Condom Residues

Latex Proteins   The water-soluble latex proteins stem only from the wall and not from the coating of the condom. Janda and Wentworth [17] proposed carrying out experimental studies using enzyme-linked immunosorbent assay (ELISA) to prove condom use through detection of these water-soluble latex proteins on vaginal swabs from alleged assault victims. Previous studies on latex gloves from diverse manufacturers had brought in different results. Therefore, it was thought that some discrimination between condom manufacturers might be possible. Unfortunately, the grant was not funded and the potential of this evidence is still unexplored.

Powdering Residues

Cornstarch   The starch particles are quite recognizable under light microscopy using hematoxylene-eosine (HE) staining (Figure 4.4a). Their diameter is from 2–32 μm. They have a polyhedric to round shape [19]. Cornstarch is very simple to identify under polarization microscopy whereby most of the particles show a typical cross (Figure 4.4b). Blackledge and Vincenti [9] have given proof of the importance of starch in establishing condom contact by presenting cornstarch collected from vaginal swabs. The authors, however, made no statement concerning the temporal validity of this detectable evidence.

Although starch particles break down biologically, they remain relatively stable in the vagina. Through examinations with volunteers using polarization microscopy, cornstarch particles could be detected in vaginal swabs up to 4 days postcoitus when a corresponding condom had been used [20]. Numerous starch particles could be detected on penile swabs taken up to 1 day after intercourse using a condom. With regard to the time frames noted, it must be taken into consideration that neither the women nor the men washed their genital areas after intercourse.

As the handling of a practical case shows (compare to Case 3 in Section 4.4), cornstarch particles captured on air-dried vaginal swabs retain an extremely stable storage life. Thus, starch particles on a 9-year-old vaginal swab could be identified beyond doubt.

An alternative procedure (used by Blackledge) for the identification of starch particles involves examination of an aqueous or methanol extract on a microscope slide with coverslip under slightly uncrossed polars. With crossed

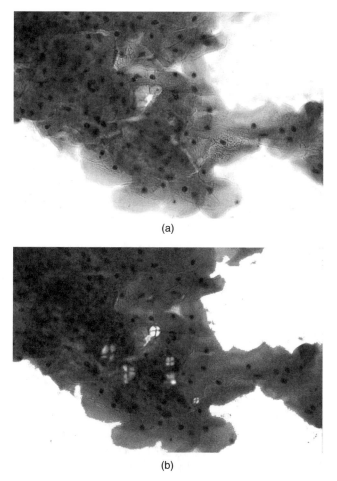

(a)

(b)

Figure 4.4 *Cornstarch particles in a vaginal swab, 18 hours postcoitus: (a) HE staining and (b) polarization microscopy.*

polars the background is black and anything that is not birefringent is not seen. With slightly uncrossed polars one can still see nonbirefringent particles while birefringent particles are still bright and the starch grains still show their characteristic interference pattern. Confirmation that the observed particles are starch is obtained by allowing a few drops of dilute Lugol's solution to flow under the coverslip while the slide is observed in brightfield. As the Lugol's solution migrates from one end of the coverslip to the other, the starch grains will slowly turn blue. If the Lugol's solution is too concentrated, the starch particles will be so strongly stained that they will appear black and will be totally opaque. One then isn't sure if they are looking at a particle that was black to begin with or a starch grain after it has been stained. Although the staining process may take longer with a dilute solution, the starch particles will just be stained a light blue and if one switches back to crossed polars the characteristic interference pattern can still be seen. Figure 4.5 shows the

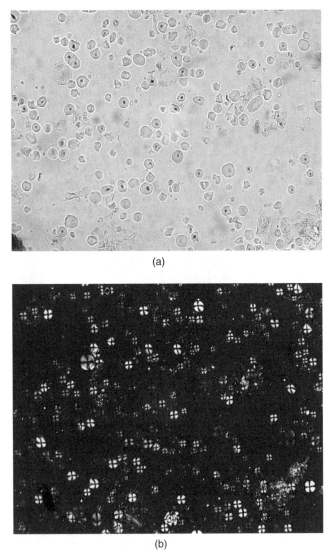

(a)

(b)

Figure 4.5 *Kava starch grains: (a) in brightfield at 100×; (b) under slightly uncrossed polars at 100×; (c) as Lugol's solution is slowly migrating from the top toward the bottom at 100×; and (d) under slightly uncrossed polars after application of dilute Lugol's solution at 400×.*

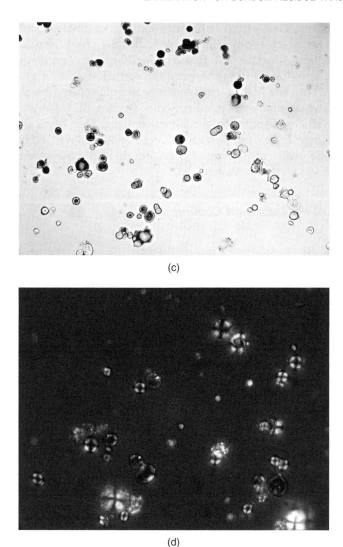

(c)

(d)

Figure 4.5 Continued

appearance of starch particles from the Kava plant (a) under brightfield, (b) under slightly uncrossed polars, (c) as Lugol's solution is migrating from top to bottom, and (d) under slightly uncrossed polars after staining with a dilute Lugol's solution.

Lycopodium Spores Spores can be detected under light microscopy. However, they cannot be presented by using customary histologic staining. Therefore, the slides need to be very carefully looked over since the spores are not recognizable according to their size and structure. Stained vaginal epithelial cell groups can considerably impede the identification of unstained spores despite their diameter of 25–40 µm. The spores indicate a characteristic tetrahedric shape and reticular surface [19]. Based on their high oil content, they are quite presentable in oil-based stains between the epithelial cells on a slide [20, 21]. Berkefeld [2] was the first to stress that lycopodium spores also carry forensic significance as condom-associated particles. Cremer [22] also presented data for the provability of spores from condom coatings. They break down biologically. Nonetheless, they can be traced in the vagina for up to 4 days postcoitus [20]. In penile swabs, the spores were evident up to 24 hours when the associated body parts were not washed.

Residues of the Lubricant Coating

Lubricants Blackledge and Vincenti [9] were the first to successfully analyze condom-associated PDMS from trace elements. The authors introduced Fourier transform infrared (FTIR) spectroscopy to forensic analysis. The matter being examined was extracted by using dichloromethane, CH_2Cl_2. Via this method not only could PDMS be identified, but divergent grades of PDMS polymerization could be recognized. In this respect, PDMS compositions of varying origins could be differentiated. It was recommended that one always follow up the results with a second procedure. In the confirmation procedure the very sensitive and highly dissolving desorption chemical ionization/mass spectrometry (DCI/MS) was applied with a previous extract of CH_2Cl_2. PDMS could be evidenced without difficulty on vaginal swabs as well as in traces found in underwear. Vaginal secretions and semen created no disturbing interferences in the PDMS depiction. The authors nonetheless supposed that larger amounts of sperm would lead to intervaginal dilution in which case the FTIR method would fall short of the minimum range of proof. Statements concerning the postcoital time frame for representation were not made. By applying DCI/MS, PDMS could be ascertained in experiments with volunteers up to 24 hours following intercourse.

Conti et al. [6] were able to represent polydimethylsiloxane taken from dilutions of the pure substance at 1:1,000,000 through proton nuclear magnetic resonance (^{1}H NMR). A carbon tetrachloride extraction was first performed. The authors found residues of silicone oil even in the vaginal swab taken immediately after forced intercourse. In the vaginal swabs of volunteers condom-associated polydimethylsiloxane was still evidenced several hours later. An exact time span was not mentioned, however. Following short-term skin contact, silicone oil was likewise detectable on skin swabs.

Maynard et al. [5] performed an extraction in two steps as proof of PDMS. The first extraction involved hexane. Then followed a diffuse reflectance infrared Fourier transform spectroscopy (DRIFTS), which proved PDMS. A confirmative investigation through pyrolysis gas chromatography in combination with mass spectrometry (PGC/MS) was necessary. After drying the swab, a second procedure was carried out with methanol. This extract was then subjected to verification of polyethylene glycol or nonoxynol-9 via infrared analysis. Confirming reactions were then carried out with liquid chromatography/mass spectrometry (LC/MS).

The authors did not share whether this procedure for condom residues of genital swabs was successful. In any case polyethylene glycol could be detected in a vaginal swab 8 hours postcoitus with a Durex brand condom. A sample taken from the vaginal opening in the same case was negative. The examiners claimed summarily that lubricants may be detected up to 12 hours after intercourse although there was no conclusive differentiation with regard to the water solubility of the lubricants under examination. The analysis of lubricants was also possible through swabs that had been stored for 3 months.

Burger et al. [23] were able to differentiate rinse solutions on condom surfaces from pure lubricants thanks to micellar electrokinetic capillary chromatography (MECC) with ultraviolet absorbance detection. The procedure was, however, not sensitive enough to be able to analyze skin swabs, extracts from pieces of clothing, and the foils of condom-associated lubricants.

Keil et al. [20] confirmed that lubricants with a silicone base could be detected quite well on condom surfaces using gas chromatography and mass spectrometry (GC/MS). Requirements were a washing process (5 mL tris buffer, pH 7.4) followed by an alkaline extraction of the material under examination (Table 4.2).

The GC method was conducted under the following conditions: GC Agilent 5890; injector temperature 270 °C; interface temperature 300 °C; gradient 100 °C, isotherm 2 min, rate 20 °C/min, 300 °C, isotherm 5 min; GC-column Varian VF-5MS, 25 m × 0.25 mm × 0.25 μm; 1 μL injection volume, splitless. MS conditions were: Agilent MSD 5972; full scan mode, m/z 50–550.

In this procedure, the rinse solutions of numerous unused condoms could be differentiated successfully based on lubrication coating. The equidistant

TABLE 4.2 Steps in Basic Extraction

Step	Basic Extraction
1	1 mL washing solution +1 mL ammonium buffer (pH 8.9) +5 mL ether/ethylacetate (1 : 1)
2	Precipitation of organic layer
3	Evaporation (N_2)
4	Derivatization (BSTFA/1% TCMS)
5	1 μL injection (GC/MS)

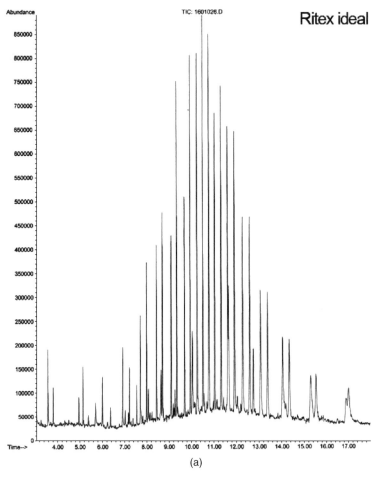

Figure 4.6   *GC/MS total ion chromatograms. Comparison of the rinse solutions of two types of condoms on the German market: (a) Ritex ideal and (b) Durex AVANTI.*

peaks in the chromatograms could be assigned to the silicone mixtures (Figure 4.6). As expected, the silicone composition was different at various manufacturers.

No reproducible relevant peaks were found in vaginal swabs where there was no preceding condom use (Figure 4.7a). Anal swabs without lubricant residues indicated numerous regular peaks, which were formed significantly different from the peaks of the silicone mixtures (Figure 4.7b).

After condom use silicone fractions could also be detected with this method (Figure 4.8). The vaginal swabs were taken from volunteer couples and were extracted at regular postcoital time intervals. Proof of lubricant residues were possible up to 12 hours postcoitus, but not in all cases [20].

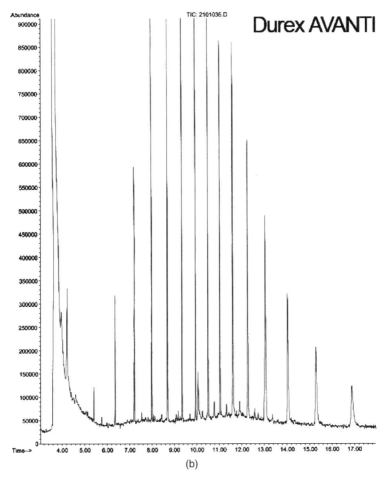

Figure 4.6 *Continued*

The addition of divergent flavorings had no noticeable influence on the total ion chromatogram of the very same condom type (Figure 4.9).

The mass spectra of certain peaks of the total ion chromatograms from 54 condoms were stored in a data bank. In this way, the spectra of unknown samples could automatically be compared with those in the data bank. Figure 4.10 offers an example of just such a comparison.

In view of the variety of available condoms, the use of a data bank is imperative in order to record positive analytical results of a certain condom type.

Spermicides Hollenbeck et al. [24] were the first to give evidence of non-oxynol-9 for investigative purposes. Liquid chromatography/electrospray ionization mass spectrometry (LC/ESI-MS), nanoelectrospray ionization mass spectrometry (NanoESI-MS), as well as matrix-assisted laser desorption/

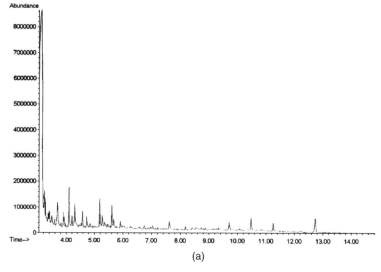

(a)

Figure 4.7 *GC/MS total ion chromatograms. Comparison of a vaginal swab with an anal swab, both without preceding condom use: (a) vaginal swab and (b) anal swab.*

ionization Fourier transform mass spectrometry (MALDI-FTMS) were applied. After pure nonoxynol-9 could be demonstrated, the spermicide was also successfully analyzed in vaginal swabs taken from test persons, as well as in the vaginal swab of a rape victim. This vaginal swab was prepared 4 hours after the crime.

Bommarito and Dougherty [25] again researched the evidence of nonoxynol-9 taken from condom lubricants. They pointed out that several routine methods leading to concrete evidence of nonoxynol-9 were not suitable. Using PGC-MS, the authors could identify nonoxynol-9 in vaginal swabs. Bommarito and Dougherty, however, could make no statement about the intravaginal stability of nonoxynol-9.

4.3 FORENSIC EVALUATION OF THE SUBSTANCES AND EXAMINATIONS

In answer to the question of whether a condom was used at all, it is important to establish which substance residues are typical of condom contact. Of significance is whether and in what time spans these substances biodegrade or are eliminated from the body.

From the forensic viewpoint, it is worth noting that condom coatings present no constant markers. Their composition is established in the end by the manufacturer who takes medical norms into consideration. Very often consistency of the end product is guaranteed by the production process. Nonetheless,

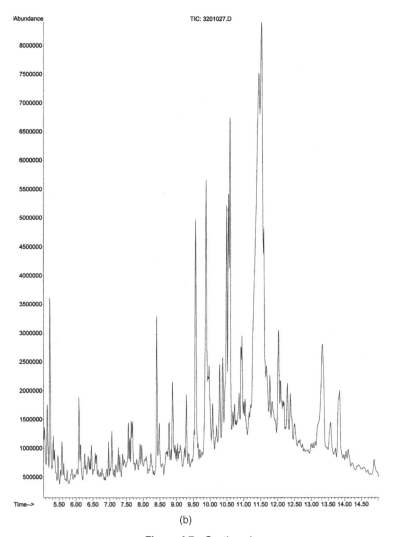

Figure 4.7 Continued

deviations can arise. This can change the composition of the original formulation. It is also worth noting that the condom industry offers the very same brand name article in great variety, namely, with or without nonoxynol-9 and with normal or heavy lubricant.

The chemical analyses are more meaningful when, in the framework of a medical examination and beside the usual swabs (proof of sperm and eventual DNA analysis), at least one separate swab is taken to register condom residues.

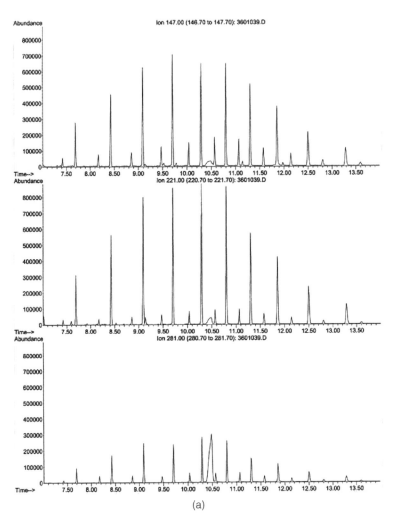

(a)

Figure 4.8   *GC/MS total ion chromatograms. Comparison of the rinse solutions of a Billy Boy brand condom with the extract of a vaginal swab. The vaginal swab was taken 12 hours after intercourse and use of a Billy Boy brand condom: (a) Rinse solutions from the Billy Boy condom and (b) extract from the vaginal swab.*

In conversation with the injured party and suspect one needs to look closely at all aspects that could play a role in the assessment of the analytical results. These are condom brand, date of last voluntary intercourse using a condom, lubricants unrelated to those on the condom, the application of a spermicide, and the use of medications and cosmetics in the genital area. By no means should the examination of the victim be carried out using powdered gloves. Practicable recommendations for a physical examination with regard to forensic analysis were published by Blackledge [26].

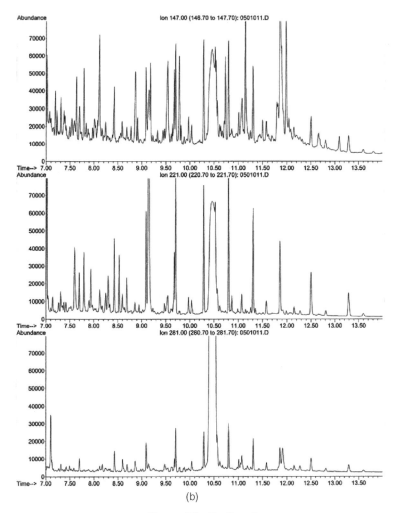

(b)

Figure 4.8 *Continued*

Residues of Rough Condoms and Vulcanization

Latex Proteins The presence of latex proteins is only possible after contact with latex-based objects. In this respect, the proof of water-soluble latex proteins from genital swabs accords the highest degree of proof of condom usage. In certain cases, one might be drawn to consider the use of other rubber objects. Attention should be paid to possible cross reactivity using ELISA.

Although a full study on the use of ELISA to detect/identify traces of soluble latex proteins from condoms has not as yet been carried out, recent

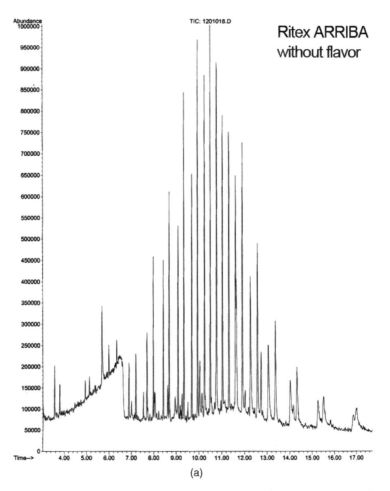

Figure 4.9 GC/MS total ion chromatograms. Comparison of the rinse solutions from two models of a Ritex ARRIBA brand condom: (a) Ritex ARRIBA without flavoring and (b) Ritex ARRIBA with flavoring (strawberry).

preliminary studies by Trochta [27] have "shown that within simulated sexual assault variables, condom proteins are transferred in amounts sufficient for immunological detection. Therefore, it is entirely feasible to develop such an assay to be included as part of a victim rape kit."

Dithiocarbamates Dithiocarbamates are used not only as accelerators in vulcanization but they can also be found in medications and in pest and weed control preparations. Evidence of dithiocarbamates on genital swabs would be a valuable sign of condom contact. Nonetheless, there has been no research so far to detect dithiocarbamates in trace elements. Only extremely small quantities of vulcanization residue can be expected in swabs.

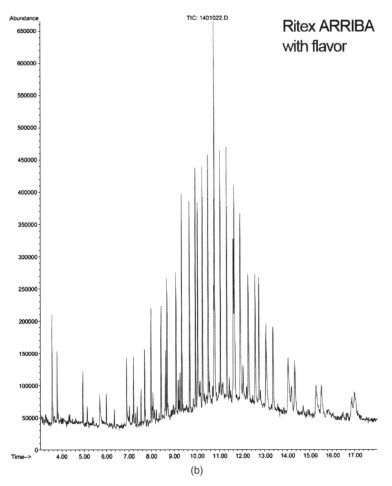

Figure 4.9 *Continued*

Nitrosamines   Nitrosamines appear as an unwanted side effect of the vulcanization process. They can thus be found in all items made of rubber. Smoking releases considerable quantities of *n*-nitrosamines. The concentrations measured in condoms lie well below those consumed in food by a person per day [28]. As a result, the nitrosamine content in swabs can also be only very low. Based on the relatively wide dissemination of nitrosamines, evidence of these substances in swabs could only be seen as a weak indication of condom use. Analytical results concerning this question do not exist.

Powdering Residues

Cornstarch   Most condoms but also some sorts of medical research gloves are powdered with cornstarch. Moreover, cornstarch is a component in numer-

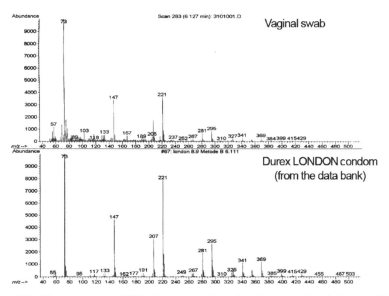

Figure 4.10 *Mass spectrum of RT 6.127. Comparison of the rinse solution of a Durex LONDON brand condom, logged into the data bank, with the extract of a vaginal swab. The vaginal swab was taken 5 hours postcoitus using a condom of the Durex LONDON brand: (a) Extract of vaginal swab and (b) rinse solution of the Durex LONDON.*

ous foodstuffs, pills, and cosmetics. Undigested cornstarch particles can also be found now and then in anal swabs. The copious presence of cornstarch granules in genital swabs can therefore be an indication of possible condom contact. In our experience, suspicion of condom use can be expressed when a smear contains more than 10 starch granules. Ample starch particles are present in the first hours following intercourse with an appropriate condom. The same applies to penile swabs.

Microscopic examination of swabs with trace elements for cornstarch is a simple but also very informative measure. It can be executed without much ado if the swabs are being appraised for sperm.

Polyethylene Polyethylene possesses very good biocompatibility. In powdered form it is not very commonplace in everyday living. The powders are produced on an industrial basis only for specific coatings. They are being increasingly used for the surfaces of condoms. Evidence of polyethylene particles on swabs would provide a high degree of proof of condom contact. The particles in pure powder compounds as well as in wash solutions of condom surfaces can be very well represented under light microscopy (Figure 4.11).

At any rate, polyethylene has not yet been detected in genital swabs. There are many reasons for this. Polyethylene is fundamentally not colorable, not even using histologic staining methods. Recognizing uncolored particles, which are often smaller than cornstarch granules, is difficult with a colored swab. Moreover, polyethylene particles are added in limited concentrations compared to cornstarch granules. Polyethylene is also not fixable with formalin or alcohol so

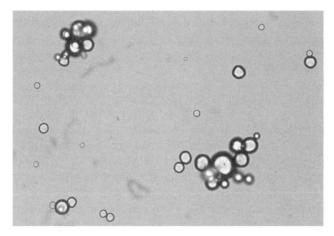

Figure 4.11 *Unstained smear of polyethylene particles as they are used in condom coating. The polyethylene powder was made available thanks to the Ritex Company.*

that the particles can be rinsed away in the additional coloring process of the slide preparation and finally are no longer contained in the smear.

Lycopodium Spores   As these spores can lead to allergic reactions and the formation of granulomes [29], they are hardly employed at all today in condom production. In a few types of condoms that still contain lycopodium spores, spores found are almost exclusively those of *Lycopodium clavatum*. Earlier, lycopodium powder was quite commonly found in pharmaceuticals. At present, it is used in very few areas of life (such as the pyrogenic and condom industries). According to our conclusions there are presently only a few odd condom types available on the German market that have minimal concentrations of lycopodium spores on their surfaces. In this sense, evidence of spores in the forensic question "Was a condom used?" is rendered meaningless. What is striking is that not a single case is reported in literary references indicating the evidence of spores to confirm condom use although lycopodium powder was used earlier as a classical coating material. A possible reason for this is the relative difficulty of identifying the spores in light microscopy.

Talc   Talcum powder is not used on most condoms anymore because it has been found responsible for the formation of granulomes [30]. Talc is comprised of anhydrous magnesium silicate that is widely distributed in various powder products. As such, it bears no significance as a "condom marker," Douglas et al. [30] showed that talc particles could be differentiated from starch granules in condom rinse solutions via scanning electron microscopy. There is no publication, however, that evidence of talc has been used to verify condom contact.

Silica   Silica is composed primarily of silicon dioxide and is only rarely used to powder condoms. It is commonly used in pharmaceuticals and cosmetics as a stabilizer. Reports on the forensic significance of silica residues do not exist.

Lubricant Coatings

Lubricants   Due to their particular qualities, silicone composites are found in numerous spheres of life. PDMS—used with priority in condom coating and being quite biologically tolerable for humans—is also used extensively in implantations, the outer coating of pills, and as a basis for salves. As a source of PDMS in genital and anal swabs, only lubricants suitable for sexual use can be considered in practice. PDMS evidence therefore has a high proof rating in cases of established condom contact when past history can exclude unrelated condom lubricants and/or the application of special salves to the skin.

In practice, the PDMS analysis involves a great deal of work since it does not belong to the routine examinations of a forensic lab. The methods of both Blackledge and Vincenti [9] and Keil et al. [20] promise success, especially since the procedure of Keil and colleagues is easier to carry out. Moreover, only average lab equipment in the form of the GC/MS is necessary. All the same, Blackledge and Vincenti were able to gather masses of up to 15,000 Da with the DCI/MS technique. In the GC/MS analysis, presentation ceases at about 500 Da, whereby usually only a portion of the PDMS mixture in application can be detected. In both techniques a sort of fingerprinting of the lubricant coating is established. The results can be compared with the contents of a data bank and associated with a specific type of condom [20].

The postcoital time frame for provability in vaginal swabs has so far been set at a maximum of 24 hours. However, in a limited controlled study involving different condom brands provided to volunteer couples, Belcher [31] used FTIR and in some cases was able to identify PDMS traces up to 48 hours postcoitus. However, swabs should be obtained from victims (and suspects) as soon as possible. PDMS is not degradable in the body. The short time span of provability can be justified through the sparse concentrations in vaginal secretion that are further reduced through natural vaginal discharge. Water-soluble lubricants were traceable in vaginal swabs for an expectedly shorter period of time postcoitus, that is, for just up to 8 hours [32].

Blackledge [33] as well as Lee et al. [1] analyzed lubricants more closely in view of fingerprinting. For this, the original substances in the lubricant coatings or the wash liquids of unused condoms were examined. Blackledge could differentiate between ten condom types on the U.S. market based on the divergent viscosity of the PDMS. For this he applied Fourier self-deconvolution (FSD) to the FTIR spectra of the coatings. Lee and colleagues constructed nuclear magnetic resonance (NMR) spectra by which all soluble organic substances on the condom surface were documented. As a result of this, numerous types of condoms could be identified. The methods were not applied to swabs in real cases, however. Blackledge rightly proved that such results are only valuable when the viscosity of PDMS does not change upon contact with body fluids. It must also be ascertained that cosmetics containing PDMS or lubricants unassociated with condoms give evidence of another composition of silicones than that on the condom coating.

Spermicides Nonoxynol-9 is a detergent that has a spermicidal effect. It is very commonly used in creams, gels, capsules, and foams as a form of contraception. Nonoxynol-9 is a lubricant component in only rare types of condoms. Evidence of this substance can only be considered as proof of condom contact when the origin has been ruled out by other preparations. Nonoxynol-9 has lost significance as a component in condom coatings because it can lead to irritations of the vaginal membrane [34]. As a result, HIV infections become more likely. Large condom manufacturers have already stopped adding nonoxynol-9 to the lubricant coating.

The methods of analyzing nonoxynol-9 in vaginal swab rinse solutions as proposed by Blackledge and Vincenti [9] led to unequivocal results. Of course, the methods (FTIR, DCI/MS) were very demanding and relatively time consuming. The extraction process took place in steps because the nonoxynol-9 spectrum could be masked by the spectrum of PDMS. Since the spermicide has a considerably lower molecular weight than that of the PDMS, there were no interferences of the mass spectra. In separating nonoxynol-9 from the condom's water-soluble lubricants, difficulties arose in that the FTIR spectra of the spermicide could not be unmistakably differentiated from those of other components in the lubricant (or from common detergents).

The serviceability of techniques applied to swab material has not been comprehensively tested probably because the necessary technical equipment is not available in most forensic labs [24]. By using FTIR, nonoxynol-9 originating from certain condom brands has been detected on swabs provided by volunteer couples as long as 72 hours postcoitus [31].

4.4 CASE STUDIES

So far, the only cases published were those limited to the question of whether condom use had even taken place [8, 9].

Case 1: From Blackledge and Vincenti [9] A female attending military training spent an evening with classmates. Alcohol was consumed. It was claimed she was raped by a classmate while she was unconscious. The man presumably used a lubricated condom that he showed his pals after the event.

As the victim regained consciousness she realized that her trousers and panties had been pulled down. The woman was medically examined and a vaginal swab was obtained. It is not known how much time passed between the assault and the swab preparation.

The suspect confessed and admitted to using a certain type of condom (Sheik Elite®). It was arranged to examine the vaginal swab for condom traces in order to verify the incident. The FTIR analysis of the vaginal swab extraction proved positive for PDMS. A DCI mass spectrum showed a typical sequence of peaks 74 daltons apart and each corresponding to the molecular weight of the sodiated PDMS ethoximer. Cornstarch particles were found under microscopic examination that occur on the surface of Sheik Elite condoms. This went to prove that vaginal penetration with a condom had taken place.

Case 2: From Blackledge and Vincenti [9]   A 17-year-old claimed to have been raped while she was using the bathroom at home. Friends and acquaintances were present in the apartment. The suspect was an HIV-positive member of the Armed Forces to whom a particular sort of condom (Prime®) was easily accessible. During a medical examination of the victim a vaginal swab was prepared. No information was given about the time of the swab preparation in relation to the time of the assault. The suspect stated that intercourse was a result of mutual compliance. He used a brand name condom (Prime) that he claimed had torn. The victim contradicted this by stating that no condom had been used during the violent sexual intercourse.

No PDMS could be proved by either FTIR or DCI/MS analysis of the vaginal swab. No cornstarch granules could be found microscopically even though the claimed condom type was coated in cornstarch. The examiners had previously ascertained through examination of a vaginal swab provided by a volunteer couple who had engaged in sex using that condom brand and simulating condom breakage, that the presence of seminal fluid in the swab extracts did not interfere with identification of PDMS by either FTIR or DCI/MS, nor did it interfere with the identification of cornstarch via PLM.

Case 3: From Keil et al. [8]   A woman was found dead in her apartment in 1992. She bled to death as a result of stab wounds and cuts on the neck. Sperm was found in the woman's mouth. The offender was discovered 9 years later through DNA analysis. He stated that he had had a secretive love affair with this woman who lived in his neighborhood. On the day in question, the woman first consented to vaginal sexual intercourse. He had used a "normal dry condom" for this. Since neither he nor the woman had reached orgasm, she proceeded to perform oral sex on him. This led to his ejaculation. The woman had made remonstrances and offended him. It escalated into an argument and he stabbed the woman out of rage. He threw away the condom after the deed. He did not remember its brand name. The sperm-negative vaginal swab had been preserved up to the arrest of the perpetrator, namely, for 9 years.

In the microscope slides of the vaginal swabs, numerous cornstarch granules could be detected (Figure 4.12).

In the chemical analysis, no silicone compositions could be established in the vaginal swab extract. The examination results did not contradict the statements made by the perpetrator. Shortly before the victim's death, he had had vaginal intercourse with her using a dry condom. Based on the analyses, the court was swayed by the defendant's portrayal. This version of the deed brought the defendant a reduction in the level of criminal charges.

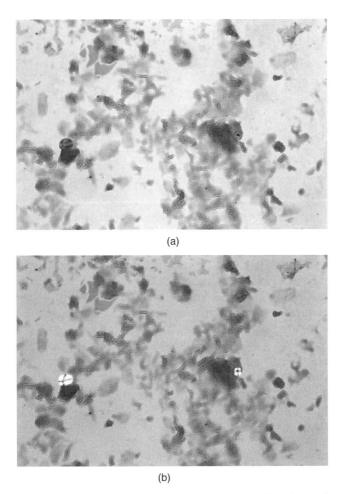

(a)

(b)

Figure 4.12 *Cornstarch granules in a 9-year-old, air-dried vaginal swab stored at room temperature: (a) HE staining and (b) polarization microscopy.*

Case 4: From Blackledge (2005, Unpublished)   Early one morning in June 1993 an employee of a car wash opened for business. He needed to use the bathroom and found that the door was unlocked and something on the inside was preventing the door from swinging. He was able to open it enough to look inside. A naked woman was lying on the floor. She was dead. Several items of evidence were found on the floor, including the decedent's clothing, a discarded condom, a foil condom package wrapper (Figure 4.13), and a metal tube like that frequently used to snort cocaine.

Although the used condom contained seminal fluid, at that time they had no suspect. However, many years afterwards the DNA profile of the seminal fluid was found to match that of an individual in a data bank. Unfortunately, attempts to extract DNA from possible vaginal epithelial cells on the outside of the condom and compare them to the victim's DNA were unsuccessful. Could a possible association be shown between (1) the used condom and the

Figure 4.13   *Empty condom packet found at crime scene. (Courtesy of Blackledge.)*

foil condom package wrapper or (2) trace evidence associated with the used condom and any trace evidence found on vaginal swabs from the victim?

The foil condom package wrapper was for a LifeStyles® brand with non-oxynol-9. This brand is made by Ansell Incorporated (Dothan, Alabama, USA). It contains a PDMS lubricant, the nonoxynol-9 spermicide, and corn-starch. The used condom was compared with a standard of the same brand and they compared in terms of dimensions, shape, and color. The used condom, the inside of the condom packet found at the crime scene, and vaginal swabs and microscope slides (obtained years earlier at autopsy) all contained numer-ous cornstarch grains. Using FTIR, PDMS was identified on all three as well. Using matrix-assisted time-of-flight mass spectrometry (MALDI-ToF MS) in the reflectron mode, nonoxynol-9 was found in the used condom and in the condom standard, but its presence on the vaginal swabs could not be proved. However, by using capillary column gas chromatography/mass spectrometry (GC/MS) methanol extracts of the used condom, the condom packet found at the scene, and vaginal swabs from the victim all showed the presence of butyl-ated hydroxytoluene (BHT), an antioxidant (Figure 4.14).

Representatives of Ansell not only confirmed that back in 1993 BHT had been part of Ansell's formulation, they were also sure that the other two major condom manufacturers (the makers of the Trojan® and Durex® brands) did not use it. In addition, based on the lot number visible on the condom packet found at the crime scene (Figure 4.13), Ansell checked its sales records and as able to state how many condoms had been produced in that lot and that the entire lot had been purchased by one customer (a municipality), and the date they were shipped.

Case 5:   (Contributed by Chris Taylor, Trace Evidence Section, U.S. Army Criminal Investigation Laboratory, Ft. Gillem, GA, USA, 2006)   When a condom wrapper is found at a crime scene, the agent knows that it needs to be collected for fingerprints, a potential lubricant standard, or possibly even for DNA. What some investigators might not realize is a simple physical fit of a used condom wrapper to a condom wrapper in the possession of the subject can associate him with the victim and/or crime scene(s). More commonly

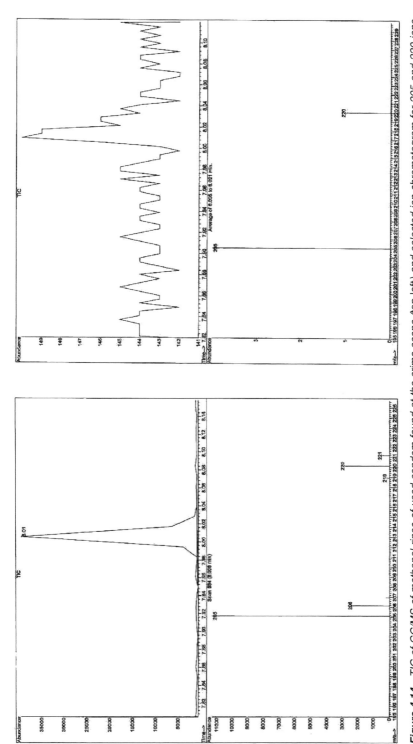

Figure 4.14 TIC of GC/MS of methanol rinse of used condom found at the crime scene (top left) and selected ion chromatogram for 205 and 220 ions for methanol rinse from a vaginal swab from the victim (top right). Bottom left: Full MS spectrum for peak at 8.01 min from GC/MS of methanol rinse from used condom (matches the antioxidant, butylated hydroxytoluene (BHT)). Bottom right: Selected ion monitoring (SIM) spectrum for 205 and 220 ions for peak at 8.01 from GC/MS of methanol rinse from a vaginal swab from the victim.

during a sexual assault, a condom wrapper may be left at the scene or discovered in the trash at the victim's residence. Because these wrappers are typically connected together in a series, upon separation of a single condom packet from its source, the wrapper can be examined at the perforation(s) or tear(s) to establish a physical fit.

Examination of condom wrappers from several manufacturing sources has revealed the sides of the wrappers shift slightly, top to bottom and side to side when being sealed around prophylactics during production. When the perforation is stamped into and between each prophylactic, the serrated perforations vary from wrapper to wrapper. This allows sufficient variability to disclose a physical fit of the separation and the offset patterns in the printing to conclude the wrappers were once a single component. This is demonstrated in Figures 4.15 and 4.16. Several cases at the USACIL have been successful in

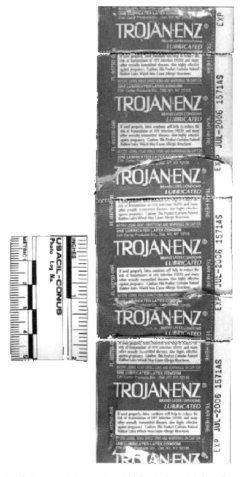

Figure 4.15   *Three condom wrappers recovered from the vicinity of the crime scene fit each other and ultimately fit the torn edge of a condom packet recovered from the suspect.*

Figure 4.16 *Close up of the physical fit of one wrapper from the scene to the wrapper in the subject's dresser. (Note the "i" and "ll" in "will" and the "h" and "l" in "help" and how the letters fit together across the separation.)*

establishing a common link between the subject and scene through the examination of condom wrappers. Upon recovering a condom wrapper from the scene, the investigator should be aware that the subject may be in possession of a critical piece of evidence that would link him to the crime. In this case, three condoms were used during the assault. Three wrappers were left and recovered near the scene. The three wrappers physically fit each other and ultimately fit a condom wrapper from the subject's top dresser drawer (Figures 4.15 and 4.16).

REFERENCES

1. Lee GSH, Brinch KM, Kannangara K, Dawson M, Wilson MA (2001). A methodology based on NMR spectroscopy for the forensic analysis of condoms. *Journal of Forensic Sciences* **46**:808–821.

2 Berkefeld K (1993). Eine Nachweismöglichkeit für Kondombenutzung bei Sexualdelikten. *Archiv für Kriminologie* **192**:37–42.

3. Blackledge RD, Cabiness LR (1983). Examination for petrolatum based lubricants in evidence from rapes and sodomies. *Journal of Forensic Sciences* **28**:451–462.

4. Yersin U, Weber M, Felber P, Maurer P, Staub R (2000). *Das Präservativ-Handbuch.* Partner & Partner AG, Winterthur, pp 6–10. See also www.bildungundgesundheit. ch/dyn/bin/82822-82707-3-manuald.pdf.

5. Maynard P, Allwell K, Roux C, Dawson M, Royds D (2001). A protocol for the forensic analysis of condom and personal lubricants found in sexual cases. *Forensic Science International* **124**:140–156.

6. Conti S, Dezzi S, Bianco A (1995). Traces of polydimethylsiloxane in case histories of rape: technique for detection. *Forensic Science International* **76**:121–128.

7. Kondomverbrauch (2005). *Kondomwelt/Zahlen & Fakten.* http://www.kondomberater.com/kondomwelt/zahlen.php?.

8. Keil W, Rolf B, Sachs H (2004). New strategies for the detection of condom residues in cases of sexual assault. Invited lecture, at PITTCON, Chicago, March 2004. See also Abstracts PITTCON, CD-ROM (Paper 7900-300).

9. Blackledge RD, Vincenti M (1994). Identification of polydimethylsiloxane lubricant traces from latex condoms in cases of sexual assault. *Journal of the Forensic Science Society* **34**:245–256.

10. Condomi AG (2004). Erfurt, Germany, personal communication.

11. Kondom Materialien. Available at http://www.lexikon-definition.de/Kondom.html. Accessed 2005.

12. Weiss J (1997). Latex allergy. Allergy Program Manager, Diagnostic Products Corporation, DPC Technical Report ZB 125-B, Los Angeles, CA, USA. See also http://www.dpc-biermann.de/allergie/immu2000_5.htm.

13. Kondom. Nitrosamine in Kondomen. Available at http://www.lexikon-definition.de/Kondom.html. Accessed 2005.

14. Firma Ritex (2004). Bielefeld, Germany, personal communication.

15. Kondom-eshop. Bunt & Duft. Available at https://kondom-eshop.de/search.jsp. Accessed 2005.

16. Durex Performa. Condoms for longer lasting pleasure. Available at http://www.durex.com/de/durexHistory.asp?intHistoryStep=226. Accessed 2005.

17. Janda K, Wentworth P (1997). Antibody-mediated latex protein recognition. Application for the National Institute of Justice, Washington, DC, December 15, cited from Blackledge [18].

18. Blackledge RD (2003). Condom trace evidence: the overlooked traces. *Forensic Nurse* **Jan/Feb**:36–38.

19. Stahl E (1962). *Lehrbuch der Pharmakognosie.* Gustav Fischer Verlag, Stuttgart, Germany pp. 44–45 and pp. 534–535.

20. Keil W, Rolf B, Sachs H (2003). Detection of condom residues by microscopic and GC/MS examination. Poster presentation (B92), at the Annual Meeting of the American Academy of Forensic Sciences, Chicago, IL, USA, February 2003. See also *Proceedings of the American Academy of Forensic Sciences* **9**:69–70.

21. Keil W, Kutschka G, Sachs H (1997). Zum Nachweis von Kondomrückständen in Vaginal- und Penisabstrichen. *Kriminalistik* **51**:439–440.

22. Cremer U (1994). Zum Nachweis von Präservativrückständen in Vaginalabstrichen. *Zentralblatt Rechtsmedizin* **42**:425.

23. Burger F, Doble P, Roux C, Kirkbride P (2003). Forensic analysis of condom and personal lubricants found in sexual assault cases by capillary electrophoresis. Poster presentation (paper MED-FP-27), at the 3rd European Academy of Forensic Science Meeting, Istanbul, Turkey, September 2003. See also Proceedings of the 3rd European Academy of Forensic Science Meeting, *Forensic Science International* **136**(Suppl. 1):247.

24. Hollenbeck TPE, Siuzdak G, Blackledge RD (1999). Electrospray and MALDI mass spectrometry in the identification of spermicides in criminal investigations. *Journal of Forensic Sciences* **44**:783–788.

25. Bommarito CR, Dougherty E (2004). Analysis of nonoxynol-9 in condom lubricants via pyrolysis gas chromatography–mass spectrometry (PGC-MS). Lecture (B121), at the Annual Meeting of the American Academy of Forensic Sciences, Dallas, Texas, USA, February 2004. See also *Proceedings of the American Academy of Forensic Sciences* **10**:83.

26. Blackledge RD (1994). Collection and identification guidelines for traces from latex condoms in sexual assault cases. *Crime Laboratory Digest* **21**:57–61.

27. Trochta AC (2004). Detection of latex condom proteins implicated in sexual assault cases: a feasibility study. Master of Forensic Sciences Thesis, National University, San Diego, CA, USA.

28. Bundesinstitut für Arzneimittel und Medizinprodukte. Nitrosamine in Kondomen. Available at http://www.calsky.com/lexikon/de/txt/k/ko/kondom.php. Accessed 2005.

29. Balick MJ, Beitel JM (1989). Lycopodium spores used in condom manufacture: associated health hazards. *Economic Botany* **43**:373–377.

30. Douglas A, Karov J, Daka J, Hinberg I (1998). Detection and quantitation of talc on latex condoms. *Contraception* **58**:153–155.

31. Belcher KL (2000). Evidentiary value of condoms: comparison of durable physical and chemical characteristics of condoms. Master of Science Thesis (Biology), University of North Texas, USA.

32. Greaves C (1998). Safe sex—is it so safe? *Contact-FSS Internal Publication* **25**:20–21. Cited from: Rogers D, Newton M (2000). Sexual assault examination. In: *A Physician's Guide to Clinical Forensic Medicine*, Stark MM (ed.). Humana Press, Totowa, NJ, pp. 82–83.

33. Blackledge RD (1995). Viscosity comparisons of polydimethylsiloxane lubricants in latex condom brands via Fourier self-deconvolution of their FT-IR spectra. *Journal of Forensic Sciences* **40**:467–469.

34. Leitlinien der Deutschen Gesellschaft für Gynäkologie und Geburtshilfe (2004). Empfängnisverhütung. Available at http://www.uni-duesseldorf.de/WWW/AWMF/ll/015-015.htm. Accessed 2004.

5

Latent Invisible Trace Evidence: Chemical Detection Strategies

Gabor Patonay

Department of Chemistry, Georgia State University, Atlanta, Georgia

Brian Eckenrode

Research Chemist, FBI Laboratory, Counterterrorism and Forensic Science Research Unit, FBI Academy, Quantico, Virginia

James John Krutak

Senior Scientist, FBI Engineering Research Facility, Operational Technology Division, Technical Operations Section, Quantico Virginia

Jozef Salon

Department of Chemistry, Georgia State University, Atlanta, Georgia

Lucjan Strekowski

Department of Chemistry, Georgia State University, Atlanta, Georgia

5.1 INTRODUCTION

Highly sensitive methods for the detection of latent trace evidence are of immense interest in the forensic science community. Due to the fact that trace evidence is often deliberately or inadvertently washed away or otherwise

Forensic Analysis on the Cutting Edge: New Methods for Trace Evidence Analysis, Edited by Robert D. Blackledge.
Copyright © 2007 John Wiley & Sons, Inc.

destroyed at crime scenes, detection of remaining latent evidence residues by sight or by using simple optical detection devices alone is usually impossible. Consequently, it is necessary to have sensitive methods for the detection of latent invisible trace evidence. Most of these methods employ some chemical or biological procedure to enhance the visibility of the trace evidence. The enhancement techniques are generally based on the principle of developing a reaction on the surfaces to be interrogated to observe changes that are visible preferably to the naked eye or via use of a simple optical device that can assist the observation process. Figure 5.1 illustrates the general procedure to visualize latent evidence. Since most often the location or even the existence of the latent evidence is not known, these tests are called presumptive tests. Any kind of presumptive test can be described with the procedure illustrated in Figure 5.1. Basically, presumptive testing can be summarized as an analytical search methodology that assumes the possible presence of trace evidence but without a priori knowledge of its location. Chemical, physical, or other methods need to be applied to the surface containing trace evidence to visualize or otherwise make it observable.

Different methodologies can be utilized to visualize trace evidence. Table 5.1 shows examples of possible chemical methods to visualize trace evidence analytes or components during presumptive testing.

This chapter will focus primarily on research performed at the FBI's Laboratory and Operational Technology Divisions and Georgia State University. Besides more recent developments, this chapter will discuss some methods already in use or published without presenting an exhaustive review. Additionally, this chapter will focus on only chemical methods. While physical methods are as important as chemical methods, they are very different from each other,

o o o o o o o o o o o o	X X X X X X X X X X X X
Surface *before* interrogation by reagent	Surface *after* application of reagent
o = Invisible trace evidence (e.g., nonfluorescent)	x = visualized trace evidence (e.g., x = fluorescence viewed or imaged with optical instruments)

Figure 5.1 *General procedure for presumptive testing.*

TABLE 5.1   Reactions for Trace Evidence Visualization and Imaging

Reactions	for	Analytes
Protonation		Pepper spray
Lewis acid complexation		Gunshot residues
Oxidation		Latent blood
		Latent prints
Hydrophobic bonding and protein bonding		Latent prints
Metal ion complexation		Specific metals (Ca^{2+}, Na^+)

and it would be difficult to discuss both strategies in the same chapter. It is important to note that although this chapter focuses on two types of latent evidence during the detailed discussions (latent blood and pepper spray residues), most of the general ideas are applicable to a much more diverse group of latent evidence visualization strategies using the chemical presumptive testing methods shown in Table 5.1.

5.2 LATENT BLOODSTAIN DETECTION

One of the most commonly used reagents for presumptive latent bloodstain detection is the chemiluminescent compound luminol (5-amino-2,3-dihydro-1,4-phthalazinedione) [1–7]. Luminol has been shown to have high sensitivity toward bloodstains and does not impair subsequent DNA analysis [2, 4, 5, 7]. However, luminol does have disadvantages: its fluorescence must be observed in a dark environment and the observation window for the reaction (i.e., the time before observation becomes impractical) is extremely short, often on the order of a few seconds. This is true for the more enhanced and much advertised BlueStar® luminol latent bloodstain reagent. This short reaction time makes photography difficult and may necessitate the use of multiple treatments that can degrade bloodstain patterns. Additionally, luminol has been classified as a possible carcinogen. For these reasons there has been considerable effort to develop alternative reagents for the detection of latent bloodstains. Some of these reagents have advantages over luminol in both the length of observation window as well as safety issues.

Fluorescein has been used in presumptive testing of latent bloodstains for several years. The major disadvantages of the early fluorescein test are that the chemistry requires significant on-site preparations (which is time consuming as well and may require special skills) and there is quickly diminishing contrast due to increasing background fluorescence. Although the time window is much longer than that of the luminol test, it is still only a few minutes and that makes photography challenging. Fluorescein and its derivatives are perhaps the most popular fluorescent dyes used in clinical diagnostics, due to their favorable photophysical properties (i.e., high quantum yields and extinction coefficients). These optical properties alone make fluorescein desirable for presumptive tests. In the past, the use of fluorescin (a reduced form of fluorescein) has been suggested [3] as an alternative to luminol. The reduced form of fluorescein exhibits negligible fluorescence. Fluorescin is oxidized to fluorescein in the presence of blood or in the presence of other reagents that oxidize or promote oxidation of the reduced form. This conversion significantly enhances fluorescence and, as a result, the fluorescence intensity is increased in the presence of blood. Hydrogen peroxide may be used to accelerate the oxidation and increase the intensity of the signal.

While using reduced fluorecein in presumptive latent bloodstain tests offers advantages over the use of luminol, it is not without disadvantages. One of the most important issues affecting latent bloodstain tests is the available time for observation and photography. To understand this problem, and to compare all previously developed methods [8], a reaction time can be defined as the time period when the bloodstain fluorescence is observable before background fluorescence develops (T_{react}). This time is often called the open-time window, for example, the time window available for the investigator to collect and photograph evidence. For the aqueous fluorescein formulation [3, 9] this time period is relatively short and is limited to several seconds or at best a few minutes. In addition, the stability of the reagent prepared by reduction with zinc in 10% sodium hydroxide as described by Cheeseman [9] is limited so that a short shelf life of the reagent is typical even when stored under refrigeration after the dry chemicals are dissolved and reduced. To alleviate the problems associated with the aqueous fluorescin reagent, a new chemistry has been developed [10].

This new chemistry was developed by KPS Technologies, LLC [10]. The advantages of the new fluorescin chemistry, as compared to the previously developed chemistries [11–20], are as follows: (1) the open time is greatly extended, (2) the shelf-life of the reagents is greatly increased, and (3) the light source is portable and permits recording of improved contrast images. In addition, use of the new reagent helps to develop evidence such as sweat and oily fingerprints at a convenient time after development of the latent blood evidence. This dual functionality of the new reagent was unexpected but is very useful in forensic crime scene examinations. The development time difference between initial visualization of bloodstains and latent fingerprint spots makes differentiation simpler.

This new fluorescin chemistry affords not just more sensitive detection of latent bloodstains but the open-time window is greatly extended. Earlier chemistries strictly focused on reducing fluorescein to its leuco form by using aqueous zinc reduction that had to be prepared on site. Such chemistry typically allows only a 30 second window to take pictures of the bloodstain. This short window creates severe difficulties and requires a great deal of preparation during the photography process. It is difficult to photograph large areas. Short exposure times are required to minimize any background fluorescence. Hence, there is a need for high powered forensic light sources. For example, one of the frequently used forensic light sources is an ISA system that utilizes a xenon arc lamp light source with an optical coupler and filter. This high intensity light source is expensive and bulky and raises portability problems. A portable power generator needs to be carried to the crime scene if the area of evidentiary interest is located in remote areas.

Fluorescein is known to exist in two forms. The lactone **1** (Scheme 5.1) is stable under low pH conditions and the open form **1′** is present in the basic solution. Leuco fluorescin **2** is prepared by a novel reduction process that requires two steps. Although a number of classical reduction methods can be used, it is important that the first step is conducted under slightly basic conditions and the subsequent final preparation of **2** is performed in the presence of acetic acid. It should be noted that the proposed structure of fluorescin **2** thus obtained has not been established unambiguously. Structure studies are currently in process in our laboratories.

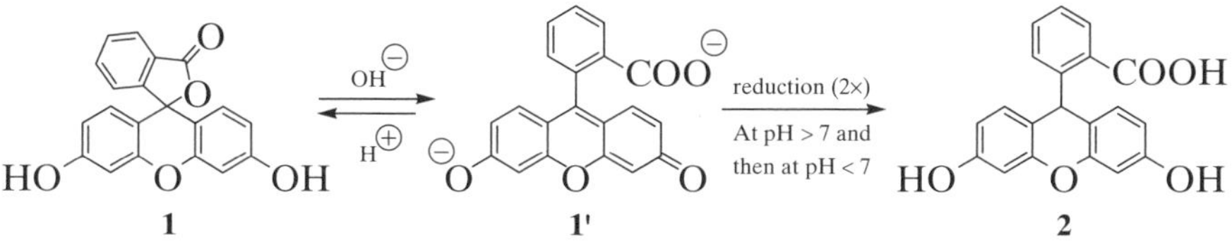

Scheme 5.1

Spectroscopic data has shown that after the first step described above, especially after drying the material under atmospheric pressure, there still exists a small percentage of fluorescein. Even if only 1% of the product obtained above is still in the oxidized state, this is enough to contribute significantly to background fluorescence. It is therefore necessary to carry out the second reduction step. The acidic solution of the final reagent **2** is stable indefinitely if stored in the dark under an inert atmosphere (e.g., nitrogen or argon).

The methodology developed represents a significant improvement over previous methods in the time and effort required by the technician for reagent preparation. The reduction procedure is new and critical to the success of the assay [10]. The presence of trace amounts of fluorescein causes the appearance of background fluorescence that significantly decreases the digital imaging contrast and overall utility of the procedure. This is the primary reason why the two-step reduction process is critical to obtain reagent **2** of high purity.

Latent bloodstains and fingerprints are detected by a relatively straightforward procedure. The area of interest is first illuminated with the light source to test for intrinsically fluorescent material that can potentially lead to a false positive result. The working solution described above is uniformly sprayed onto a surface suspected of having latent bloodstains. Immediately following the treatment with the working solution, the surface is sprayed with a 15% hydrogen peroxide solution. Both solutions are applied as a fine mist via aerosol propellant. The area of interest is then illuminated with a forensic light source. If bloodstains are present in the area of interest, fluorescin **2** is converted to fluorescein and a fluorescent spot indicating the presence of blood is observed. After 10–20 minutes on most surfaces, any sweat or oily fingerprints deposited in the area sprayed also develop. The chemical mechanisms responsible for this are currently under investigation. The excitation light is filtered by the use of appropriate wavelength filters. Documentation is usually performed by photographic means.

Two factors affect the utility of the described method as a means of detecting latent bloodstains. First, it has been observed that bloodstains and fingerprints on surfaces that are the same color as fluorescein emission are difficult to detect, due to the fact that the emission is difficult to discern or record versus background color. Light colored wood is a potentially problematic matrix. Although it is possible to detect blood and fingerprints on this matrix, the emission is not as distinct as it is on darker matrices. However, these problems can be overcome through the use of a different leuco dye system whose emission wavelength is distinctly different from that of the background matrix. Second, the composition (i.e., porosity) of the matrix has been observed to affect bloodstain detection. While porous substrates are more difficult to clean, reducing the likelihood that all blood present is removed, detection of latent bloodstains is still more difficult on porous substrates than on nonporous surfaces.

The use of an ethanol solution containing acetic acid improves the duration of signal, that is, the time before the signal is lost in the background. The acidic environment enhances stability of the reduced form of the dye to aerial oxidation, prolonging the time before the background fluorescence appears. The nonaqueous solvent increases the quantum yield of the developed dye, thereby increasing the fluorescence intensity of the dye. Through the use of the acidic ethanol environment, the background is minimized while the fluorescence of the dye, when in contact with blood, is enhanced. The composition of the working solution allows for at least an order of magnitude improvement in the signal duration time in comparison to that of the method based on luminol chemiluminescence. A second advantage of the ethanol-based formulation is that fast evaporation of ethanol makes it less susceptible to smearing or running relative to aqueous solutions, when applied on vertical surfaces. Importantly, bloodstain patterns and fingerprints are not degraded by the method. A second leuco dye system was also prepared by reduction of rhodamine 6G and found to work well for the detection of latent bloodstains (vide infra) on light colored wood.

The results of the blood detection and fingerprint experiments are illustrated in Figure 5.2. Figure 5.2a is an exposure taken 5 minutes after treating a bloodstain (1:100 dilution with pure water) deposited on white paper with the new fluorescin product. Illumination was accomplished through use of a forensic light source (commercial Crime Scene Scope) and an LED lamp. After 5 minutes no background fluorescence is present, only the specular reflection is visible. Figure 5.2b was taken under the same conditions as Figure 5.2a, with the exception of the light source. The light source used for Figure 5.2b was a 20 blue LED light source. The spot intensities are comparable in both Figures 5.2a and 5.2b. Figures 5.2c and 5.2d are exposures of the same bloodstain taken after 10 minutes with the forensic light source and the LED light source, respectively. Again, very little background is present. Figures 5.2e and 5.2f are exposures of the bloodstain taken after 3 hours with the forensic light source and the LED source, respectively. Some background has appeared, as can be seen through comparison between Figures 5.2e and 5.2a. However, the latent blood spots are still clearly visible. It was observed that several of

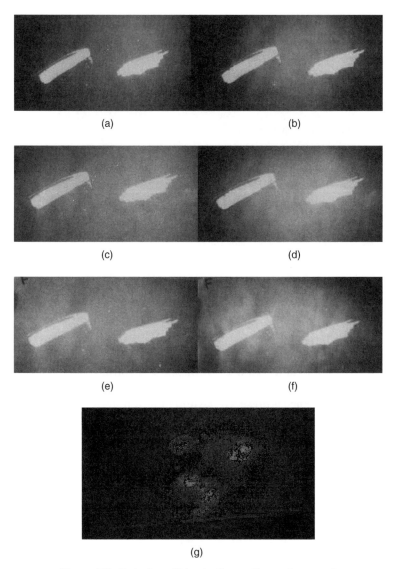

(a)

(b)

(c)

(d)

(e)

(f)

(g)

Figure 5.2 *Detection of blood with new fluorescin reagent.*

the treated latent bloodstains were clearly observable even months after spraying with leuco fluorescein. Similar results were obtained after treating the surface with a reduced rhodamine 6G dye. While the rhodamine dye did not perform as well as fluorescein regarding background suppression; on the substrates selected, the blood spots were clearly visible. If the blood spots are deposited on a light yellow wood or fabric substrate, the rhodamine dye reagent yields enhanced contrast over the fluorescein dye. Figure 5.2g is a digital image of the fluorescence observed from a fingerprint that developed

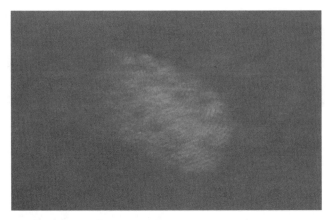

Figure 5.3 *Latent (nonbloody) fingerprint details 15 minutes after treatment with fluorescin using an MCS-400 MiniCrimeScope light source.*

about 15 minutes after the initial detection of bloodstains using the fluorescin reagent. The digital camera was operated in black/white mode so the fluorescence is revealed as a white on black contrast image instead of the yellow on black. By appropriate magnification (see Figure 5.3), it is possible to record detailed images of latent prints.

There are additional advantages of this new chemistry when compared to earlier methods [21–29]. Bloodstain patterns are fixed by this composition so that additional evidence is preserved. Another advantage of the ethanol-based working solution is that the low vapor pressure of ethanol makes it less susceptible to running and smearing of bloodstains and latent prints compared to aqueous solutions, when applied on vertical surfaces. Additionally, the presence of ethylene glycol in the solution improves adhesion to nonporous surfaces. The use of polymeric leuco dye compositions further improves adhesion especially when used on vertical surfaces. Importantly, bloodstain patterns and fingerprints are not degraded by the method [30, 31]. The chemicals for processing trace evidence scenes in this manner are available in a kit form that is extremely simple to use. No knowledge of the chemical reactions occurring is required of the technicians. This is an important criterion since not all persons involved in trace evidence determinations will have a working knowledge of chemistry. The reduced form of the dye is provided as a ready-to-use formulation. Another very important advantage is that the organic formulation also allows for fluorescin to penetrate latex or similar paints. In the event that the evidence was washed and covered by latex paint, use of the fluorescin reagent permits detection of latent bloodstains without removing or damaging the painted surface.

Latent blood can be detected on several different surfaces at very low concentrations. Figures 5.4 and 5.5 are illustrations for 1 : 1,000,000 dilution of blood on different materials.

Figure 5.4 *Leuco fluorescein detection of 1:1,000,000 dilution of blood on PVC floors.*

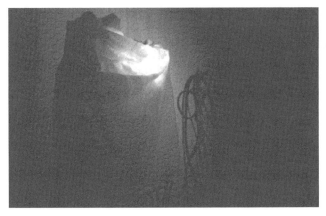

Figure 5.5 *Leuco fluorescein detection of 1:1,000,000 dilution of blood on garment (T-shirt).*

Method Overview Latent bloodstains or blood spatter patterns at crime scenes can be detected by using the following procedure:

1. The area of interest is illuminated with an LED light source or other forensic crime scene light source to test for intrinsically fluorescent material that can potentially lead to a false positive result.
2. The working solution is uniformly and lightly sprayed onto a surface suspected of having latent bloodstains or patterns. Immediately following treatment with the working solution, the surface is sprayed with a 15% hydrogen peroxide solution. Both solutions are applied as a fine

mist via aerosol spray bottle. Light spraying is important to preserve bloodstain patterns and retard oxidation, hence improving contrast and the open-time window.

3. The area of interest is then illuminated with the LED light source or other forensic crime scene light source using the appropriate filter for fluorescent dye excitation.

4. If bloodstains or blood patterns are present in the area of interest, fluorescin is converted to fluorescein and fluorescent spots or patterns indicating the location of blood are observed. The excitation light is filtered out through the use of appropriate wavelength filters (e.g., orange colored eyeglasses for visual observation or orange filters on the photographic or digital camera selected for recording the bloodstain pattern). Alternatively, under dark conditions, the fluorescent spots or patterns are observed by using a modified night vision pocket scope containing an LED illuminator of appropriate output wavelength excitation range (e.g., from 430 to 470 nm for fluorescein) and an appropriate interference filter placed in front of the pocket scope lens for the fluorescent dye emission (e.g., 500–530 nm for fluorescein). Other leuco dye reagents require LED light sources and interference filters optimized for the reagent.

5. Additional tests can follow (DNA, blood type, fingerprints, etc.). The reagent was shown to not interfere with the usual DNA analytical protocols [32].

Sulforhodamine B, λ_{max} = 554 nm

In certain cases background fluorescence may interfere with fluorescein fluorescence. This can happen if the surface contains highly fluorescent materials (e.g., carpet). One solution to the problem is to move to a longer wavelength part of the electromagnetic spectrum. This can be achieved using longer wavelength leuco dye (e.g., leuco rhodamine). Leuco rhodamine chemistry works very similarly to fluorecein.

A significant disadvantage with leuco rhodamine is that the background fluorescence develops faster than that of leuco fluorescein. Even so this method is superior to other methods [1–9]. The background development after 1 hour is illustrated in Figure 5.6.

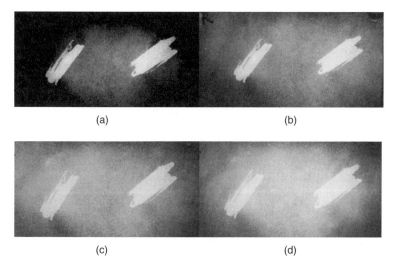

(a)   (b)

(c)   (d)

Figure 5.6 *Latent bloodstain 1 minute after leuco rhodamine treatment: (a) MiniCrimeScope and (b) blue LED illumination. Latent bloodstain 60 minutes after leuco rhodamine treatment (c) MiniCrimeScope and (d) blue LED illumination.*

5.3 FINGERPRINT DETECTION WITH NEAR-INFRARED DYES

A new method to detect latent prints directly utilizes near-infrared (NIR) fluorescent dyes [33, 34]. To understand the potential advantages of using NIR dyes for detection of trace evidence, it is important to first consider important properties of these dyes. A general structure of polymethine cyanine or carbocyanine dyes is shown below.

Cyanine dyes

They absorb light strongly with measured extinction coefficients in the range 200,000–300,000 $M^{-1}cm^{-1}$. Quantum yields of fluorescence are large and often exceed 40% with Stokes shifts in the range 30–50 nm. Due to these unique properties, carbocyanine dyes have found numerous applications. Many new uses are being developed or proposed for the carbocyanines ($n \geq 1$) that absorb and fluoresce at NIR wavelengths (>700 nm). Intense research in the area of NIR dyes is being stimulated by the fact that virtually all organic materials including biological tissues, natural products, and man-made products are penetrated by NIR radiation.

Since the NIR region (700–1200 nm) is inherently a region of low background light interference, it is well suited for analytical work using highly complex samples without any preseparation and for fast identification of materials that are tagged with invisible NIR tracers. The NIR cyanines that absorb and fluoresce in the region 700–800 nm have already found applications in protein labeling including immunoassays, DNA sequencing, electrophoretic determination of functional groups in polymers, determination of metal ions, and NIR photography, among others. The wavelength advantage of NIR dyes is best illustrated by Figure 5.7. Background fluorescence is minimal in that region. Trace evidence detection using NIR fluorescent dyes has been achieved by design and synthesis of novel near-infrared fluorescent (NIRF) dye systems and integration of NIR imaging systems for detection of these dyes when used to identify trace evidence. For this approach the following important factors in new analytical methods have to be considered: safety of reagents, sensitivity and selectivity, ease of application and implementation, open-time windows for observation and digital imaging, price, and nondestruction of analyte prior to other analytical methods for positive identification (e.g., DNA in blood, LC/MS of organic components in pepper spray).

Fingerprint detection has been quite successful using cyanoacrylate fuming methods. In many cases, however, reflection of illuminating light from the tested surface hinders contrast development and hence visual observation and digital imaging of super glued fingerprints. In such cases NIR dyes can enhance detection by moving the detection into the NIR range. A super glued fingerprint sprayed with an alcohol solution of an NIR fluorescent dye can easily be viewed with a night vision pocketscope modified to contain an appropriate filter and NIR excitation LED. NIR illumination of the sprayed surface can be achieved using a conventional MiniCrimeScope with an NIR interference filter and the white light settings. Figure 5.8 depicts an ISA/SPEX MiniCrimeScope and a homemade NIR LED light source.

Visualization is convenient using any night vision equipment. The investigator can observe directly through the eyepiece of the night vision scope, or by using an attached CCD camera, it can be observed in real time using a monitor

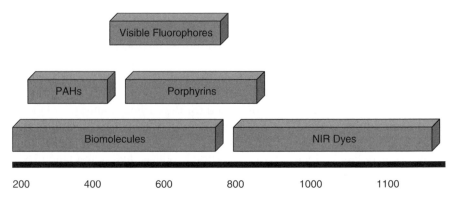

Figure 5.7 *Spectral regions of various molecules.*

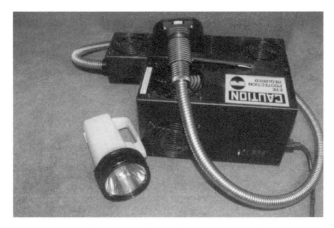

Figure 5.8 *ISA/SPEX MiniCrimeScope and NIR LED light source.*

Figure 5.9 *Night vision scope attached to a CCD camera.*

or a laptop computer. A night vision scope CCD camera setup is shown in Figure 5.9.

Figure 5.10 shows a digital image of a NIR fluorescent enhanced fingerprint developed on a white surface.

A new class of NIR fluorescent dyes, bis(carbocyanines), has been designed for direct fingerprint detection without any other enhancement techniques. These dyes form nonfluorescent closed clamshell-like structures when not hydrophobically bound to compounds present in oily and sweat prints. This property permits fingerprint detection because, as shown in Figure 5.11, the

Figure 5.10 *NIR fluorescence enhanced super glued fingerprint.*

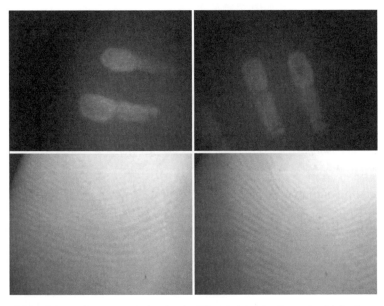

Figure 5.11 *Schematic representation (top) of fingerprint detection using bis(carbocyanines) and actual fingerprints detected using NIR fluorescence excitation and night vision goggles (bottom).*

open clamshell structure hydrophobically bound to fatty esters in oily prints or proteins in sweat prints is highly fluorescent.

As shown in Scheme 5.2 the bis(carbocyanine) molecule consists of two cyanines linked by a flexible polymethylene chain that can be varied in length. The abbreviation of a particular dimer is **BHmC** [**Bis**(**H**epta**m**ethine **C**yanine)] followed by the number of methylene units in the linker. Their synthesis involves three distinct steps. First, an indolium substrate is quaternized by the reaction with an α,ω-dibromoalkane to give a bis(indolium) salt **3**. Second, a mono(indolium) salt is allowed to react with an excess of the Vilsmeier–Haack reagent **4** to furnish a half-dye **5**. Dye **BHmC** is produced by condensation of **1** and **5** under standard conditions.

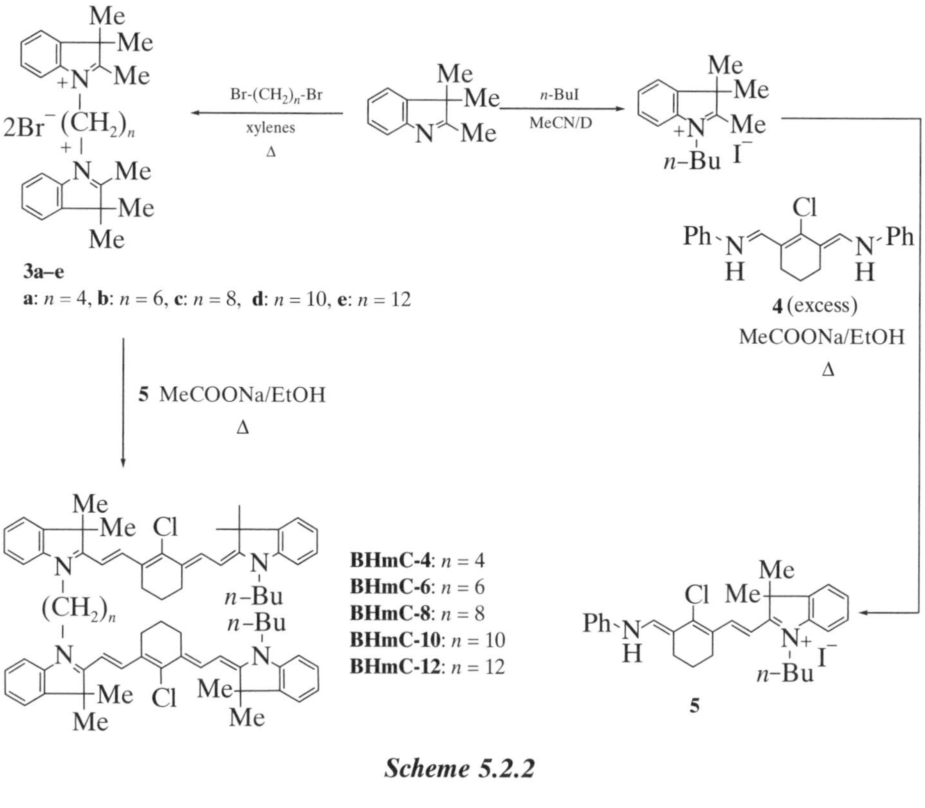

Scheme 5.2.2

An interesting application of **BHmC-12** reflects its ability to bind lower molecular weight biologically important molecules, such as fatty acids. One possible use of this phenomenon is latent fingerprint detection. The use of a monomeric counterpart shows no discernable interaction to latent fingerprints. By contrast, **BHmC-12** exhibits strongly enhanced fluorescence upon binding to fatty acids that are present in latent fingerprints and results in a clear fluorescence image of the latent fingerprint. Figure 5.11 illustrates this phenomenon.

5.4   PEPPER SPRAY DETECTION

The terms *capsaicinoid* and *oleoresin capsicum (OC)* refer to the family of molecular homologs and analogs from the *Capsicum* fruits having the basic structure shown below [35–40].

capsaicin: R = $CH_2NHCO(CH_2)_4CH{=}CHCH(CH_3)_2$
dihydrocapsaicin: R = $CH_2NHCO(CH_2)_6CH(CH_3)_2$

The most abundant members of this family are capsaicin and dihydrocapsaicin, which account for an estimated 80–95% of naturally occurring capsaicinoids [36, 37]. Subsequent references to capsaicinoids in this chapter will be to these two forms unless otherwise indicated.

Since the main component in pepper spray is capsaicin and this compound contains a phenolic group with the generalized structure shown above, this moiety can be exploited. Capsaicin can be reacted in at least two different ways for detection of the presence of pepper spray residues. One type of detection would directly utilize near-infrared (NIR) absorbing dyes for this purpose. To understand the reasons for using NIR dyes, it is important to first discuss the properties of these dyes.

Pepper sprays containing A, B, or analogs are readily available to both law enforcement personnel and the general public for a variety of uses, including riot control and self-defense. In those instances, the presence or absence of pepper spray on an evidentiary garment may help determine the facts of the incident. Previous laboratory analysis of pepper sprays at the FBI Laboratory relied on visually inspecting the garments for a colored dye, as found, for example, in CAP-STUN® (Zarc International, Inc., Gaithersburg, MD, USA) or examining the garments under a UV light source for pepper sprays that contained a UV fluorescent marker. For example, Fox Labs Pepper Sprays (Fox Labs International, Clinton Township, MI, USA) contain 2,5-bis(5′-*tert*-butyl-2-benzoxazol-2-yl)thiophene (CAS# 7128-64-5), a general use fluorescent optical brightener, to permit UV visualization of the product. Once located, suspect stains were cut from the garment, extracted with methanol, and analyzed for capsaicin and dihydrocapsaicin via gas chromatography/mass spectrometry (GC/MS).

There are now a number of products on the market that do not contain a colored or UV dye, for example, MK4 First Defense® pepper spray (Defense Technology Corporation of America, Casper, WY, USA). When no stains were visible on evidentiary garments submitted for pepper spray analysis, testing was limited to random sampling of areas of the garment for extraction and subsequent GC/MS analysis. Negative results from this type of blind sampling were always inconclusive.

5.4.1 Pepper Spray Detection Using Near-Infrared Fluorescent Dyes

The weak acidic nature of the phenolic moiety may be used for detection. Several NIR pH sensitive dyes have been synthesized that may be suitable for this purpose [39]. Due to the weak acidity of the phenolic moiety, only NIR pH sensitive dyes that show a keto–enol transition were evaluated. The absorption profiles for the cyanine NIR dye **6** is shown in Figure 5.12. The two forms of the dye, **6=O** and **6—OH**, are shown in Figure 5.13. The keto form **6=O** and the enol form **6—OH** show absorption at 541 nm and 706 nm, respectively.

This protonation detection approach for pepper spray was tested on difficult surfaces as well, for example, denim, black leather, and dark red artificial

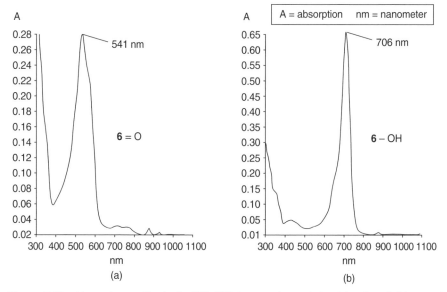

Figure 5.12 *Absorption profiles in the VIS–NIR for experimental dye **6** in ethanol: (a) neutral pH and (b) acidic pH.*

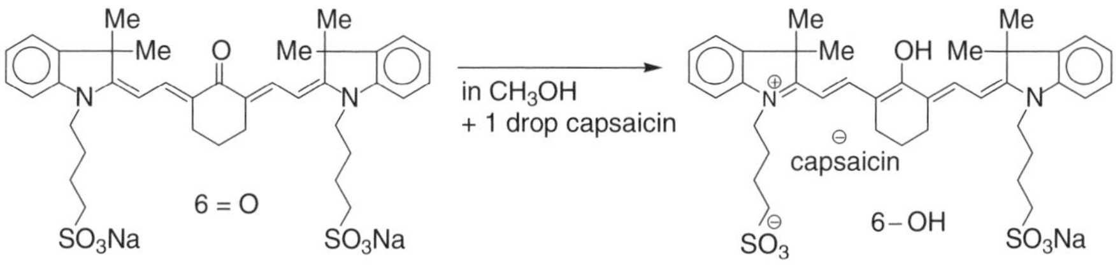

Figure 5.13 *Experimental NIR dye **6=O** in ethanol and its protonated form **6—OH** in the presence of capsaicin.*

Figure 5.14   *T-shirt (left) sprayed with MK4 pepper spray (right).*

leather. Pepper spray was applied and allowed to dry. After drying, a 1 mM solution of an NIR dye **6** in ethanol was sprayed as a fine mist (about $2 \, mL/ft^2$) and allowed to dry (a few seconds drying time). The surface was observed using a MCS-400 MiniCrimeScope. A 710 nm interference filter was inserted in the light source for excitation. An NIR night vision pocket scope was used for observation. The pocket scope was equipped with a double 800 nm cut-on filter. The area where pepper spray was present was clearly indicated by higher fluorescence intensity. The advantage of this approach is that the NIR dye does not chemically react with the pepper spray active content; hence, positive identification using mass spectrometry (MS) methods are not hampered. The investigator may collect swab samples under NIR light observation and then analyze these samples by liquid chromatography and mass spectrometry (LC/MS) for positive identification.

NIR fluorescence spectroscopy was shown to have advantages over visible fluorescence spectroscopy [33, 34]. Its main advantage is the virtual elimination of background interference. A new NIR reagent was developed for pepper spray detection using a keto–enol system protonation mechanism. The method was demonstrated on dark blue colored denim clothing. Dark colors pose no difficulty to this method. NIR fluorescence can conveniently be observed using NIR excitation sources, for example, MiniCrimeScope equipped with appropriate filter as well as NIR LED flashlights. NIR sensitive digital cameras and night vision equipment are suitable for observation and documentation of the NIR fluorescence signal. The major challenge in developing new NIR presumptive tests for crime fighting applications is the development of new chemistry that utilizes a turn-on mechanism in the NIR dye, and that can be quite challenging.

Figures 5.14–5.16 illustrate the method of detecting pepper spray using NIR dyes.

Figure 5.15   *T-shirt (left) with pepper spray after treatment with the NIR reagent (right).*

Figure 5.16   *Digital images of NIR fluorescence on a T-shirt (left) and on blue denim (right). A MiniCrimeScope with a 710 nm filter and a pocket scope with a 760 nm filter were used along with a Nikon D1X camera.*

The night vision scope arrangement is very similar to that depicted in Figure 5.9 except the filter is replaced according to the excitation and fluorescence wavelength of the dye (Figure 5.12). Since this wavelength is shorter than in the fingerprint application discussed earlier, a sensitive digital camera may be used instead of a night vision scope, such as a Nikon D1X (Figure 5.17).

5.4.2   Pepper Spray Detection Using Chemical Derivatization

A very different approach for pepper spray detection was described by Cavett et al. [35]. Others described pepper spray detection based on different chemistries [36–39]. A presumptive and confirmatory method, based on a chemical derivatization of capsaicinoids to azo dyes, has been developed for the visualization of colorless, ultraviolet (UV) dye-free pepper sprays on textiles. Identification of both the capsaicinoids and their derivatives can be confirmed via extraction of the derivatized capsaicinoids followed by liquid

153 mm (6.1")

157 mm (6.1")   85 mm (3.1")

Figure 5.17 *Nikon D1X camera.*

chromatography/mass spectrometry (LC/MS) analysis. This method is qualitative. Visible detection of as little as 50 μL of a 0.2% pepper spray (equivalent to ~0.1 mg) on a variety of garments was shown possible. The authors describe a presumptive method for the visualization of the capsaicinoids present in pepper spray and a rapid LC/MS method for analytical confirmation.

This method was developed primarily for rapid qualitative assessments followed by definitive MS confirmation. Limitations of the method include difficulty observing the colored derivative stain generated by the visualization procedure on certain fabrics, including those that have been treated for outdoor use (e.g., water-repellent garments). Experiments were conducted to test the visibility of the derivative by spotting various fabrics with measured amounts of the actual pepper spray product. Visualization was successful with amounts as small as 50 μL of MK4 (0.2% capsaicinoid content) on all of the following fabrics: 100% cotton (white, yellow, blue denim), 100% acrylic (neon pink), 100% Dacron polyester (multihued and patterned), and a blend of 50/50 cotton/polyester (patterned white) as illustrated in Figure 5.18. The size of the spot associated with 50 μL of MK4 First Defense pepper spray varied due to the chemical characteristics of the fabric but was generally approximately the size of a quarter.

After laundering, most of the bleached fabrics showed no indication of the presence of pepper spray; controls showed clear evidence of the violet stain produced on derivatization. Faint stains were produced on a few of the bleached fabrics, indicating that the reaction proceeds, provided that a small amount of the capsaicinoids remains on the fabric. Results of the visualization reaction performed after bleaching fabrics treated with pepper spray are summarized in Figure 5.18 and Table 5.2.

A false positive is generated if the violet color indicator develops on derivatization in the absence of capsaicinoids. Because the diazonium coupling reaction used in the visualization procedure is nonspecific, it is possible that

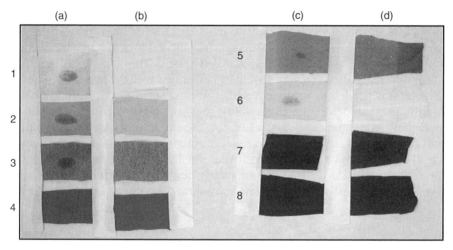

Figure 5.18 *(See color insert.) Fabric swatches spiked with 50μL MK4 brand pepper spray. Columns (a) and (c): swatch of each column after visualization reaction; columns (b) and (d): swatch of each column with no visualization reaction. (a1, b1) white cotton,(a2, b2) yellow cotton, (a3, b3) blue cotton denim, (a4, b4) pink acrylic; (c1, d1) multihued polyester, (c2, d2) 50/50 cotton/polyester, (c3, d3) brown leather, (c4, d4) red nylon.*

TABLE 5.2   Results of Visualization Reaction Performed after Bleaching Fabrics Previously Treated with Pepper Spray

Fiber Type	MFF Warp Stripe Fiber	Color Indicator After Bleaching
Synthetic fibers	Spun diacetate	Faint
	Filament triacetate	None
	Nylon 6.6 polyamide	Faint
	Dacron 54 polyester	Faint
	Dacron 64 polyester	None
	Creslan 61 acrylic	None
	Orlon 75 polyacrylic	None
	Spun viscose (rayon)	None
	SEF modacrylic fire-retardent fiber	None
	Polypropylene	None
Natural fibers	Bleached cotton	Faint
	Spun silk	None
	Worsted wool	None

other common garment stains may yield a false positive during the presumptive analysis. Positive confirmation of capsaicinoid presence is performed via the LC/MS analysis previously described. Color development on derivatization did occur on several samples (human saliva, perspiration, and urine); however, no violet stains were generated. A faint reddish color was developed on derivatization of the single lot of human urine tested. On derivatization, three of the six human perspiration samples produced faint orange-to-red "speckles."

Due to the presumptive nature of this test, confirmatory analysis is necessary using MS methods. A typical chromatogram produced from an injection of 3 μL is shown in Figure 5.19. The mass spectra for the capsaicinoids are consistent with previously published spectra [40] and logical fragment ions are obtained for the capsaicinoid derivatives (Figure 5.20).

The derivatization reaction should be performed with the minimum amount of solutions needed to lightly cover the garment surface. Oversaturation can lead to gross discoloration of the fabric and may make the interpretation of color development more difficult. Limiting the amounts of base and diazonium salt solution that are applied also reduces potential salt buildup in the heated capillary of the mass spectrometer. Additionally, it is preferable to show the presence of both derivatized and unreacted capsaicinoids during LC/MS analysis. Atomizers or bottles with fine mist spray dispensers are the preferred

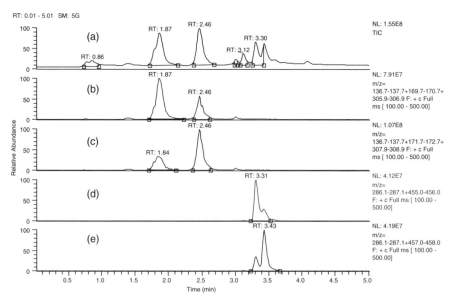

Figure 5.19 *Chromatogram generated from pepper spray derivatized with 4-nitrobenzene-diazonium tetrafluoroborate: (a) total ion current (TIC). Extracted ions of interest: (b) capsaicin, (c) dihydrocapsaicin, (d) capsaicin derivatized with the diazonium reagent, and (e) dihydrocapsaicin derivatized with the diazonium reagent.*

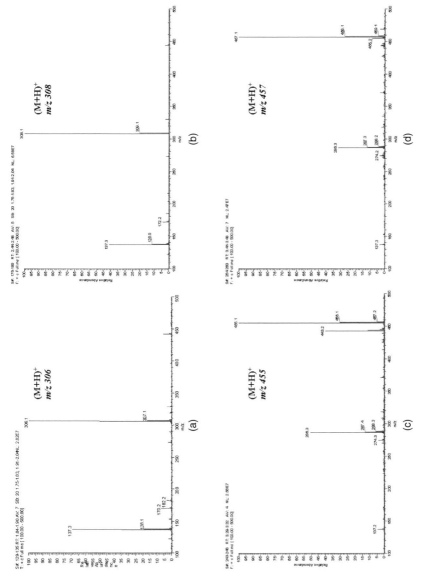

Figure 5.20 Electrospray mass spectra of (a) capsaicin, (b) dihydrocapsaicin, (c) capsaicin derivatized with a diazonium reagent, and (d) dihydrocapsaicin derivatized with a diazonium reagent. The [M + H]⁺ ion is present for each compound of interest.

means of applying the solutions used in the visualization reaction. Analysts are strongly urged to perform this reaction on test samples to gain familiarity with and proficiency in the method prior to performing evidence examinations. The described method is robust and colorless pepper sprays may be visualized via derivatization of capsaicinoid molecules under basic conditions with a diazonium salt solution to produce a violet chromophore. Interferences from common garment stains are negligible, except in cases where the stain is overly dark (e.g., blood) and overwhelms the violet color indicator that is developed at the location of the pepper spray stain. None of the interference tests yielded false positives. If false positives are obtained via the visualization procedure, they will be eliminated by the confirmatory LC/MS analysis.

REFERENCES

1. Gaensslen RE (1983). In *Sourcebook in Forensic Serology, Immunology, and Biochemistry*, ZS89-94, United States Department of Justice, National Institute of Justice, United States Government Printing Office, Washington, DC, pp 112–114.
2. Lee HC, Gaensslen RE, Pagliaro EM, Buman MB, Berka KM, Keith TP (1989). The effect of presumptive test, latent fingerprint and some other reagents and materials on subsequent serological identification, genetic marker and DNA testing in bloodstains. *Journal of Forensic Identification* **39**:331–350.
3. Grispino RR (1990). The effects of luminol on the serological analysis of dried human bloodstains. *Crime Lab Digest* **17**:13–23.
4. Hochmeister MN, Budowle B, Baechtel FS (1991). Effects of presumptive test reagents on the ability to obtain RFLP patterns from human blood and semen stains. *Journal of Forensic Sciences* **36**(3):656–661.
5. Gross AM, Harris KA, Kaldun GL (1999). The effect of luminol on presumptive tests and DNA analysis using the polymerase chain reaction. *Journal of Forensic Sciences* **44**(4):837–840.
6. Hochmeister MN, Budowle B, Sparks R, Rudin O, Gehrig C, Thali M, Schmidt L, Cordier A, Dirngofer R (1999). Validation studies of an immunochromatographic 1-step test for the forensic identification of human blood. *Journal of Forensic Sciences* **44**(3):597–602.
7. Fregeau CJ, Germain O, Fourney RM. Fingerprint enhancement revisited and the effects of blood enhancement chemicals on the subsequent Profiler Plus™ fluorescent short tandem repeat analysis of fresh and aged bloody fingerprints. *Journal of Forensic Sciences* (2000), **45**(2):354–380.
8. Viscasillas M (2002). In: *Proceedings of the American Academy of Forensic Science Annual Meeting*, Vol. VIII, February 11–16, 2002, Paper B53, p 54.
9. Cheeseman R (1999). Fluorescein blood stain detection method, US Patent, 5,976,886.
10. Sowell J, Patonay G, Strekowski L, and Krutak J. Non-aqueous leuco dye bloodstain detection technology, US Application Number 10/366,289.
11. Marie C, Boan T, Di Benedetto J. *Detection of Latent Blood. A New Fluorescein Formulation*, IABPA, Houston TX.

12. Budowle B, Leggitt JL, Defenbaugh DA, Keys KM, Malkiewicz SF (2000). The presumptive reagent fluorescein for the detection of dilute bloodstains and subsequent STR typing of recovered DNA. *Journal of Forensic Sciences* **45**(5): 1090–1092.

13. Quickenden TI (2001). Increasing the specificity of the forensic luminal test for blood. *Luminescence* **16**(3):251–253.

14. Arnold RR, Gallant JR, Kaufman RP (1988). Two latent prints developed and lifted from homicide victim identify perpetrator. *Journal of Forensic Identification* **38**(6):295–298.

15. Burt JA, Menzel ER (1985). Laser detection of latent fingerprints: difficult surfaces. *Journal of Forensic Sciences* **13**(2):364–370.

16. Christman DV (2004). A study to compare and contrast animal blood to human blood product. Virginia Department of Forensic Science, *Bloodstain Pattern Analysis Training Manual*, Section 4.2.3.7, Oct.

17. Cox M (1991). A study of the sensitivity and specificity of four presumptive tests for blood. *Journal of Forensic Sciences* **36**(5):1503–1511.

18. Dalrymple BE (1982). Use of narrow-band-pass filters to enhance detail in latent fingerprint photography by laser. *Journal of Forensic Sciences* **27**(4):801–805.

19. Dalrymple BE, Duff JM, Menzel ER (1976). Inherent fingerprint luminescence—detection by laser. *Journal of Forensic Sciences* **21**(1):107–115.

20. DeForest P (1990). A review of interpretation of bloodstain evidence at crime scenes. *Journal of Forensic Sciences* **35**(6):1491–1495.

21. Doherty P, Mooney D (1990). Deciphering bloody imprints through chemical enhancement. *Journal of Forensic Sciences* **35**(2):457–465.

22. Eckert W, James S (1989). *Interpretation of Bloodstain Evidence at Crime Scenes.* Elsevier Science Publishers, New York.

23. Hawkes V (1996). The OPP: leaders in bloodstain technology. *The OPP Review* (Ontario Provincal Police), December.

24. Wolson TL (2001). DNA analysis and the interpretation of blood stain patterns. *Journal of Canadian Society of Forensic Sciences* **34**(4):151–157.

25. Laux DL (1990). The effects of luminol on the subsequent analysis of bloodstains. Paper presented to the Annual Meeting of the American Academy of Forensic Sciences, February.

26. Ristenbatt RR, Shaler RC (1995). A blood stain pattern interpretation in a homicide case involving an apparent stomping. *Journal of Forensic Sciences* **40**(1):139–145.

27. Neureiter F, Pietrusky F, Schutt E (1940). Dictionary of forensic medicine and biologic criminalistics. In *Forensic Examination of Blood*. Hirschwaldsche Buchhandlung, Berlin, pp 220–222.

28. Pascual FA, Verd MS, Grifo G (1995), Investigation of bloodstains: false negative results of the benzidine test. *Forensic Science International* **71**:85–86.

29. Raymond M, Smith E, Liesegang J (1996). The physical properties of blood—forensic considerations. *Science & Justice* **36**(3):161–171.

30. Everse KE, Menzel ER (1987). Blood print detection by fluorescence. *SPIE* **743**: 184–190.

31. Thornton JI, Heye CL (1987). Fluorescence detection of bloodstain patterns. *SPIE* (*Fluorescence Detection*) **743**:190.

32. Seubert HJ, Craig R, Leggitt J, Patonay G, Hoskins KA, Watkins TG, Miller KWP, Bartholomew R, Guerrieri R (2005). A comparison of KPS fluorescein to other presumptive blood identification techniques and its effects on PCR based DNA analysis methods. Paper presented at the American Academy of Forensic Sciences, 57th Annual Meeting, New Orleans, LA, February 21–26.

33. Krutak JJ, Eckenrode BA, Patonay G, et al. (2005). Fluorescent heterocyclic dye systems for visualization of trace evidence at crimes scenes. Paper 1-PO57, 20th International Congress of Heterocyclic Chemistry, July 31 to August 5, Palermo, Italy.

34. Krutak JJ, Eckenrode BA, Patonay G, et al. (2005). Novel near-infrared fluorescent dye systems for visualization of trace evidence at crime scenes. Paper A0027, 17th International Association of Forensic Sciences Meeting, August 20–26, Hong Kong, China.

35. Cavett V, Waninger EM, Krutak JJ, Eckenrode BA (2004). Visualization and LC/MS analysis of colorless pepper sprays. *Journal of Forensic Sciences* **49**(3): 469–476.

36. Dong MW (2000). How hot is that pepper? *Today's Chemist at Work* (ACS Publication) **May**:17–20.

37. Govindarajan VS, Sathyanarayana MN (1991). Capsaicin—production, technology, chemistry, and quality. *Food Science Nutrition* **29**:435–474.

38. Reilly CA et al. (2001). Determination of capsaicin, dihydrocapsaicin, and noni-vamide in self-defense weapons by liquid chromatography–tandem mass spectrometry. *Journal of Chromatography A* **912**:259–267.

39. Streitwieser A Jr, Heathcock CH (1976). *Introduction to Organic Chemistry*. Macmillan Publishing, New York, p 1010.

40. Reilly CA et al. (2002). Detection of pepper spray residues on fabrics using liquid chromatography–mass spectrometry. *Journal of Forensic Sciences* **47**(1):37–43.

6

*Applications of Cathodoluminescence in Forensic Science**

Christopher S. Palenik†

Visiting Scientist, FBI Laboratory, Counterterrorism and Forensic Science Research Unit, Quantico, Virginia

JoAnn Buscaglia

Research Chemist, FBI Laboratory, Counterterrorism and Forensic Science Research Unit, FBI Academy, Quantico, Virginia

6.1 INTRODUCTION

The phenomenon of cathodoluminescence (CL) refers to the emission of visible (or near-visible) light from a sample that has been bombarded by an electron beam (Figure 6.1). This analytical technique, often confused with fluorescence or chemiluminescence, is most commonly recognized by the brilliantly colored images that result (see Color Plate 6.1). While the technique has been explored for applications in the forensic analysis of paint and glass

---

* This is Publication No. 06-01 of the Laboratory Division of the Federal Bureau of Investigation (FBI). Names of commercial manufacturers are provided for information only, and inclusion does not imply endorsement by the FBI.
† Research Microscopist, Microtrace LLC, Elgin, IL.

Forensic Analysis on the Cutting Edge: New Methods for Trace Evidence Analysis, Edited by Robert D. Blackledge.

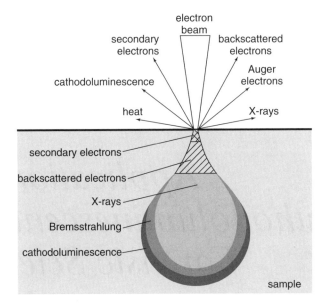

Figure 6.1 *Schematic illustration of various sample–electron interactions that can occur and the relative volume of these probes in a typical sample.*

[1, 2], CL has been applied to only limited types of forensic evidence. The recent emphasis placed on obtaining provenance or source information from unknown samples (i.e., determining the path a sample has traveled, or the geographic origin of a sample) in forensic examinations has sparked a new interest in CL due to the strong dependence of luminescence on the geologic history of a sample.

In contrast to the presently limited number of applications of CL in forensic science, applications of CL in academic science are well established. CL was first reported by Crookes in 1879 [3] when studying discharge phenomenon under vacuum. Since that time, CL has developed into a standard technique in physics and materials science for studying trace element and defect incorporation in semiconductors and insulators. In the geological sciences, CL has been applied to the study of trace element zoning in various minerals and to obtain information for provenance studies. It is this body of research, which uses CL as a tool to reconstruct ancient geologic environments, that will be drawn from to provide similar constraints upon the history of a modern forensic sample. In addition to sand and soil, luminescent minerals are also commonly used in a processed or synthetic form as pigments or filler/extenders. These anthropogenically modified minerals are encountered in trace evidence in a variety of materials such as glass, paint, and duct tape.

With this range of materials applications and the recent developments in analytical instrumentation, CL has the potential to become a more widely used analytical tool in the trace evidence section of a forensic science laboratory.

The goal of this chapter is to present, in a manner accessible to both students and experienced forensic scientists, a basic introduction to the theory, instrumentation, and range of applications of CL in a forensic laboratory.

6.2 THEORY

To develop and appreciate applications of CL in forensic science, a basic understanding of the processes that cause and influence luminescence is necessary. This section aims to present an overview of the mechanisms associated with CL. Textbooks with more detailed explanations are readily available and should be consulted. Texts by Marshall [4] and Götze [5] provide applied descriptions of the theory, while Yacobi and Holt [6] provide a more theoretical (but still accessible) treatment of the topic.

6.2.1 Luminescence Terminology

To begin, it is necessary to define cathodoluminescence and its distinction from other types of luminescence. The term luminescence is general and is used to refer to any type of photon emission (typically visible) resulting from the relaxation of an electron from an excited state to a ground state. Within this category, there are many types of luminescence such as bioluminescence, chemiluminescence, photoluminescence, and cathodoluminescence. In these cases, the prefix denotes the source of excitation energy.

Cathodoluminescence refers specifically to sample excitation via an energized electron beam. Photoluminescence, also known to many forensic scientists colloquially as fluorescence (i.e., excitation of a sample by ultraviolet light followed by visible emission) is analogous to CL in many respects, with the largest difference being in the energetics of the excitation source. CL excitation uses a higher energy (typically 10–15 kV) and higher power density than photoluminescence, which results in a greater penetration depth, sampling volume, and photon emission. As a result, many materials that do not photoluminesce efficiently enough to detect can be observed by CL (e.g., quartz and feldspar).

The terms fluorescence and phosphorescence also require clarification. These terms refer only to the lifetime of an emission after excitation has ceased. Therefore, depending on the material, CL can be termed either fluorescence ($<10^{-8}$ s) or phosphorescence ($>10^{-8}$ s), though the distinction is rarely quantified. In this chapter, the most general term, luminescence, will be used to refer to the observed emission.

6.2.2 Electron Source

The interaction of an electron beam with a sample (Figure 6.1) results in a variety of emissions that include electrons (i.e., secondary and backscattered electrons used for SEM image formation), X-rays (i.e., energy collected by

X-ray spectrometers to identify the elemental composition of a sample or to provide structural information in electron diffraction), and photons (i.e., visible, infrared, and ultraviolet light). An additional property of CL that can be noted in Figure 6.1 is that luminescence arises not from a point on the surface of the sample, but from an excitation volume that can extend as far as several micrometers into the sample surface [4]. This penetration depth increases with the excitation voltage.

In CL, the electron beam is supplied by either a cold or hot cathode. The electron beam from a cold cathode (similar to the cathode ray tube in older televisions) results from the discharge of an ionized gas formed between a cathode and anode. In a hot cathode instrument, such as an electron microscope, electrons are emitted from a heated filament. The benefits of each of these electron sources will be discussed in Section 6.3 of this chapter.

6.2.3 Cathodoluminescence

Luminescence occurs through a process of excitation, energy transfer, and relaxation. Initially, an energy source (i.e., the electron beam in CL) excites ground state electrons to a higher energy level (Figure 6.2a). In CL, relaxation energy is released in the form of photons with energies in the visible, ultraviolet, and infrared regions of the spectrum (~1–5 eV or 200–900 nm). Cathodoluminescence can be categorized as intrinsic or extrinsic based on the type of luminescence center (i.e., the activator or source of the luminescence in a material). Intrinsic luminescence refers to the direct recombination of electron–hole pairs (Figure. 6.2b), a process that is characteristic of the structure of the material being studied (e.g., blue luminescence observed in pure calcite). The energy of intrinsic luminescence is equal to the energy of the band gap (E_g). Direct recombination of electron–hole pairs has a small probability of occur-

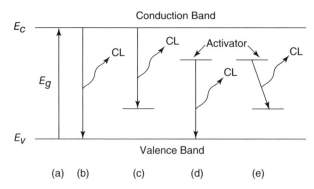

Figure 6.2 *Schematic illustration of luminescence mechanisms: (a) excitation of an electron from the valence to conduction band; (b) emission of intrinsic luminescence due to direct recombination; (c)–(e) extrinsic luminescence that occurs as a result of recombination involving energy levels in the forbidden zone (i.e., activators). Extrinsic luminescence is typically stronger than intrinsic luminescence due to a higher probability of occurrence.*

rence; therefore, intrinsic luminescence is typically weak and often obscured by extrinsic luminescence [6].

Extrinsic luminescence describes emissions that result from the presence of defects or impurities in the structure of the material being examined (e.g., the orange luminescence observed in calcite due to ppm levels of Mn^{2+}) (Figure 6.2c–d). In this case, defects and trace elements in the crystal lattice responsible for luminescence, known as activators, with energy levels in the band gap (forbidden zone), provide additional energy levels for excitation and relaxation processes. Due to the nature of geologic samples and the dominance of extrinsic processes, most CL emissions of interest in this chapter are extrinsic.

The exact energy and peak width are functions of temperature, pressure, and interactions between the activator ion and the surrounding atoms of the crystal (i.e., crystal field). Increases in temperature cause thermal broadening of the peaks due to the population of excited vibrational states. Cooling samples to liquid helium temperatures provides sharper peaks and is used in fundamental studies to assign peaks to particular luminescence centers. However, forensic applications of CL typically involve minerals and activators that have been characterized (e.g., calcite and feldspar) or treat uncharacterized systems without assigning peaks (e.g., quartz luminescence has not been completely characterized).

Of more relevance to CL in forensic science are peak broadening and shifts that occur due to the crystal field strength (D_q). Crystal field strength is calculated by assessing the influence of an activator's ligands (coordinating atoms) or the surrounding crystal structure on the energy levels of a given atom (e.g., an activator atom). Activators that have a strong interaction with the surrounding crystal structure will lose energy to the structure through phonons (i.e., vibrational energy) (Figure 6.3). This energy transfer between the

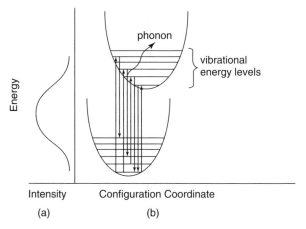

Figure 6.3 *Schematic illustration of (a) peak broadening that results from (b) occupation of higher vibrational energy levels in samples containing activators that interact strongly with the surrounding crystal structure.*

activator and crystal structure occurs in the time between the excitation of an activator by an electron and the subsequent photon emission. The loss of vibrational energy to the crystal structure results in a decrease in the emission energy relative to the excitation energy (Stokes shift). Therefore, the peak position of activator ions will shift based on their interaction with the surrounding crystal structure. This shift is observed in the Mn^{2+} peak in calcite, which shifts to higher wavelengths as a function of increasing amounts of Mg^{2+} (Figure 6.4a). Similarly, the peak width at room temperature will be influenced by the surrounding crystal structure.

In contrast, some activators, such as the rare earth elements (REEs), have only weak interactions with the crystal structure due to shielding by 5p and 6s electrons. As a result, the peak position and peak width of these activators remain relatively independent of the surrounding crystal structure, making their identification in minerals such as zircon or apatite relatively straightforward (Figure 6.4b).

In addition to activators, other atoms termed sensitizers and quenchers also play a role in luminescence. Sensitizers are atoms that, upon being excited by the electron beam, transfer their excitation energy to an activator atom, resulting in an increase in the intensity of a weakly luminescing activator (e.g., Nd^{3+} in the presence of Sm^{3+}, Dy^{3+}, or Eu^{3+}). In other cases, this can cause a non-

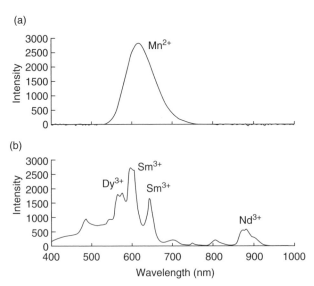

Figure 6.4 (a) CL spectrum of a calcite crystal. The broad orange peak is due to the strong interaction between trace amounts of the activator Mn^{2+} and the crystal structure. Variations in the Mg^{2+} content result in peak shifts. (b) CL spectrum of a purple luminescent apatite crystal with sharp peaks caused by rare earth elements that interact only weakly with the surrounding crystal structure.

luminescing atom to become activated and luminesce [7]. Quenching atoms cause a decrease in luminescence with increasing concentration. For example, the intensity of Mn^{2+} luminescence in calcite is decreased by increasing amounts of Fe^{2+}. Additionally, at sufficiently high concentrations, it appears that Mn^{2+} results in self-quenching [4].

6.2.4 Limitations

Despite an increasing amount of research into CL theory, applications, and instrumentation, significant aspects of CL remain to be fully explained. The defects or impurities responsible for luminescence in many materials are still a matter of controversy. The effects of quenching and sensitizing are not particularly well understood and illustrate why CL intensities cannot be treated quantitatively. As a result, much of the theory that has been developed for other spectroscopic techniques, such as Beer's law for spectrophotometry or ZAF-related corrections for wavelength dispersive spectroscopy, does not exist for CL. Thus, at present, quantification of activator concentration can only be applied under controlled conditions.

6.3 INSTRUMENTATION

CL instruments are typically assembled from components that include an electron source, microscope, camera, and spectrometer. Currently, improvements in both technology (hardware and software) and price are being made continually to most of these components, particularly with regard to cameras and spectrometers. Due to these technological improvements and the presently developing state of application in forensic science, there is currently no single ideal configuration. Instead, each component and each configuration comes with a particular set of strengths and weaknesses. This section aims to present the current state-of-the-art in CL by introducing the range of instrumentation available along with the accompanying benefits, limitations, and considerations for forensic application. These points are summarized in Table 6.1.

6.3.1 Electron Source

The electron beam that supplies the energy for CL can be provided by either a cold or hot cathode (explained in Section 6.2.2). The cold cathode operates with a broad beam that is controlled by a focusing coil and two magnets physically oriented on the top of the chamber. To operate, the beam requires a specific gas pressure in the chamber, which is regulated by a leak valve that can be controlled manually or, preferably, through a feedback loop. This configuration results in a low beam density, and hence low luminescence intensity,

TABLE 6.1 Comparison of Cold and Hot Cathode CL Configurations

Cathode Type	General Considerations	Stereo Microscope	Research Microscope	Scanning Electron Microscope
Cold cathode	• Broad beam • Low current density • Current density not reproducible • Quantitative CL not possible • No coating required • Low vacuum • Less expensive	• Permits overview images of sample • Good depth of field allows analysis of unpolished grains • Photo ports not always available • Least expensive viewing option	• Spectrometer available • Spectroscopy limited to 400–1100 nm unless quartz optics are used • Large working distance limits maximum magnification • Poor depth of field makes high magnification study of grains difficult	N/A
Hot cathode	• Higher beam energy • Regulated energy • Coated samples required • High vacuum required • More expensive	N/A	• Beam comes from below sample; thin sections required	• Best spatial resolution; allows study of zoning in individual crystals and pigments in paint • Easily combined with EDS for sample identification • Most imaging is B&W • New color systems available • Spectroscopy covers UV range • Most expensive option

relative to hot cathode instruments. Furthermore, intensity measurements can only be considered semiquantitative at best. The advantages of a cold cathode system are the ability to work with a broad beam (up to 2.5 cm) on uncoated samples in a relatively low vacuum (~10–50 mTorr). This is extremely useful in characterization of soils or in the study of concrete blocks, where overviews of sample heterogeneity can provide quick diagnostic information.

CL instruments based on a hot cathode include scanning electron microscopes (SEMs) or standalone instruments that can be attached to the stage of a light microscope. Hot cathode instruments require higher vacuums and coated samples. However, as previously mentioned, a hot cathode beam is more intense and can be focused to a finer point, and the current can be regulated. The latter point makes quantitative measurement of CL intensity possible. In terms of applications, the hot cathode on a light microscope provides higher resolution images, which can be useful when studying mineral zoning or fine features. In addition, the SEM-CL can provide even higher resolution and elemental analysis when the SEM is also equipped with an X-ray detector.

6.3.2 Microscope

When choosing a light microscope on which to mount a CL chamber, the most important feature to consider is the objectives. Extra-long working distance objectives are required due to the distance between the sample and the top of the chamber window (~1 cm in a cold cathode instrument). Additionally, the CL chamber contains a leaded glass viewing window, which is approximately 5 mm thick, to absorb emitted X-rays. This window is considerably thicker than a standard 0.17 mm microscope coverslip, resulting in aberrations in the final image. Low magnification objectives (4× and 10×) are most useful for imaging in forensic examinations, because samples are typically grain mounts, which suffer from depth of field issues at magnifications higher than 10×. For spectroscopy, an objective with the highest numerical aperture (N.A.) is recommended in order to capture the largest solid angle of light possible. Other considerations include a rotateable polarizer and analyzer (since chambers are not designed to rotate) and two phototubes, one for the spectrometer and the other for a camera. Polarized CL has been examined and orientation effects on luminescence have been reported [4]; however, the details of these investigations are beyond the present level of forensic CL applications. Finally, a stereo microscope with a phototube has proved to be a useful accessory for collecting extremely low (0.6–4×) magnification survey images. This setup has been found to be particularly valuable when studying minerals in soils, concrete, or even cross sections of duct tape.

6.3.3 Camera

Virtually any digital camera can be used to collect images of brightly luminescing samples such as calcite or feldspar. However, cameras with exposure times on the order of 1–2 minutes can be required to collect images of faintly luminescing samples on a cold cathode instrument. To accommodate these long exposure times, cameras that are Peltier cooled are recommended to reduce noise. Monochromatic cameras with an RGB filter wheel are more difficult to use for low intensity luminescence photography because exposure times must be tripled, and the intensity of various luminescence centers can change over this period.

6.3.4 Spectrometer

Spectrometers are typically attached to a microscope phototube via a fiberoptic cable. The sampling area is defined by the diameter of the fiberoptic cable and the magnification of the objective used. The imaged area sampled by the fiber optic is then projected onto a grating and dispersed onto a detector. Traditionally, photomultiplier tube (PMT) detectors have been used and remain the cost-effective option. However, CCD detectors (typically Peltier or liquid nitrogen cooled) have become the preferred, if not standard, detector. The benefits of a CCD detector are a higher sensitivity and signal-to-noise ratio than a PMT, resulting in decreased spectral collection time. As opposed to a PMT, which collects one data point at a time, a CCD captures a solid angle of the dispersed spectrum with each pixel representing a channel. The result is that a single exposure can capture a portion of the spectrum anywhere from a few nanometers to 1000 nm wide, depending on the dispersion of the grating used. Typically, a 300 g/mm grating has been used to collect spectra in research studies. This resolution is generally adequate for CL as peaks tend to be broad. However, higher resolution gratings (e.g., 1200 g/mm) provide additional resolution that is advantageous when characterizing the narrow and often overlapping bands of the REEs.

6.3.5 SEM-CL

Until recently, SEM-CL had been largely limited to panchromatic black and white images or spectra collected by a PMT. Over the last several years, improvements in detectors and software, as well as the standardization of external beam control on electron microscopes, have been combined to produce CCD-based SEM-CL spectrometers as well as live-color SEM-CL accessories [8]. Hyperspectral imaging using a PMT detector has also been applied to SEM-CL to produce calculated-color images based on a grid of spectra collected over the area of a rastered SEM beam. Subsequent spectral processing converts each spectrum to an RGB value, which together can be combined to form a calculated-color image [9]. Advances in the field of SEM-CL will continue and promise improved analytical characteristics.

6.4 TECHNIQUES AND FORENSIC CONSIDERATIONS

Most forensic scientists do not have hands-on experience or even secondary experience with CL, while scientists more experienced with applications of CL in the geological or semiconductor industry may not be aware of the practical limitations inherent to forensic science (e.g., variety of samples, sample size, available analysis time, and limitations on the destruction and alteration of evidentiary samples). This section is written to present both audiences with basic information regarding techniques of sample preparation, as well as the analytical considerations and limitations that become relevant to the interpretation of luminescence.

6.4.1 Instrumental Conditions

The FBI Laboratory Counterterrorism and Forensic Science Research Unit's CL system (Figure 6.5) utilizes a Luminoscope ELM-3R mounted on either a Nikon E800 polarized light microscope (Figure 6.5a; 4×/0.2 N.A./15.7 mm W.D.; 10×/0.3 N.A./16.0 mm W.D.) or a Nikon stereo microscope (Figure 6.5b). The E800 is equipped with transmitted and reflected light. The stereo microscope utilizes a fiberoptic ring illuminator. An oblique fiberoptic light source is also used in both setups. The Luminoscope is operated at beam conditions rang-

(a)   (b)

Figure 6.5 *Images of a cold cathode CL instrument mounted on (a) a polarized light microscope equipped with a fiberoptic cable for spectroscopy and a camera and (b) a stereo microscope with a camera used for collecting low magnification overview images.*

ing from 5 to 15 kV, beam currents of 0.5 to 1.5 mA, and a vacuum of ~10 to 60 mTorr. Spectra are collected using an Ocean Optics HR2000 CCD-based spectrometer and/or a Horiba Jobin-Yvon Triax 320 spectrometer with PMT and CCD detectors. The Ocean Optics system is equipped with a 5 μm slit and a 300 g/mm grating. This results in a measured full width at half maximum (FWHM) of the Hg 546.07 nm line of 6 nm for the Ocean Optics CCD. The Triax 320, which utilizes a 1200 g/mm grating and slit widths that range between 0.1 and 1.5 mm, has a measured $\text{FWHM}_{\text{Hg 546.07 nm}}$ of 0.1–2.0 nm for the CCD detector and 0.3–2.0 nm for the PMT detector. Images are collected using either a QImaging Micropublisher 5.0 megapixel camera or the Optronics Magnafire. For broad beam experiments, the Luminoscope is placed on the stage of a Nikon stereo microscope with a photoport. Spectral measurements and deconvolutions are made using PeakFit v4.12.

6.4.2 Sample Preparation and Preservation

Under ideal circumstances, grains or slabs of material are analyzed only after samples are polished and mounted. These samples are prepared for CL in the same manner used to prepare thin sections or polished slabs for light microscopy (i.e., the final polish uses ≤0.1 μm paste). A final polish reveals a cross-sectional view that provides additional information over that observed in grain mounts such as zoning (i.e., compositional changes in a crystal that occur during growth), microcracks, or surface alteration (e.g., coatings or recrystallized rims on altered minerals). Additionally, polished samples luminesce more efficiently, eliminating geometric effects in the electron–sample interaction.

Despite the benefits of polishing, in forensic casework, they are often outweighed by the additional time required. Therefore, applications regarding the analysis of grain mounts will be stressed in this chapter. Grain mounts can be prepared for CL by scattering a given amount of sample onto a slide, or specific minerals can be plucked using a tungsten needle and mounted on double-sided carbon tape on an SEM stub for CL and later examination by EDS. One disadvantage to the study of grain mounts by CL is that a limited depth of field in a typical research microscope limits the useful maximum magnification to approximately 10× for CL imaging (which is adequate for most particles). While glass slides can be used, most glass slides show a certain amount of luminescence and/or reflect the violet ionization color of the electron beam, resulting in a higher background than is desirable in CL images and spectra. To avoid this, polished aluminum slides can be used because they do not contribute to background luminescence that would interfere with long-exposure images or spectroscopy. The disadvantage is that samples mounted on aluminum slides cannot be observed by transmitted light.

For softer materials that luminesce, such as duct tape and paint, cross sections are useful for the study of internal structures. Cross sections can be cut with a microtome, but hand-cut sections using a single edged razor blade typically prove to be a faster and satisfactory alternative. For most types of microscopic examination, cross sections on the order of 1–10 μm are necessary to provide optical or infrared transparency. Because CL is a surface technique, these cross sections only have to be thin enough that they can be mounted in an on-end orientation.

When using a hot cathode instrument, samples are typically made conductive to eliminate charging by coating with a layer of carbon. Because a cold cathode operates at a lower current density on the sample, coating is not necessary. In both instruments, sample heating as a result of buildup can melt or discolor softer samples, such as paint, or cause samples to become brittle (e.g., duct tape). Finally, intense heating of some materials resulting from a focused beam can cause the sample to incandesce, emitting blackbody radiation, which can obscure the fainter luminescence.

6.4.3 Image Collection

As mentioned in Section 6.3.3, a color CCD camera is preferred for CL imaging; however, digital images face a variety of hurdles including color balancing, exposure selection, and correct postprocessing of images. To accurately reproduce color, digital cameras require the relative exposures of the red, green, and blue pixels to be adjusted. Typically, this is done by white balancing; however, color adjustment remains subject to judgment. An interlaboratory test showed that among various CL laboratories, color reproduction varied dramatically. The Society for Luminescence Microscopy and Spectroscopy circulated a calcite standard to 11 laboratories for photography [10]. The results showed similar levels of detail in the images, but a range of colors were recorded that varied from yellow to orange. Beyond adjustment of camera settings to produce an image that corresponds to a reference image, no standard for color exists.

A further complication of CL imaging is that minerals luminesce over wide ranges of intensities. When studying a single mineral, this is not a problem; however, in soils, the range of luminescence intensities among the minerals present can make exposure selection difficult. For example, grains of feldspar (a relatively strongly luminescent mineral) are often observed in the same field of view with grains of quartz (a relatively weakly luminescent mineral). Selection of an exposure time that will adequately expose the luminescence of both minerals without overexposing the feldspar or underexposing the quartz is, at times, not possible. Ideally, two images can be collected with different exposures. However, changes in the luminescence with time may preclude this option. In these cases, adjustment of the image level using postprocessing software can bring out underexposed color and details not observable in the original sample. Analysts should be mindful that manipulation of digital images can have implications for court acceptance. Therefore, all processing should be documented and the original image should be maintained with the case.

6.4.4 Spectral Collection

Although the collection of spectral data has become commonplace in research studies, very little standardization exists with regard to data reporting. As a result, interlaboratory comparison of luminescence data is often not possible. Marshall and Kopp [10] discuss five performance characteristics that must be considered when reporting data: resolution, wavelength accuracy, transmission, signal-to-noise, and spectral acquisition time.

Wavelength accuracy is the primary (as well as easiest) characteristic to calibrate and can be corrected within any laboratory using a Hg lamp or HeNe laser [11]. CCD detectors, which have a nonlinear pixel-to-wavelength relationship, make this calibration more difficult, particularly when a high resolution grating requires that the visible spectrum be collected in segments that are later glued together by software. Manufacturers include algorithms to correct for this problem; however, it is necessary to check the wavelength accuracy of known peaks across the visible spectrum, as well as the accuracy of a single known peak when it is projected onto different pixels of the CCD chip.

Spectral response is a more complicated parameter to correct because all optical components in a system influence the transmission efficiency at different wavelengths. These include lenses, mirrors, gratings, and detectors, which means that the detected intensity of emission at a particular wavelength for a standard sample will be different for different systems. This point is illustrated in Figure 6.6, which shows the luminescence spectrum for the same sample collected by three different detectors. Without a correction for spectral

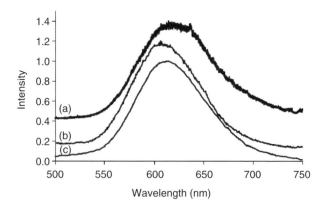

Figure 6.6 *Illustration of the varying spectral responses of different detector/spectrometer configurations for a given microscope. Shown are spectra from a luminescing calcite crystal collected by the (a) Triax CCD, (b) Triax PMT, and (c) Ocean Optics CCD detectors. Note that the peak position varies among the different configurations due to the spectral response of the detector, not the wavelength accuracy. The intensity scale has been normalized and offset.*

response, these spectra show differing peak intensities and peak widths, which prevent direct comparison between the systems. The differences are due to the differing sensitivities of the three detectors as a function of wavelength. The simplest way to correct for spectral response is to measure the spectrum of an intensity-calibrated light source. Light sources calibrated from an absolute intensity standard maintained by the National Institute of Standards and Technology (NIST) are commercially available.

Other problems also plague transmission measurements. Over time, hydrocarbon (e.g., epoxy) or other contamination in or around a sample are vaporized and redeposited on the chamber walls, sample, and viewing window. This gradual buildup of contamination will decrease the transmissivity at a rate that is not noticeable by visual inspection. Over time, this buildup can reduce intensity and/or shift the spectral response significantly. The amount and rate of buildup will depend on the samples being studied.

Finally, the beam intensity of a cold cathode instrument cannot be regulated well enough at the sample surface to make consistent measurements between samples. The sum of these spectral collection issues is illustrated in the results of an interlaboratory comparison that showed a ratio of peak heights measured from a standard sample that varied from 0.6 to 3.9 [10].

6.4.5 Luminescence Fading

Upon initial observation of the mineral in a typical continental soil sample, a 0.5 cm field of view will be filled with vivid colors ranging from blue to green to yellow to red, as seen in Plate 6.1a,b. Over a period of seconds to minutes of electron exposure, this luminescence fades to the extent that luminescence in many minerals is difficult to observe visually. Longer exposure photographs will reveal that the vivid colors have faded to darker shades of blue, yellow, and brown. Fading is a result of defect migration, annealing, or heating (Plate 6.1c–e). Fading occurs in almost all quartz samples. In some minerals, such as feldspar, fading is not as pronounced, but can be confirmed through spectroscopy (see Section 6.5.2).

6.4.6 Sample Alteration

For forensic purposes, CL, like electron microscopy, can be considered nondestructive. Potential changes in a sample due to electron beam irradiation include defect and impurity migration in a sample. As with electron microscopy, the electron beam does not typically affect the elemental composition of a sample. However, the electron beam can damage the organic component of a sample when the luminescent properties of the inorganic components are being studied. Potential alteration and consumption of evidence by a specific analytical technique must be considered carefully before any technique is applied so as to ensure that additional tests or analysis by other experts are not excluded.

TABLE 6.2 Commonly encountered luminescent minerals and the range of colors encountered[a]

Mineral		None	Violet	Blue	Green	Yellow	Orange	Brown	Red
Quartz	SiO_2			w					
Opal	SiO_2			w					
Calcite	$CaCO_3$						s		s
Aragonite	$CaCO_3$					s			
Dolomite	$CaMg(CO_3)_2$						s		s
Albite	$NaAlSiO_4$				m	m		m	m
Anorthite	$CaAl_2Si_2O_8$			m	m				
Orthoclase	$KAlSiO_4$			m					
Zircon	$ZrSiO_4$			s	s	s			
Apatite	$Ca_3(PO_4)_3(F,Cl,OH)$		s	s		s			
Monazite	$(REE,Th)PO_4$							m	m
Spinel	$MgAl_2O_4$	x			m				
Anatase	TiO_2			m					
Rutile	TiO_2	x							
Talc	$Mg_2Si_4O_{10}(OH)_2$	x		m					
Zincite	ZnO				s				
Kyanite	Al_2SiO_3			m					
Andalusite	Al_2SiO_3								m
Sillimanite	Al_2SiO_3								m
Gypsum	$CaSO_4 \cdot 2H_2O$			m					

Legend: x = observed; w = weak; m = moderate; s = strong.
[a]This table lists commonly observed colors; however, other colors may be observed.

6.5 LUMINESCENT MINERALS

Luminescence has been observed and characterized in many classes of minerals including silicates, carbonates, oxides, sulfides, hydroxides, halides, nitrates, borates, sulfates, and phosphates [4]. Table 6.2 is a partial list of the more commonly encountered luminescing minerals and the range of colors observable for these minerals. This section discusses the origin of luminescence in some of the more commonly encountered groups of carbonate, quartz, feldspar, and accessory minerals. Although not inclusive of all minerals, the luminescence mechanisms and related provenance information discussed should provide a basis for understanding the general applications of mineral luminescence in forensic science examinations.

6.5.1 Calcium Carbonate Group

In both geology and forensic science, calcium carbonates represent one of the most frequently encountered mineral groups, of which calcite ($CaCO_3$), aragonite ($CaCO_3$), and dolomite ($MgCa(CO_3)_2$) are the most commonly encountered members. Carbonate minerals are found, for example, in sedimentary rocks in limestone, in metamorphic rocks in marbles, and in igneous carbonatites. In addition, carbonate minerals are formed by solution precipitation in voids of preexisting rocks, form the structural material of shells, and can be a major component of sand and soil. Carbonates are also used in a large number of industrial applications, including cement and building materials, and as filler/extenders in paints and polymers.

Calcite has an intrinsic faint blue luminescence due to peaks around 410 and 520 nm that result from structural defects. Due to the presence of activator elements (impurities), the intrinsic luminescence is typically overshadowed and rarely observed in natural samples. This intrinsic luminescence can sometimes be observed in biogenic calcite, which often contains relatively low levels of Mn^{2+} [12]. More commonly, carbonates have a bright luminescence that ranges from orange to orange-red in calcite (tending to red with increasing Mg content) (see Color Plates 6.1e, f). In carbonates, Mn^{2+} is the dominant activator, while Fe^{2+} is the most important quencher [4]. In calcite, this substitution occurs in the Ca^{2+} site in the crystal, which has an emission around 575–610 nm (Figure 6.7a). With increasing Mn^{2+}, the peak position shifts to higher wavelengths and the peak broadens [12]. Self-quenching becomes important as the Mn^{2+} exceeds ~3000 ppm, while quenching by Fe^{2+} becomes significant above ~1000 ppm. For samples with low Fe content (<200 ppm), the Mn^{2+} concentration has been shown to be linearly related to the CL intensity [12].

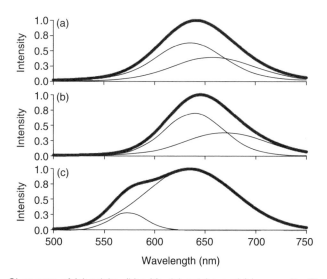

Figure 6.7 *CL spectra of (a) calcite, (b) a Mg-rich calcite, and (c) aragonite. The spectrum of each sample (thick line) has been deconvoluted into component peaks to illustrate that the luminescence activator Mn^{2+} can structurally substitute for both a Ca^{2+} site around 570–610 nm and a Mg^{2+} site around 630–670 nm. As the amount of Mg in the sample increases, substitution of Mn^{2+} into the Mg^{2+} site becomes more common and luminescence color becomes redder. Also note that the overall peak position and peak width vary with composition and structure.*

In dolomite, Mn^{2+} and Fe^{2+} can also occupy the Mg^{2+} site, which has an emission line around 650 nm [13]. Therefore, with increasing Mg^{2+}, the luminescence in carbonates becomes redder (Figure 6.7b). Dolomite exhibiting yellow luminescence has been reported and is believed to result from microscopic domains of calcite (yellow to orange luminescence) that are dispersed in the dolomite [14]. Aragonite, a polymorph of calcite found commonly in shells, luminesces green or yellow (see Color Plate 6.1h). The exact cause of green luminescence in aragonite (Figure 6.7c) is not known.

In addition to the broad luminescence bands of Mn^{2+}, trace incorporation of REEs (e.g., $Eu^{2+/3+}$, Dy^{3+}, Sm^{3+}, Tb^{3+}, Pr^{3+}) in calcite can be detected in the spectrum. The trivalent REEs that substitute are shielded from the crystal structure and, as a result, show sharp peaks with a relatively constant emission energy. In rare cases, the spectrum and/or color of the luminescence is dominated by rare earth emissions. More frequently, the concentration of rare earth activators is low enough that the sharp emission lines of the REEs do not contribute significantly to the spectrum. In this case, the REE content can be analyzed by subtracting the spectrum of a pure Mn^{2+} calcite spectrum [15].

Zoning (i.e., spatial changes in the chemical composition of a single crystal) and textural data can also provide information about crystal formation conditions in calcite and other minerals. Zoning can be broadly categorized as either concentric zoning or sector zoning [16]. In the first, compositional changes occur during crystal growth, which are often related to changes in fluid composition. In sector zoning, compositional variations are tied to specific crystallographic orientations and represent a preferential incorporation of the activator into certain growth faces of the crystal, rather than a change in the fluid composition. In a forensic context, the mechanism for zoning is less important than the observation and classification of zoning, which can be used to correlate or disassociate various samples. Plate 6.1i shows an example of concentric zoning observed by CL in marble. The zoning observed by CL is commonly not observable by backscatter electron (BSE) imaging in an SEM, because the BSE detector is not sensitive enough to observe the low concentrations of activator elements responsible for the zoning. Color Plate 6.1j shows a green-luminescing feldspar that contains an interior zone with blue luminescence, which is not observable by SEM-BSE or EDS.

Finally, surface texture information can provide additional clues to the origin of a sample. Color Plate 6.1k is a CL image of a fragment of biogenic calcite. Although this fragment is easily recognizable, texture, combined with a typically weak blue luminescence, is indicative of a relatively unaltered (fresh) biogenic calcite. With alteration, the calcite increases in Mn^{2+} content and the characteristic orange begins to dominate. This effect can be observed in Color Plate 6.1l, which is a CL image of biogenic sand particles consisting mainly of calcite with different intensities of luminescence. Finally, fossilized biogenic materials can be observed in polished sections of rocks (see Color Plate 6.1m).

6.5.2   Feldspar Group

Feldspar minerals represent another of the most commonly occurring groups of minerals, particularly in many younger soils. Due to the relatively fast weathering rate of feldspars (which weather to clay minerals), mature soils often contain only a small or even nonexistent feldspar component. The feldspar group is compositionally described by three end-member components— albite ($NaAlSiO_4$), anorthite ($CaAl_2Si_2O_4$), and orthoclase ($KAlSiO_4$). Feldspar minerals with intermediate compositions, or solid solutions, are observed between the K and Na end-members (known as alkali feldspars) and Na and Ca end-members (known as plagioclase feldspars). Structurally, feldspar minerals can accommodate various cations, particularly those of a similar size and valence (e.g., CL activators Mn and Fe). Additionally, feldspars can accommodate REEs in their crystal structures.

Luminescence from feldspar is typically moderate to bright and includes blue, violet, green, yellow, and red hues. Upon initial exposure to an electron beam, some varieties of quartz can be confused for feldspar minerals; however, quartz luminescence fades quickly in intensity and color to a brown luminescence. The strong luminescence makes classification and estimation of the relative abundance of feldspar minerals by CL possible. Alkali feldspars typically exhibit blue or red luminescence (see Color Plates 6.2a–c). Microcline, a polymorph of orthoclase, often shows a deep yellow luminescence. Plagioclase feldspars can luminesce in a range of colors but are most commonly blue and violet (see Color Plates 6.2d–f). Experiments have shown that the composition of feldspars can be correlated to the position of the red peak [17]. Authigenic (i.e., formed in situ) feldspar has low concentrations of trace elements and, as a result, is weakly luminescent.

The perceived visual color of luminescence in feldspar minerals can be spectroscopically attributed to several activators. The contribution of these activators varies as a function of exposure to the electron beam and as a function of the composition of the feldspar. Typically, feldspar luminescence is dominated by three to four components. A blue luminescence centered in the region of 400–475 nm has been associated with an oxygen hole center that forms as a result of a structural substitution of Al^{3+} for Si^{4+}. With increasing electron exposure, this defect can migrate, causing a decrease in the overall intensity of this activator's intensity. Another defect, possibly associated with an oxygen atom adjacent to a 2+ charged impurity atom, occurs around 500 nm in many alkali feldspars. A green luminescence center attributed to Mn^{2+} occurs in the range of 540–700 nm. Finally, Fe^{3+} has been associated with a band occurring within the range of 690–740 nm [17]. The exact energy of the red Fe^{3+} peak can be correlated with the bulk composition of the feldspar [18]. The contributions of these four peaks to the overall spectrum of an orthoclase feldspar are shown in Figure 6.8. Figure 6.9 shows the CL spectra of the feldspar minerals shown in Color Plate 6.2 and their sensitivities to varying lengths of electron beam exposure.

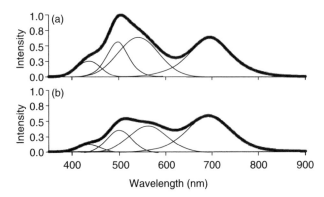

Figure 6.8 *CL spectra of an orthoclase ($KAlSi_3O_8$) feldspar (a) immediately after exposure to the electron beam and (b) after 10 minutes exposure. The spectra (thick lines) have been deconvoluted into individual peaks (thin lines) to illustrate the contributions from various activators. Note that after 10 minutes of electron beam exposure, the overall intensity of the spectrum has decreased, and the contributions of certain defects to the overall spectrum have changed.*

Similar to REE substitutions in calcite, REEs that have substituted in feldspar can be measured by CL. In certain instances, the rare earth peak will visually alter the luminescence color (e.g., Eu^{2+}); however, rare earth emissions are generally weak relative to those of the Mn^{2+} and Fe^{3+} luminescence centers. Nonetheless, peaks from REEs (Ce^{3+}, Sm^{3+}, Dy^{3+}, Eu^{2+}, Tb^{3+}, and Nd^{3+}) in the low ppm range can be resolved in high resolution spectra [19].

6.5.3 Quartz

Quartz (α-SiO_2) is one of the most durable of all detrital minerals. Sand samples freshly disaggregated from a parent rock collected close to source rock will contain a suite of minerals representative of the original rock. Over time, weathering processes will cause less durable minerals to react to form new minerals or dissolve. Eventually, quartz will represent the principal original mineral in a mature silicate soil or sand. Even quartz does not remain completely unreacted, as the surface layer can be weathered and reprecipitated.

In comparison to feldspars and carbonates, quartz is not able to structurally accommodate large concentrations of impurity elements. The interconnected framework of SiO_4 tetrahedra that make up the structure of quartz limits the size and quantity of impurities that can be included structurally. Therefore, quartz typically contains structural impurities (e.g., Ti and Al) on the order of only a few hundred parts per million. As a result, quartz from different environments has only subtle chemical differences. The commonality of quartz and these minor differences between quartz from different sources means that it is often ignored in most current forensic casework.

The lack of elemental impurities means that quartz luminescence is typically weak relative to the feldspar and carbonate minerals. The observed

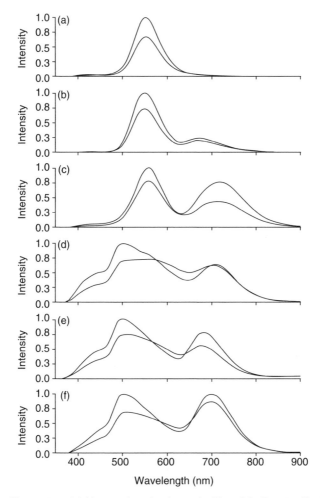

Figure 6.9 *CL spectra of feldspar minerals shown in Plate 6.2. Spectra illustrate that the luminescence is in part tied to the bulk chemistry of the mineral and include different activators. Each plot contains a spectrum collected immediately after electron excitation began (more intense spectrum) and after 10 minutes of electron excitation (less intense spectrum). Note that the decrease in intensity with time for various peaks (i.e., activators) varies among the minerals: (a) albite (NaAlSi$_3$O$_8$), (b) anorthite (CaAl$_2$Si$_2$O$_8$), (c) anorthite, (d) albite, (e) microcline (KAlSi$_3$O$_8$), and (f) orthoclase (KAlSi$_3$O$_8$).*

luminescence in quartz and amorphous silica (e.g., opal, agate) results mainly from structural defects and changes with time. Luminescence colors include dull shades of red, blue, violet, and green. The visually observed luminescence color of quartz depends on the relative proportion of the major peaks. Color Plates 6.2g–l shows images of several detrital quartz grains, as well as opal and chert. At least 10 emission bands have been reported in quartz, which range from the ultraviolet to the near infrared. A spectrum of quartz luminescence

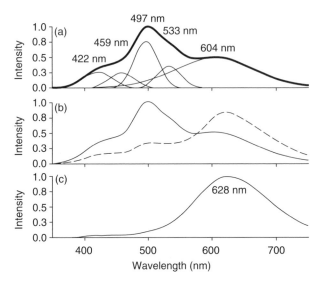

Figure 6.10 *CL spectroscopy of quartz. (a) Spectrum of a blue quartz grain immediately after exposure to the electron beam (thick line). The spectrum has been deconvoluted into component peaks signifying the contribution of various activators (thin line). (b) The spectrum of the same quartz grain collected immediately after exposure to the electron beam (solid line) and after 10 minutes. of electron beam exposure (dashed line). (c) The brown spectrum of a typical detrital quartz grain is dominated by the peak at 628 nm attributable to a nonbridging oxygen defect.*

is shown in Figure 6.10a and has been deconvoluted to illustrate contributions from five of the most common quartz activators. These defects are related to oxygen vacancies (~600–630 nm), interstitial impurities (460, 500, and 530 nm), and intrinsic lattice defects (420 nm). The exact assignment of many of these activators is still the subject of debate [20, 21]. With increasing electron beam exposure and temperature, the locations of various activators change, as does the color of luminescence (see Section 6.4.4 and Color Plates 6.1a–e). Figure 6.10b illustrates the spectral change in color from blue to brown that occurs during 10 minutes of exposure to the electron beam.

In the geological sciences, the resiliency of quartz has been the motivation for a significant amount of research into the potential application of quartz as a geologic or environmental indicator. For example, quickly fading green or blue quartz is associated with hydrothermal solutions, while more time-stable quartz luminescence intensities have been associated with high temperature quartz formed from a melt [21]. In general, various luminescence colors have been associated with quartz from different environments: plutonic quartz with blue to violet luminescence; matrix quartz in volcanic rocks with red luminescence; quartz from regional metamorphic rocks with brown luminescence; and authigenic environments with no or weak luminescence [14]. Color shifts observed during electron irradiation (e.g., changes from blue to brown or blue to violet) have been used in provenance studies as a measurable characteristic of a given environment [14]. Despite these efforts, the conclusions of most provenance studies are able to make only general associations of quartz with either a metamorphic, volcanic, or authigenic source. An additional complication is that quartz with identical CL properties can be formed in different environments. A more recent study demonstrated that provenance correlations using SEM-CL can be improved when they are combined with features observed by light microscopy (e.g., internal microcracks) and included a diagnostic flowchart that can be used to assign provenance (Figure 4 in Reference 22). Unfortunately, even these additional associations are subjective. In the end, it should be noted that luminescence of quartz can provide another constraint on the provenance for a sample, but must be checked for consistency with other indicators in a sample.

In addition to measurement of color, other features observable in quartz by luminescence can provide additional information. For example, internal zoning can reveal changes in melt chemistry during precipitation. Quartz overgrowths, cementation, and grain dissolution–reprecipitation features in detrital grains can also be recognized through examination by CL.

6.5.4 Accessory Minerals

Quartz, carbonates, and feldspar represent the most commonly encountered minerals in soils and are arguably the most characterized luminescent minerals. However, there are a variety of other luminescent minerals that will likely be encountered in forensic situations. Information analogous to that discussed in previous sections can be applied to these other minerals, including the use of CL as a tool in mineral identification, the detection and identification of trace elements, and the correlation of luminescence with composition.

Less common, but potentially equally important, minerals in forensic CL examinations include zircon ($ZrSiO_4$), apatite ($Ca_5(PO_4)_3OH$), and monazite ($(La,Ca,REE)PO_4$). These are most easily recovered and concentrated for study by CL using heavy mineral separations [23]. Zircon (see Color Plates 6.2m–o) luminesces in shades of bright green, yellow, and blue. Often internal zoning can be resolved even in grain mounts (see Color Plate 6.2n). Apatite typically luminesces in shades of violet, blue, and yellow (see Color Plates 6.3a, b). Monazite, another heavy mineral, exhibits a relatively weak brown luminescence when observed visually (see Color Plate 6.3c). Spectroscopy reveals that this luminescence results only from the sharp emission lines of the REEs (Figure 6.11a). All three of these minerals typically accommodate large and different concentrations of REEs (Figures 6.4 and 6.11), which suggest that they would contain useful information for differentiation among sources. Similar to other minerals, the physical geological processes that control the incorporation of trace elements into the structure of these minerals is not particularly well understood on a practical level.

Other minerals that can exhibit luminescence include diopside ($Ca(Mg,Fe)Si_2O_6$), a pyroxene that luminesces blue; spinel ($MgAl_2O_3$), which exhibits a green luminescence; and gypsum ($CaSO_4 \cdot 2H_2O$), which shows a dull

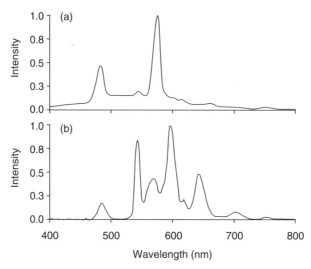

Figure 6.11 *CL spectroscopy of heavy minerals (a) monazite and (b) zircon, which are dominated by the luminescence of REEs.*

blue luminescence. Kyanite, andalusite, and sillimanite are three polymorphs of Al_2SiO_5 that can be distinguished on the basis of CL. The CL of the above-mentioned minerals are illustrated in Color Plates 6.3g–i. More specific information about these and other luminescent minerals can be found in references such as Marshall [4], Pagel et al. [24], and Gorobets and Rogojine [25].

6.6 FORENSIC APPLICATIONS

To this point, this chapter has discussed the theory, instrumentation, and type of analytical information that can be obtained from luminescing minerals. The remainder of this chapter will discuss the ways in which this information can be applied advantageously in forensic casework. As discussed to this point, the major mineral components in soil and sand are luminescent, as are a significant number of minor, but geologically important, minerals. In addition, many of these minerals are used as natural resources in manufactured materials commonly encountered as evidence, including dust, brick, concrete, slag, glass coatings, paint, and tapes [26]. Due to the widespread use of these materials, excellent analytical techniques already exist for the discrimination and sourcing of these materials. CL is an additional analytical method that, in certain investigations, may provide a benefit in speed or, more significantly, in the level of source discrimination obtained. As discussed earlier, CL provides information directly related to the trace element and defect content of materials. Presently, no analytical technique routinely utilized in forensic examinations provides information about the defect structure of a material. Furthermore, the presently applied range of analytical techniques found in a typical forensic

Body Gel Holo Glam

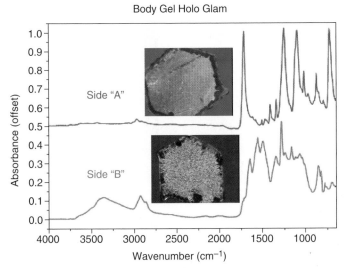

Figure 1.1 *Glitter particle morphology. Two sides of the same particle (700× original magnification) with corresponding ATR FTIR spectra. (From Reference 5.)*

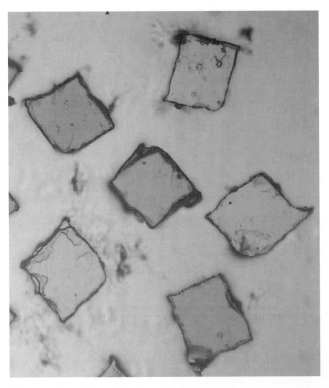

Figure 1.4 *Glitter particles showing cutting machine anomalies (one rounded corner and protrusion on adjacent corner. (From Reference 6.)*

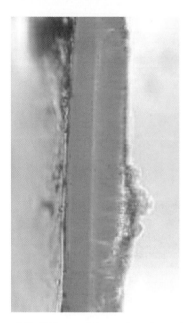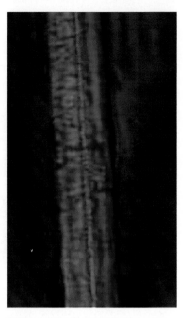

Figure 1.7 *Glitter particle as viewed on edge at 400×. Transmitted light (left) and transmitted light with crossed polars (right). Meadowbrook Inventions, Inc., Crystalina 300 Series, #326, C3625HX, 0.025 in. hexagonal. (Photomicrographs by Michelle Siciliano (Intern) and Gene Lawrence (Criminalist Supervisor), San Diego County Sheriff's Crime Laboratory.)*

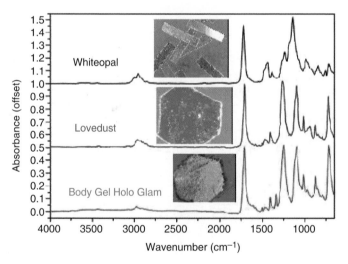

Figure 1.8 *ATR FTIR spectra of individual glitter particles in three different commercial cosmetic products. (From Reference 5.)*

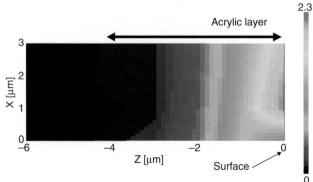

Figure 1.14 *Crystalina 421 peak (2 vs. 1) ratio x-z confocal map. (From Reference 11.)*

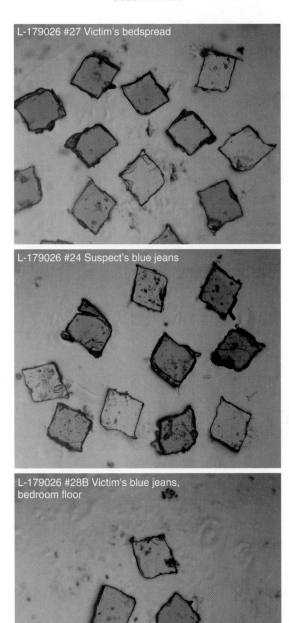

Figure 1.17 *Glitter particles recovered from victim's bedspread (top), suspect's blue jeans (middle) and from Victim's blue jeans (bottom). (From Reference 6.)*

Figure 2.18 Discolored dot pattern on the bottom front of the shirt from the suspected driver of a crashed 1992 Cadillac Deville.

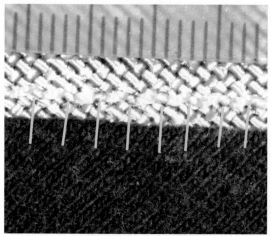

Figure 2.19 Comparison of the discolored dot pattern on the shirt with the seam of the deployed driver side airbag from a crashed 1992 Cadillac Deville.

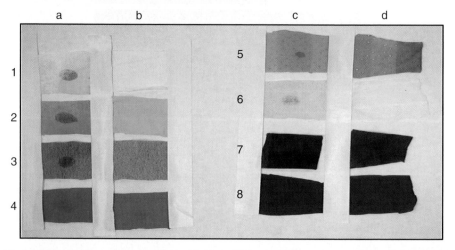

Figure 5.18 Fabric swatches spiked with 50μL MK4 brand pepper spray. Columns (a) and (c): swatch of each column after visualization reaction; columns (b) and (d): swatch of each column with no visualization reaction. (a1, b1) white cotton, (a2, b2) yellow cotton, (a3, b3) blue cotton denim, (a4, b4) pink acrylic; (c1, d1) multihued polyester, (c2, d2) 50/50 cotton/polyester, (c3, d3) brown leather, (c4, d4) red nylon.

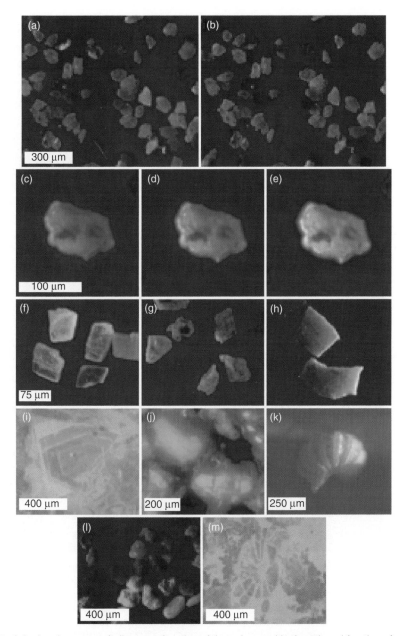

Plate 6.1 *Luminescence fading as a function of time observed in the mineral fraction of a soil from Rockford, IL after (a) 30 s and (b) 120 s of exposure to the electron beam and in a quartz grain from Sistersville, WV exposed to the electron beam for (c) 30 s, (d) 60 s, and (e) 120 s. Carbonate minerals typically luminescence in the range of orange to red as seen in the (f) calcite and (g) dolomite. Aragonite (h), a polymorph of calcite, shows a green luminescence. (i) CL zoning in this polished marble sample likely occurred due to changes in the solution chemistry during crystal growth. (j) Yellow feldspar grains with blue interior zoning caused by unknown activators. This zoning is not observable by SEM-BSE. (k); Faintly blue luminescing biogenic carbonate from a tropical soil sample with some surface precipitation of calcite. (l) CL image of sand from a beach made up largely of biogenic and altered biogenic carbonates that show varying luminescence intensities. (m) Fossilized biogenic carbonate observed in a polished marble sample.*

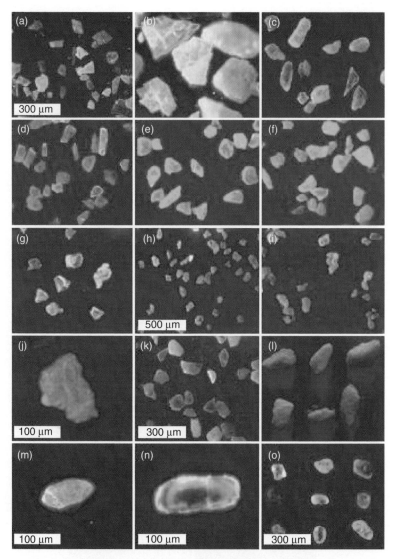

Plate 6.2 *CL (a–f) images of feldspar materials of different compositions that show some of the variety of colors observed in feldspar minerals. Nominally these feldspars are (a) albite, (b) anorthite, (c) anorthite, (d) albite, (e) microcline ($KAlSi_3O_8$), and (f) orthoclase ($KAlSi_3O_8$). Quartz and other silica minerals: (g) violet quartz, (h–j) three quartz samples from the mineral fraction of soil samples, (k) opal, and (l) chert. Varieties of luminescenct zircon ($ZrSiO_4$): (m) yellow zircon, (n) zoned green zircon, and (o) variety of blue zircon crystals.*

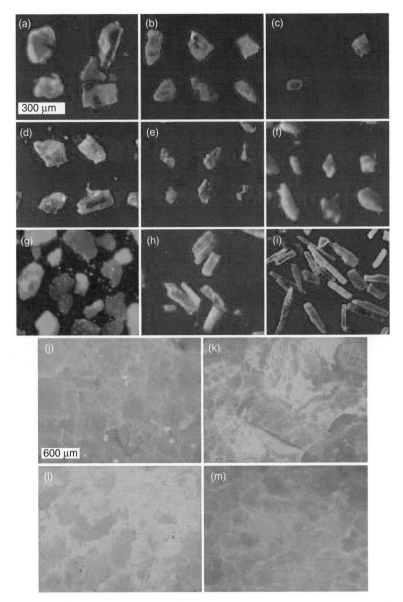

Plate 6.3 Luminescence of accessory minerals of forensic interest: (a,b) apatite ($Ca_5(PO_4)_3OH$), (c) monazite (($Ce,La,Th)PO_4$), (d) diopside ($CaMgSi_2O_6$), (e) spinel ($MgAl_2O_4$), and (f) gypsum ($CaSO_4 \cdot 2H_2O$). Varying luminescence of Al_2SiO_5 polymorphs: (g) andalusite, (h) kyanite, and (i) sillimanite. (j)–(m) Polished marble samples from various sources.

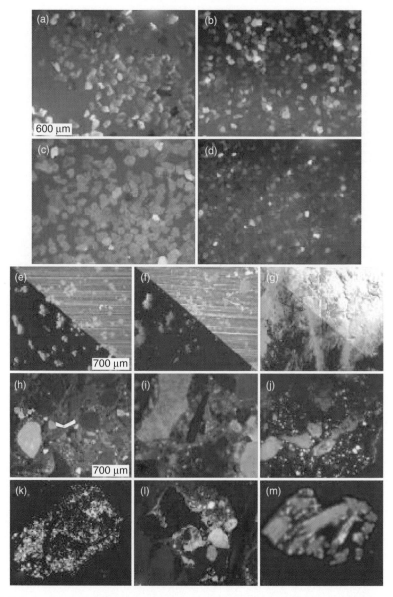

Plate 6.4 *Luminescence from the separated mineral fraction of soils from different environments: (a) Hilo, HI; (b) Juarez, Mexico; (c) Jacksonville, FL; and (d) Alton, IL. In addition to calcite, luminescent minerals such as (e) anatase, (f) talc, and (g) zincite are used as fillers, extenders and pigments (shown in reflected light (upper right) and CL (lower left)). Polished slabs of various concrete blocks (h–j) show that CL can be used to discriminate among concrete on the basis of mineralogy and grain size and shape. The mineralogy of slag can be extremely complicated. CL provides a means to quickly identify various components such as the blue, green, and yellow feldspar grains in (k); the purple quartz, orange calcite, and blue and yellow feldspar in (l); and the orange calcite and yellow feldspar in (m).*

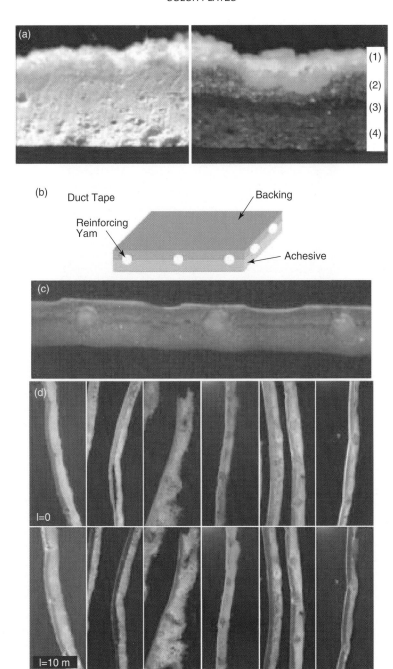

Plate 6.5 Luminescence of paint and duct tape. (a) CL image of a layered white chip of architectural paint. By reflected light (left) the chip appears homogeneous. CL (right) shows that the paint consists of four layers, three of which use calcite as the dominant filler/extender and one that uses anatase (layer 3). (b) A schematic diagram of duct tape construction and a CL image of duct tape (c), which shows that each layer contains a luminescent component. (d) Six duct tape samples are shown to illustrate the range of features and textures that can be observed by CL. The upper row of (d) shows the duct tapes immediately after exposure to the electron beam. The lower row shows the same samples after 10 minutes of electron beam exposure.

Figure 12.7 Fluorescence of scrim fibers under long wave UV.

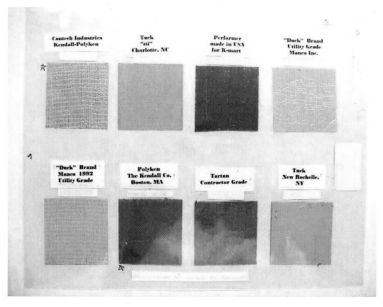

Figure 12.8 Variation in adhesive color of eight different duct tapes.

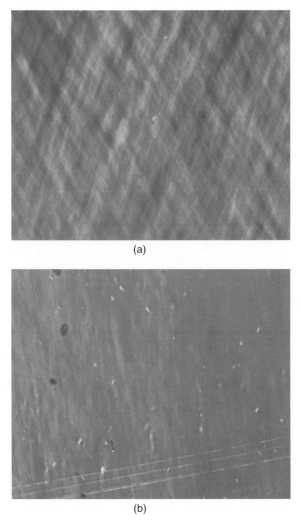

(a)

(b)

Figure 12.19 (a) Biaxially oriented polypropylene (BOPP) film. (b) Monoaxially oriented polypropylene (MOPP) film.

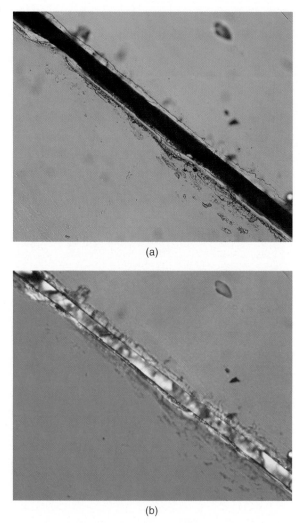

(a)

(b)

Figure 12.20 *(a) A threatening note was taped to a car window using a polypropylene office tape. The questioned piece and known piece were mounted edge to edge. Under cross-polarized light, a slight difference in retardation is noted. (b) The same field as (a) but with a quarter-wave plate (147.5 nm) inserted in the path accentuates the difference in retardation and shows that these two tapes have a significant difference in thickness.*

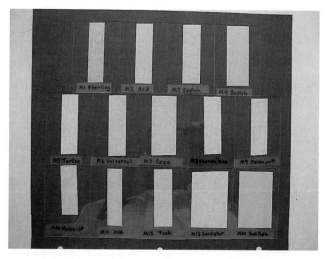

Figure 12.23 *Masking tape color and width are variable.*

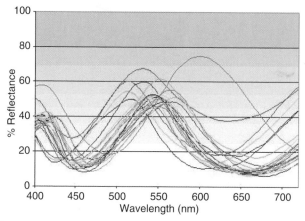

Figure 14.3 *Reflectance scans of 25 individual flakes for Silver-to-Green OVP. Note: Effective angle of incidence is 35° (50× objective).*

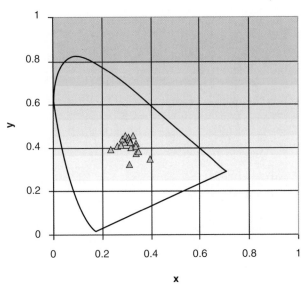

Figure 14.4 *1931 x, y, CIE chromaticity diagram of the color of 25 individual flakes of Silver-to-Green OVP (calculated from the spectra in Figure 14.3).*

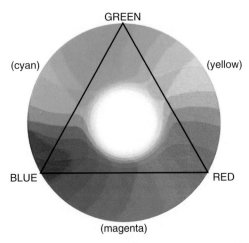

Figure 14.5 *Primary colors for additive color mixing shown in capital letters. The lower case letters in parenthesis are the secondary colors.*

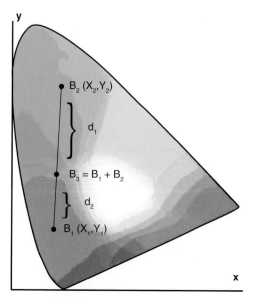

Figure 14.6 *1931 CIE chromaticity diagram showing "straight line" and "center of gravity" rules for additive color mixing.*

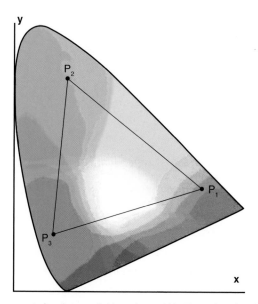

Figure 14.7 *Color gamut showing available colors within the color triangle bounded by three primary colors.*

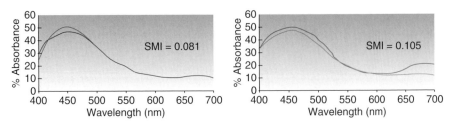

Figure 14.9 *Examples of different spectral profiles and associated spectral matching indices. An SMI of zero indicates an exact spectral match.*

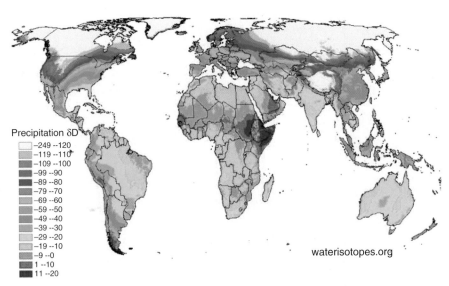

Precipitation δD

- −249 --120
- −119 --110
- −109 --100
- −99 --90
- −89 --80
- −79 --70
- −69 --60
- −59 --50
- −49 --40
- −39 --30
- −29 --20
- −19 --10
- −9 --0
- 1 --10
- 11 --20

waterisotopes.org

Figure 15.5 *Predicted long-term annual average precipitation hydrogen isotope ratios for the land surface. This continuous layer is produced with a combination of empirical relationships between measured precipitation $\delta^2 H$ and latitude and elevation, and a geostatistical smoothing algorithm for variation not explained by that relationship. Measured precipitation values are those maintained in the International Atomic Energy Agency (IAEA) water isotope database. Methods and grids are available at http://waterisotopes.org.*

laboratory either cannot measure trace element chemistry or cannot measure trace elements at the dimensional scale (~1 μm) or sensitivity (~1 ppm) of which CL is capable. In fairness to other analytical techniques, it is should be noted that only a limited number of trace elements act as activators and activator concentrations cannot be analyzed practically by CL. Therefore, CL offers a relatively inexpensive, complementary technique that can be used in forensic investigations.

6.6.1 Screening and Comparison

Cathodoluminescence provides a visual, quick, and relatively nondestructive means to screen and discriminate among many types of samples. Screening provides qualitative information over a length scale range similar to light microscopy (~1 μm to ~25 mm), with the added benefit of being able to visualize information directly related to trace element composition. In addition, the trace elements are observed in situ, where luminescence can be related to surface textures (see Color Plate 6.1m) and internal zoning (see Color Plate 6.2n). In some anthropogenically modified samples, such as paint or duct tape samples, luminescence can be coordinated with structural information.

6.6.2 Identification

Once a familiarity with luminescent materials has been achieved, a microscopist can identify the majority of luminescent components in a particular sample with only a cursory visual or spectroscopic examination. Furthermore, in geologic samples, multiple populations of a given mineral can be identified (e.g., the class of feldspar) and relative abundances of various components can be estimated. In anthropogenic samples, inorganic components such as pigments and fillers extenders are often too small (<1 μm) for identification by light microscopy or X-ray diffraction. Although individual particles can not be resolved by light microscopy, their luminescence can be observed and measured with CL.

In all cases, visual discrimination can be confirmed by spectroscopic examination of the luminescence or by other techniques (e.g., EDS, LA-ICP-MS). Examiners inexperienced with CL must use caution when making an identification based solely on color, as color is not always unique to a given mineral. For example, calcite typically shows bright orange luminescence. Biogenic calcite, however, can show virtually no luminescence or a faint blue luminescence, which could be mistaken for quartz. Similar to the cautions given to assigning peaks in an EDS spectrum, mineral identifications should be made only after consideration of the context in which a potential identification is made. The potential confirmation of CL identification by EDS is a situation in which an SEM-CL detector is advantageous. However, such confirmation can also be done by isolating the particle after analysis by cold cathode CL and transferring it to a variable pressure (VP) SEM with an EDS detector.

6.6.3 Authentication

Various types of luminescence are presently used by gemologists to aid in making distinctions between natural and synthetic gemstones or gemstones from various locations. Similarly, luminescence could be applied to source authentication for various materials. Marbles, for example, which are metamorphosed calcite, can be distinguished on the basis of luminescence color, grain size, internal zoning, and the presence of other minerals. Color Plate 6.3j–m shows the CL images of four marble samples from different sources. The first marble (see Color Plate 6.3j) includes yellow luminescing feldspar grains. The second marble (see Color Plate 6.3k) is distinguishable by fossils that can be observed. The remaining marble samples can be differentiated on the basis of their luminescence colors.

6.6.4 Provenance

The concept of a provenance exam originated from geology, where geologic samples that may be thousands to millions or even billions of years old are studied to determine their original geological source or to place constraints on the environment that existed when they formed. In forensic science, a provenance exam, also known as geographic sourcing, has developed to identify the origin of a sample by placing constraints on the environment from which a sample originated [27]. These exams differ from authentication examination in that no original reference material is available for comparison. In these cases, every type of information available in a sample is used to constrain the geographic source of a sample (e.g., mineral, biological, anthropogenic). CL adds another complementary technique that can contribute to the pool of information that must be interpreted to locate a sample's origin, and can provide corroborating information or confirmation of hypotheses developed from the results of other techniques.

One of the motivations for the research into forensic applications of CL was to apply CL as a provenance indicator. For example, the luminescence colors of quartz can be indicative of a particular environment of formation (e.g., metamorphic or authigenic) or the trace element signature of a zircon crystal may be consistent with a particular region of formation. Similarly, the manganese content, which is directly correlated to the luminescence intensity in calcite, has been linked to various environmental constraints [28]. Unfortunately, while the luminescence signal is a function of a sample's original environment and subsequent history, the CL signature is not necessarily unique to any given environment. Furthermore, the geologic research into these topics is still limited and additional research is necessary. While CL cannot yet stand independently as an indicator of provenance, the evidence provided by CL *can* provide evidence indicative of a particular environment, and this value should not be overlooked as an investigative tool in forensic examinations.

6.7 GEOLOGICAL SAMPLES: SOIL AND SAND

Soil and sand are routinely encountered as evidence in forensic laboratories; however, at present, many forensic laboratories either do not accept or are unable to adequately test soil evidence. One reason for this is that the mineralogical and analytical knowledge required to examine and interpret evidence requires a large amount of training and experience. CL provides a means to supplement the information obtained from traditional mineralogical examinations or, in certain cases such as screening, to simplify the examination.

In a screening or comparison examination, low magnification CL microscopy can be used to visually compare samples via bulk mineralogy. Color Plates 6.4a–d shows CL images of samples from four distinct geologic and geographic environments (Hawaii, Mexico, Florida, and Illinois). The luminescence in the Hawaiian sample is attributable mainly to orange luminescing carbonates and fragments of volcanic glass with blue luminescent coatings of TiO_2. The Mexican soil is composed of brightly luminescing blue and green grains of feldspar surrounded by weakly luminescing brown and grey grains of quartz.

Soil samples from within the continental United States typically show less variation due to more uniform weathering processes across the continent that produce soils with more similar bulk mineralogical compositions. A purely visual CL soil comparison survey conducted by the Smithsonian Institute of 20 splits from 10 soil samples found that while soils from more exotic geologic environments could easily be distinguished using CL, soil samples consisting mainly of feldspar and quartz were considerably more difficult to associate [29, 30]. The samples from Florida (see Color Plate 6.4c) and Illinois (see Color Plate 6.4d) are typical of luminescence in many soils from the continental United States. These samples consist mainly of varying proportions of authigenic quartz and feldspar, both exhibiting blue luminescence. The Florida sample, which has been highly weathered, now consists almost exclusively of blue-grey luminescing quartz with only a small feldspar content relative to the Illinois sample. The Florida sample also contains a small component of orange luminescing sillimanite. In these types of samples, which will likely comprise the majority of the samples encountered as evidence, distinction between soils using low magnification CL becomes less obvious. Nonetheless, many of these soils can be differentiated using the full suite of luminescing minerals, their CL spectra, and the relative proportions of the major mineral types.

The significance of soil examinations using CL remains tied to many of the same issues that soil examinations have always faced: Over what distance do CL activators vary? Can specific environments be assigned a "CL signature"? What size sample is representative of an environment? The answer to all these questions is that the significance of geological (and CL) examinations is heavily dependent on the geology of the environment being studied and the question being asked. Therefore, the significance of CL in soil evidence must be evaluated on a case-by-case basis.

6.8 ANTHROPOGENIC MATERIALS

Some of the most heavily utilized industrial minerals, such as calcite and anatase, are luminescent. Color Plates 6.4e–g shows CL images of three inorganic materials commonly used as pigments or fillers/extenders: anatase (TiO_2), talc, and zincite (ZnO). It should be noted that rutile (TiO_2) does not luminesce. This category of mineral evidence is used in a variety of products encountered as forensic evidence including paint (automotive and architectural), plastics, adhesives, cleansers, and foodstuffs. Depending on the end-use, industrial minerals undergo varying amounts of processing. When larger particle sizes can be used, crushing of the mined mineral is adequate. In this case, the mineral retains the trace element chemistry of the mined mineral. In cases where a finer quality ingredient is needed, as is commonly the case for automobile pigments, the inorganic compound is synthesized. In this case, the resulting pigment will contain an altered trace element chemistry and typically be more pure than the original mineral (unless the material is doped to achieve specific properties). For forensic applications of CL, this also implies that the synthetic form of a mineral from different sources generally will not be as distinguishable. Fortunately for the forensic scientist, the use of these industrial products is governed by economics and the least expensive suitable material available is generally used.

6.8.1 Cement and Concrete

Concrete is a mixture of cement and aggregate. Cement consists of calcium silicates, calcium and iron aluminates and gypsum, which react with water and the aggregate to form various hydration phases. The luminescence properties of these phases have yet to be characterized; however, luminescence of specific cement phases might provide rapid insight into the conditions of formation or state of degradation for a given sample. Color Plates 4h–j show CL images of three different polished slabs from concrete blocks and illustrates some of the variation that may be encountered. Here the luminescence shown is dominated by differences in the aggregate mineralogy, which can consist of quarried materials, recycled concrete or slag. A combined study of aggregate and cement luminescence could provide source discrimination or constraints on sample origin.

6.8.2 Slag, Fly Ash, and Bottom Ash

Slag, fly ash, and bottom ash are the waste products of industrial processes, such as iron and power production. The high temperatures of these processes result in the formation of unusual phase assemblages. Color Plates 6.4k–m shows examples of three samples from a steel plant slag. Color Plate 6.4k is a conglomerate consisting mainly of yellow, green, and blue luminescing feldspar particles. Color Plate 6.4l shows another conglomerate with two purple quartz grains and blue and yellow feldspar minerals in a calcite matrix. Color Plate 6.4m shows orange calcite laths with a yellow pyroxene (diopside). At the very least, these unusual CL textures would give some indication that the particles recovered were characteristic of an anthropogenic process. Further research may facilitate the identification of a specific process or even manufacturing source.

6.8.3 Glass

Glasses are routinely examined by fluorescent light to identify float surfaces or distinguish between bulk glass samples. The cathodoluminescence of glass was first studied for forensic applications by Stoecklein and Göbel [1] to show that the Sn^{2+} layer of the float surface can be distinguished from the nonfloat surface. Furthermore, they concluded that Sn^{4+} in SnO_2 layers can be distinguished from Sn^{2+} of the float surface. Our results confirm that the spectrum of the bulk glass can be spectroscopically distinguished from that of the float surface (Figure 6.12).

The nonfloat surface, which can be left uncoated or coated with one or more nanolayers of a coating such as SnO_2, TiO_2, or Cr-spinel, is more complex. The analytical difficulty is that the coating is thinner than the electron penetration depth. As a result, the CL signal also originates from the underlying bulk glass. Additionally, the nonfloat surface may also contain traces of Sn^{2+} from tin vapor over the bath in the float process. Finally, none of the coatings studied (identified as containing Sn and Cr-spinel by literature, laser ablation inductively coupled plasma mass spectrometry (LA-ICP-MS), or total reflection X-ray fluorescence spectrometry (TXRF)) appear to luminesce. The nonfloat (i.e., coated) surface of these glasses instead showed a weak luminescence that appeared to be a combination of a weak Sn^{2+} signal and the bulk glass (Figure 6.12a). The identification of a distinct Sn^{4+} luminescence could not be confirmed.

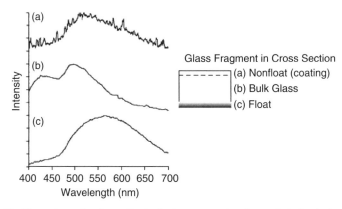

Figure 6.12 *The schematic shows the typical components of a coated sheet glass fragment (a) the nonfloat side to which coatings are typically applied, (b) the bulk glass, and (c) the float side into which Sn^{2+} has diffused. Adjacent are spectra from a spinel-coated sheet glass.*

6.8.4 Paint

Paints were one of the first forensic materials examined by CL [1, 2, 31]. CL is most useful in the study of architectural paints, which often consist of multiple layers of paint containing white pigments that are difficult to identify and discriminate. Color Plate 6.5a shows a cross section from a wall with several layers of white paint. Note that when studying paints by CL, one advantage is that a true thin section is not necessary as in the case for light microscopy or infrared spectroscopy. From this image, four layers of white paint can be identified, including one layer based on an anatase filler (layer 3) rather than calcite, which dominates the other layers. The luminescence in automobile paints is typically much weaker and due to the use of organic and largely nonluminescing inorganic pigments. The exception is the use of anatase as a white pigment, which results in a blue luminescence in automobile paint layers that use this pigment.

6.8.5 Duct Tape

Duct tape consists of three components: an adhesive layer, a colored backing layer and a network of reinforcing yarns that provide strength (see Color Plate 6.5b). The reinforcing yarn is typically cotton along the length of the tape, which tears easily, and polyester along the width of the tape to guide the tear perpendicular to the length of the tape. Duct tape is typically characterized first by physical properties such as color, texture, and fiber count, followed by elemental and chemical analysis using FTIR spectroscopy, X-ray diffraction, and SEM-EDS to identify the polymers and mineral components in each layer.

CL provides an alternative or supplemental means by which to examine duct tapes. Color Plate 6.5c shows the CL image of a duct tape sample in cross section. The backing layer shows a green luminescence, which is possibly attributable to zincite. The reinforcing fibers, which can be cotton or polyester, have a blue luminescence due to anatase or an organic optical brightener. The adhesive layer in this sample shows a gray to blue luminescence, possibly from the natural rubber and additives such as anatase and talc, along with some larger particles of calcite, which is used as a filler/extender. Within duct tape, the layered structure, voids, and pigment distribution can be observed visually. For example, the intricate structure of the green backing layer is visible in the fourth sample (see Color Plate 6.5d).

Over time, luminescence fading results in an overall change in luminescence from green and blue colors to brown and orange colors. Color Plate 6.5d shows CL images of six duct tape samples collected immediately after exposure to the electron beam (top row) and after 10 minutes of electron beam exposure (bottom row). Useful information can be obtained from each time period, but the examiner must be certain that all images being compared are collected after the same length of exposure to the electron beam. However, after approximately 10 minutes, the luminescence remains fairly constant and stable. While these results were obtained solely with duct tape, it is likely that other types of tape and polymers that use inorganic fillers/extenders could be studied advantageously using CL.

6.9 CONCLUSIONS AND OUTLOOK

This chapter has outlined the fundamental causes for cathodoluminescence and the types of information that can be obtained through CL analysis. It has also discussed applications of CL to a variety of categories of forensic evidence. In the most basic applications, CL provides a complementary method to commonly applied microanalytical techniques. For example, CL can provide a simple and quick means to screen soils or discriminate among feldspar minerals. CL can also provide information that other commonly applied forensic methods do not provide, such as visualization of trace elements and structural defects in a material. Finally, in provenance examinations, the most challenging question in forensic geology, CL can provide additional environmental or geologic constraints on the origin of a sample.

As with many of the forensic science disciplines, the role of forensic soil, mineralogical, and trace evidence examinations is expanding beyond their traditional role of trial support to be more investigative. Additional research into the correlation of luminescence with specific environments, as well as further research into the applications of CL for the comparison and sourcing of anthropogenic materials, will facilitate the use of CL for both traditional and investigative trace evidence examinations in the future and provide greater understanding of the significance of evidentiary associations using CL.

ACKNOWLEDGMENTS

The authors gratefully acknowledge the support of our colleagues and friends during this project. Very special thanks are extended to skip Palenik of Microtrace for providing several of the samples used throughout this study and, more importantly, for providing the first author with his introduction to CL. The loan of a spectrometer to us by Linda Lewis of Oak Ridge National Laboratory during the upgrade of our system was invaluable. Don Marshall has continually provided kind and timely advice regarding practical CL. Additionally, the authors wish to thank Maureen Bottrell, Roger Keagy, Robert Koons, David Korejwo, Kim Mooney, Sorena Sorenson, Wilfried Stoecklein, Ed Suzuki, and Jodi Webb for their discussions about CL and forensic materials examinations. The FBI Laboratory is acknowledged for providing the funding for a postdoctoral fellowship, administered by Oak Ridge Institute for Science Education, which supported this work.

REFERENCES

1. Stoecklein W, Göbel R (1992). Application of cathodoluminescence in paint analysis (AMF O'Hare, Chicago). *Scanning Microscopy International* **6**: 669–678.
2. Hopen T (1997). Cathodoluminescence of paint samples. Inter/Micro-97, Chicago, IL.
3. Crookes W (1879). *Contributions to molecular physics in high vacua. Philosophical Transactions of the Royal Society* **170**:641–642.
4. Marshall DJ (1986). *Cathodoluminescence of Geological Materials.* Unwin Hyman, Boston.
5. Götze J (2000). *Cathodoluminescence Microscopy and Spectroscopy in Applied Mineralogy.* Technische Universitat Bergakademie, Freiberg, Germany.
6. Yacobi BG, Holt DB (1990). *Cathodoluminescence Microscopy of Inorganic Solids.* Plenum Press, New York.
7. Götze J (2002). *Potential of cathodoluminescence (CL) microscopy and spectroscopy for the analysis of minerals and materials. Analytical and Bioanalytical Chemistry* **374**:703–708.
8. Gatan (2005). www.gatan.com/sem/cl.html.
9. Lee MR, RW Marting, Trager-Cowan C, Edwards PR (2005). Imaging of cathodoluminescence zoning in calcite by scanning electron microscopy and hyper-spectral mapping. *Journal of Sedimentary Research* **75**:313–322.
10. Marshall DJ, Kopp OC (2000). The status of the standards program of the Society for Luminescence Microscopy and Spectroscopy. In: *Cathodoluminescence in Geoscience*, Pagel M et al. (eds.). Springer Verlag, New York.
11. Remond G, Balk L, Marshall DJ (eds.) (1995). *Luminescence.* Scanning Microscopy International, Chicago.
12. Habermann D, Neuser RD, Richter DK (1998). Low limit of Mn^{2+}-activated cathodoluminescence of calcite: state of the art. *Sedimentary Geology* **16**: 13–24.
13. Gillhaus A, Richter DK, Meijer J, Neuser RD, Stephan A (2001). Quantitative high resolution cathodoluminescence spectroscopy of diagenetic and hydrothermal dolomites. *Sedimentary Geology* **140**:191–199.
14. Richter DK, Götte Th, Götze J, Neuser RD (2003). Progress in applications of cathodoluminescence (CL) in sedimentary petrology. *Minerology and Petroloy* **79**:127–166.
15. Habermann D, Neuser RD, Richter DK (1996). REE-activated cathodoluminescence of calcite and dolomite: high-resolution spectrometric analysis of CL emission (HRS-CL). *Sedimentary Geology* **101**:1–7.
16. Reeder RJ (1991). An overview of zoning in carbonate minerals. In: *Luminescence Microscopy and Spectroscopy—Qualitative and Quantitative Applications*, Baker CE, Koop OC (eds.). SEPM (Society of for Sedimentary Geology) Short Course, No. 25, pp. 77–82.
17. Götze J, Krbetschek MR, Habermann D, Wolf D (2000). High-resolution cathodoluminescence studies of feldspar minerals. In: *Cathodoluminescence in Geoscience*, Pagel M et al. (eds.), Springer Verlag, New York.

18. Krbetschek MR, Götze J, Irmer G, Rieser U, Trautmann T (2002). The red luminescence emission of feldspar and its wavelength dependence on K, Na, Ca—composition. *Minerology and Petrology* **76**:167–177.

19. Götze J, Habermann D, Neuser RD, Richter DK (1999). High-resolution spectrometric analysis of rare earth elements-activated cathodoluminescence in feldspar minerals. *Chemical Geology* **153**:81–91.

20. Kalceff MAS, Rhillips MR, Moon AR, Kalceff W (2000). Cathodoluminescence microcharacterisation of silicon dioxide polymorphs. In: *Cathodoluminescence in Geoscience*, Pagel M et al. (eds.). Springer Verlag, New York.

21. Götze J, Plötze M, Habermann M, Habermann D (2001). Origin, spectral characteristics and practical applications of the cathodoluminescence (CL) of quartz—a review. *Mineralogy and Petrology* **71**:225–250.

22. Bernet M, Bassett K (2005). Provenance analysis by single-quartz-grain SEMCL/optical microscopoy. *Journal of Sedimentary Research* **75**:492–500.

23. Mange MA, Heinz FWM (1992). *Heavy Minerals in Colour.* Chapman and Hall, London.

24. Pagel M, Barbin V, Blanc P, Ohnenstetter D (eds.) (2000). *Cathodoluminescence in Geosciences.* Springer, Berlin.

25. Gorobets BS, Rogojine AA (2002). *Luminescent Spectra of Minerals: Reference Book.* All-Russia Institute of Mineral Resources (VIMS), Moscow.

26. Palenik CS, Buscaglia J (2005). Applications of cathodoluminescence in forensic geology. *Geochimica et Cosmochimica Acta*, Goldschmidt Conference Abstracts Volume, A591.

27. Palenik SJ (1979). The geographic orgin of dust. In: *The Particle Atlas*, Volume X, Mcrone WC (ed.). Ann Arbor Scientific Publishers, Ann Arbor, MI.

28. Machel HG (2000). Application of cathodoluminescence to carbonate diagenesis. In: *Cathodoluminescence in Geoscience*, Pagel M et al. (eds.). Springer Verlag, New York.

29. Avery V (1995). Use of cathodoluminescence in the FBI forensic geology study. Smithsonian Report.

30. *Smithsonian* (1995). FBI forensic soil/sands final report. Department of Mineral Sciences. Smithsonian Institute.

31. Stoecklein W, Franke M, Goebel R (2001). Cathodoluminescence in forensic science. In: *Problems of Forensic Science*, Volume XLVII, Wojcikiewics J (ed.). Institue of Forensic Research Publishers, Krakow, Poland, pp. 122–136.

7

Forensic Application of DART^{TM} (Direct Analysis in Real Time) Mass Spectrometry

*James A. Laramée

Principal Chemist, EAI Corporation, a Subsidiary of SAIC, Inc., Abingdon, Maryland

Robert B. Cody

Product Manager, JEOL USA, Inc., Peabody, Massachusetts

J. Michael Nilles

Senior Scientist, Geo-Centers, Inc., a Subsidiary of SAIC, Inc., Abingdon, Maryland

H. Dupont Durst

Research Chemist, Edgewood Chemical Biological Center, Edgewood, Maryland

7.1 INTRODUCTION

DART^{TM} (Direct Analysis in Real Time) is a novel mass spectrometer sample ionization technique that has demonstrated utility for the analysis of both gaseous and condensed-phase samples. When combined with a high-resolution

* Author to whom correspavdance should be directed.

Forensic Analysis on the Cutting Edge: New Methods for Trace Evidence Analysis, Edited by Robert D. Blackledge.
Copyright © 2007 John Wiley & Sons, Inc.

mass spectrometer, it enables the analyst to rapidly detect and unambiguously identify a wide variety of analytes. Solid, liquid, and gas samples can be analyzed directly, under ambient conditions, with minimal or no sample preparation. This decreases analysis time, increases productivity, and prevents the numerous problems associated with the generation, handling, and disposal of hazardous waste streams [1, 2]. Unlike other desorption ionization techniques, DART does not expose the sample, or the instrument operator, to high voltages, ionizing radiation, or sprays of environmentally burdening toxic solvents. The prototype DART device was constructed and tested in 2002, and a patent application filed in 2003; since then, DART has been used increasingly in forensic applications. DART has been commercially available for two years. Now we report additional applications that may be of interest to the forensic community.

7.2 EXPERIMENTAL

Mass spectra were acquired with an AccuTOF (JEOL USA, Inc., Peabody, MA, USA) high-resolution time-of-flight mass spectrometer equipped with an atmospheric pressure ionization interface and a DART ionization source. The scientific principles underlying the operation of DART have been described in detail previously [1, 2]. Briefly, helium was excited in a 5 keV electrical discharge [2], and subsequent electrodes remove most ionic species, 150 and 250 VDC for electrodes 1 and 2, respectively (Figure 7.1). The gas stream was warmed to 200 °C. After exiting the DART device, excited-state species undergo chemical reactions with other reagents, such as atmospheric water. In this case, hydronium (H_3O^+) ions are formed, which react with the analyte (M) to produce protonated [MH]$^+$ cations. These enter the mass spectrometer for mass analysis and detection.

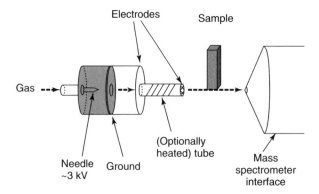

Figure 7.1 *Schematic diagram of the DART ion source.*

7.3   DRUG AND PHARMACEUTICAL ANALYSIS

More than two dozen licit and illicit drugs were successfully analyzed by DART. Basic compounds typically produce protonated [MH]⁺ cations with excellent detection sensitivity. Examples include amphetamines, opiates, cocaine, and lysergic acid diethylamide (LSD). Inert ingredients give lower instrument responses, thus simplifying the identification of the active components. High-fidelity mass spectra of a steroid (methandrostenolone), a tranquilizer (diazepam), and a narcotic analgesic (acetaminophen and oxycodone) are easily obtained without breaching the tablet (Figure 7.2). Acidic compounds were also successfully analyzed as the ammoniated neutral, [M+NH₄]⁺, in the presence of dilute ammonium hydroxide.

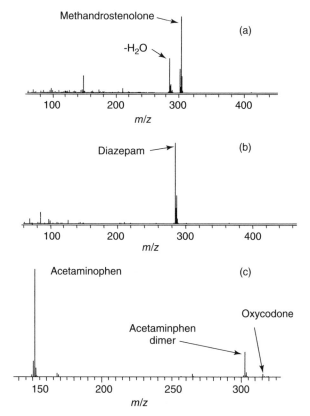

Figure 7.2 *DART analysis of drugs in pill form showing protonated molecules for (a) methandrostenolone, (b) diazepam, and (c) oxycodone "512" (325 mg acetophenone, 5 mg oxycodone). A fragment corresponding to water loss from methandrostenolone is also observed for (a), and a proton-bound dimer is also observed for the acetaminophen in (c).*

7.3.1   Confiscated Samples

An early case study analyzed four confiscated, unmarked gelatin capsules. DART identified cathinone (a banned stimulant), caffeine, and magnolia bark (a Chinese herbal medicine) in the first capsule. The next two capsules contained only magnolia bark, and the remaining capsule contained conjugated linoleic acid, an over-the-counter weight-loss aid.

7.3.2   Endogenous Drugs

Marijuana is easy to identify using DART. Two tiny (1.5 mm^2) leaf particles showed different cannabinoid contents (Figure 7.3a,b). "Leaf Sample A" (Figure 7.3a) contained a small amount of cannabinol as indicated by a peak at m/z 311.201 and a trace of tetrahydrocannabinol (m/z 315.231). By contrast, "Leaf Sample B" (Figure 7.3b) contained more tetrahydrocannabinol, with little or no cannabinol.

A single poppy seed is sufficient sample to detect morphine and codeine by the DART method (Figure 7.3c). The maximum concentration of morphine and codeine per seed is 20 and 10 nanograms, respectively [3]. Cooked food

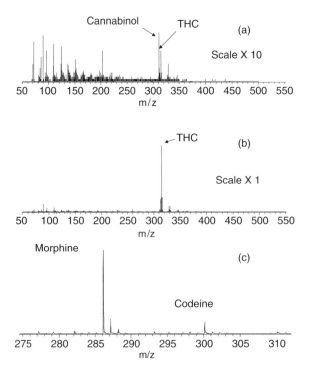

Figure 7.3 *Drugs in vegetable materials: (a) Confiscated marijuana sample containing a small amount of cannabinol and tetrahydrocannabinol (THC), (b) confiscated marijuana sample containing larger amounts of THC and only trace amounts of cannabinol, and (c) opiates in a single poppy seed.*

products containing poppy seeds (bagels, crackers, etc.) showed lower amounts of these opiate alkaloids.

7.3.3 Drug Residues on Surfaces

Detection of drug residues on a variety of surfaces was achieved with minimal sample preparation and instantaneous results. Cocaine residues present on a confiscated facial tissue and an aluminum foil smoking pipe were detected within seconds, simply by placing them in front of the DART beam (Figure 7.4). An empty glass vial that had previously contained a 250 µg/mL standard of fentanyl was analyzed in a similar fashion (Figure 7.5a) confirming the pres-

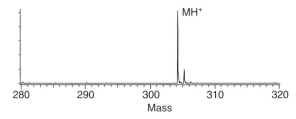

Figure 7.4 *Cocaine residue detected as [M+H]⁺ (m/z 304.1549) on a confiscated tissue.*

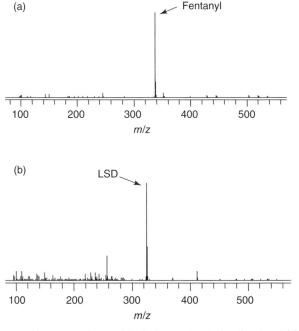

Figure 7.5 *Drug residues on surfaces, detected as protonated molecules: (a) fentanyl residue on "empty" glass vial containing a 250 ppm solution, and (b) mass spectrum of blotter paper spiked with 1 µg of LSD.*

ence of residual fentanyl. A piece of blotter paper spiked with only 1/250th of a typical street dose of LSD gave the expected mass spectrum, with an excellent signal-to-noise ratio (Figure 7.5b).

The advantage of an exact mass measurement in forensic science and other disciplines is that it reduces the occurrence of false positive test results. A false positive occurs when two or more chemicals occupy the same spectral position. When that happens, it is not possible to uniquely assign the instrument response to a particular analyte. For example, the compound triethanolamine, is present in many cosmetic and personal hygiene products. As a consequence it is commonly found on currency. Methamphetamine is also occasionally found on currency, although usually for less innocent reasons. Both of these compounds have the same nominal mass of 150 Da. Without additional analytical selectivity, a false positive occurs. Accurate mass measurement by the AccuTOF mass spectrometer easily distinguishes between the ubiquitous triethanolamine (m/z 150.113) and the illicit methamphetamine (m/z 150.128).

7.4   SAMPLES FROM THE HUMAN BODY

7.4.1   Fingerprints

Fingerprints typically contain a wide variety of chemicals from many different compound classes. This high chemical background increases the probability of false positives when testing for trace-level analytes. Exact mass measurements go a long way toward mitigating this problem. One example is a study in detecting low levels of explosives transferred from a fingertip to a secondary material. One microgram of 2,4,6-trinitrotoluene (TNT) was applied to a fingertip and then successively fingerprinted onto an ABS plastic strip. The TNT molecular radical anion at m/z 227.018 was successfully discriminated from the presence of myristic acid at m/z 227.201, a common component in cooking oils (Figure 7.6). Even the 52nd fingerprint showed the presence of TNT with a signal-to-noise ratio of 20. The minimal sample preparation required, combined with ease-of-use and speed, makes DART a very attractive screening system.

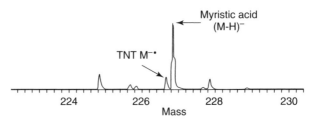

Figure 7.6   *Fingerprint on ABS plastic after depositing 1 µg of TNT on the fingertip. The mass spectrum shows that both TNT M⁻ and myristic acid [M-H]⁻ are clearly separated and identified by their exact masses.*

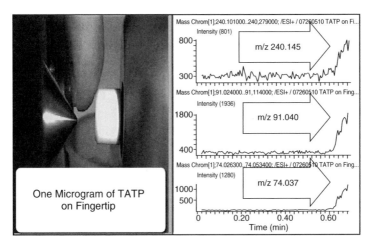

Figure 7.7 *Identification of TATP on a fingertip. The left panel shows the fingertip placed in front of the DART source and the right panel shows the ion current traces for the TATP ammonium adduct (top) and the characteristic TATP fragments $C_3H_7O_3^+$ (middle) and $C_3H_6O_2^+$ (bottom).*

Recent world events have unfortunately seen the use of triacetone triperoxide (TATP). The ability to rapidly identify the presence of TATP on a fingertip or some other matrix, without the need for solvent extractions or time-consuming chromatography steps, is a significant DART capability (Figure 7.7). Identification of TATP was achieved by monitoring three characteristic signature ions. These were the ammonium adduct of TATP at mass 240.145, a tris(-oxygen) dimethyl fragment ion at mass 91.040, and the bis-(oxygen) dimethyl fragment ion at mass 74.037.

Another example involved the analysis of a fingerprint made on a glass vial after the subject had briefly held an oxycodone tablet. Oxycodone, aspirin, and human sweat components were found (Figure 7.8).

7.4.2 Bodily Fluids

DART can rapidly screen large numbers of body fluid samples for expected and unexpected substances. This can be achieved without chromatographic separation or flow-injection systems. Quantitative analysis has been demonstrated with the addition of an internal standard. Laboratories responsible for analysis of time-critical samples could benefit from the DART method.

Urine Positive-ion urinalysis by DART gave simple and easily interpretable mass spectra. Creatinine, urea, amino acids, drug metabolites, and other nonendogenous compounds were found. One of the first applications

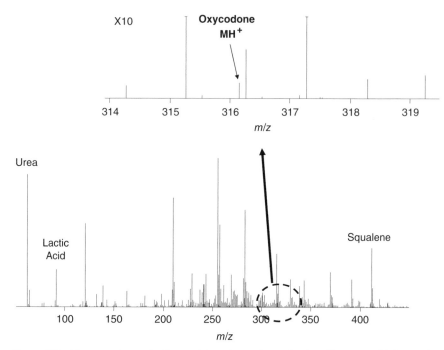

Figure 7.8 *Mass spectrum of a fingerprint made on glass after handling an aspirin/oxycodone tablet. The aspirin and oxycodone are detected together with compounds present in sweat including lactic acid, urea, fatty acids, and squalene.*

of DART was the detection of ranitidine (Zantac®) (Figure 7.9a) and atenolol (Figure 7.9b) in urine. Negative-ion mass spectra of urine were more complex than the positive-ion spectra, with a wide variety of compounds detected including organic acids, metabolites, vitamins, and nucleotides (Figure 7.9c).

Quantitation is demonstrated using the date rape drug, γ-hydroxybutyrate (GHB) [4]. A calibration curve was linear from 5 to 800 parts per million. Good analytical technique includes the use of internal standards, in this case GHB-d$_6$ (Figure 7.10). A correlation coefficient of greater than 0.999 was obtained for five replicate analyses on five successive days.

7.5 CONDOM LUBRICANTS

Condoms are increasingly used by sexual assailants [5]. Thus, the rapid detection of the condom lubricant nonoxynol-9 from evidence items should aid forensic investigators. Nonoxynol-9 is a nonylphenol-ether polymer with ethylene oxide repeat units, $C_{15}H_{24}O(C_2H_4O)_n$. It is well suited for DART analysis.

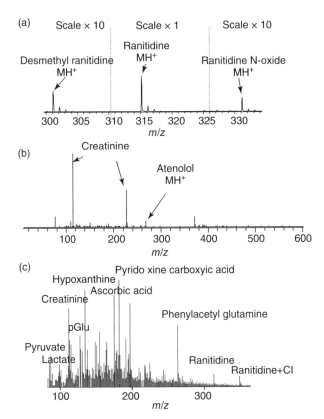

Figure 7.9 *Urinalysis of subjects taking prescription drugs. (a) Positive-ion mass spectrum of urine from subject "A" taking ranitidine, (b) positive-ion mass spectrum of urine from subject "B" taking atenolol, and (c) negative-ion mass spectrum of urine from subject "A" showing many common compounds present in urine (for readability, only selected abundant peaks are labeled).*

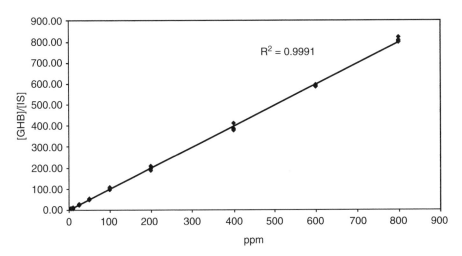

Figure 7.10 *Standard curve for urine spiked with γ-hydroxybutyrate (GHB) in the range 15–800 ppm. Internal standard (GHB-d$_6$) was added at a concentration of 50 ppm. Five replicates were measured on five successive days.*

The positive-ion DART mass spectrum obtained from the swab of an unused condom showed a series of peaks corresponding to the protonated neutrals (n = 3 to 12); less abundant $[M+NH_4]^+$ peaks were also observed. (Figure 7.11a). A vaginal swab, taken 3.5 hours after sexual intercourse with a subject using a condom, likewise shows the presence of nonoxynol-9 (Figure 7.11b).

7.6 DYES

7.6.1 Self-Defense Sprays

Self-defense sprays contain a lachrymatory agent and sometimes a dye. Baseball caps were spiked with different sprays [6] and aged for several weeks before analysis. The mass spectrum (Figure 7.12) showed the presence of capsaicin and dihydrocapsaicin, the fluorescent dye BBOT (2,5-bis(5-*tert*-butylbenzoxazol-2-yl)thiophene), and *o*-chlorobenzaldehyde, a degradation product of CS (2-chlorobenzalmalononitrile). Even the small mass spectral peaks were clearly resolved from isobaric interferences. All elemental compo-

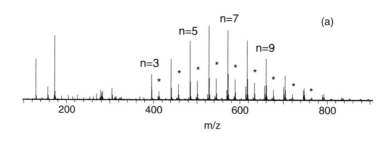

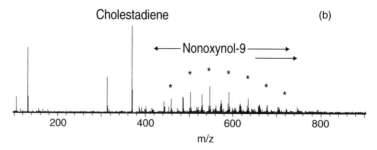

Figure 7.11 *(a) Swab from a condom showing nonoxynol-9 (as [M+H]⁺). (b) Vaginal swab taken after intercourse showing nonoxynol-9 detected as [M+NH4]⁺.*

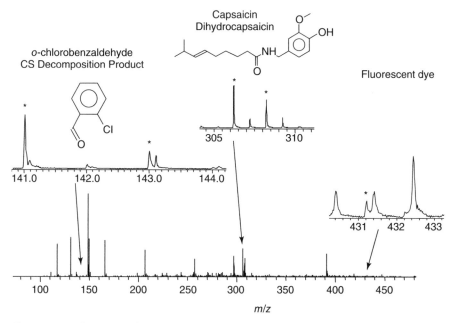

Figure 7.12 *Detection of capsaicin ("pepper spray"), the fluorescent dye BBOT, and o-chlorobenzaldehyde (a CS decomposition product) on a baseball cap that had been sprayed with self-defense spray.*

sition assignments were confirmed by exact mass measurements with excellent agreement between the expected and measured isotopic abundances.

7.6.2 Currency-Pack Dye

Exploding bank security devices in currency packs contain a lachrymator and the dye 1-methylaminoanthraquinone (MAAQ). The bright red dye marks the money as stolen, rendering it worthless. Current analytical methods for MAAQ require extensive sample preparation[6–9]. DART unambiguously confirmed the presence of MAAQ on samples of currency and clothing contaminated with MAAQ without any sample preparation whatsoever (Figure 7.13a). The abundant [MH]+ peak at m/z 238.086 readily distinguishes MAAQ from other red dyes that might be present, such as rhodamine from red marker pens (Figure 7.13b).

7.7 EXPLOSIVES

No explosive has yet been found that DART cannot directly and instantaneously detect, and this includes the military significant compounds RDX and HMX, which have very low vapor pressures. Various explosives have been

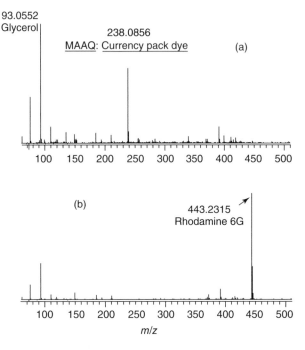

Figure 7.13   *DART mass spectra of red dyes on currency: (a) currency-pack red dye (MAAQ) and (b) rhodamine 6G from a red marker pen.*

detected, and unambiguously identified, in or on a wide variety of matrices, including ditch water, luggage, clothing, airline boarding passes, and living tissue (human and plant).

Peroxide-based explosives are making a comeback after first being studied by the U.S. Army in the 1920s. These explosives can easily be synthesized in one-pot reactions at room temperature, but have undesirable characteristics to render them unsuitable for military and industrial applications. Recent world events have unfortunately highlighted the use of both hexamethylene triperoxide diamine (HMTD) and triacetone triperoxide (TATP). One hundred nanograms of HMTD were spiked onto a DC Metro rail card and the card placed in the DART beam. A peak corresponding to the $[M-H + NH_3]^+$ cation for HMTD was immediately observed (Figure 7.14a).

Nitroaromatic explosives such as trinitrotoluene, trinitrobenzene, and isomers of dinitrotoluene were DART detected as molecular radical anions with excellent sensitivity. Their mass spectra are similar to those observed in electron capture negative-ion mass spectra. More energetic operating conditions produce fragment ions such as $[M-H]^-$, $[M-NO]^-$, $[M-OH]^-$, NO^-, and NO_2^- that provide isomer differentiation. A detonated improvised explosive device (IED) was examined for traces of TNT. High resolution and accurate

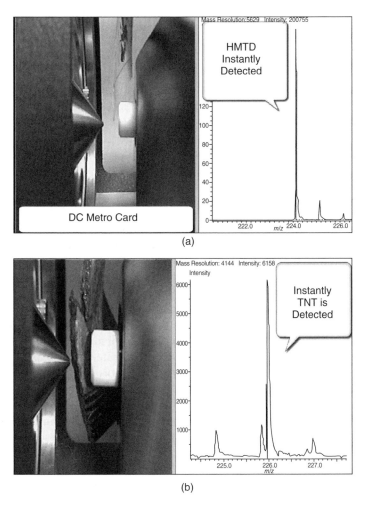

Figure 7.14 *Explosives detection by DART: (a) 100 nanograms of HMTD spiked on a DC Metro rail card, detected as [M+H]⁺, and (b) TNT detected on an exploded improvised explosive device (IED).*

mass measurement permitted the TNT analyte to be distinguished from an interference at the same nominal mass (Figure 7.14b).

Nitramine and nitrate esters did not produce molecular radical anions or deprotonated molecules but did form adduct anions in the presence of a dopant. These explosives include ethylene glycol dinitrate (EGDN), nitroglycerin (NG), 2,4,6-*N*-tetranitro-*N*-methylaniline (tetryl), pentaerythritol tetranitrate (PETN), hexahydro-1,3,5-trinitro-1,3,5-triazine (RDX), and octahydro-1,3,5,7-tetranitro-1,3,5,7-tetrazocine (HMX). Suitable dopants are Lewis bases that can fit into the electron-accepting cleft of the polynitro moieties. Dopants that produce chloride anions are acceptable candidates. Thus,

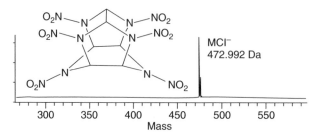

Figure 7.15   *Mass spectrum of CL-20 (hexanitro-hexaazaisowurtzitane), detected as [M+Cl]⁻.*

an open vial of dichloromethane in the vicinity produced [M+Cl]⁻ anions characteristic of the explosive's molecular weight.

CL-20 (2,4,6,8,10,12-hexaazaisowurtzitane) is an extremely high-energy polycyclic nitramine compound with a rigid caged structure [10]. Due to its superior explosive properties, it may replace conventionally used explosives in the future as an environmentally more benign material. Chloride attachment to CL-20 produced an abundant [M+Cl]⁻ anion (Figure 7.15). Without chloride, CL-20 gave a somewhat less intense molecular ion.

DART offers a significant advantage when compared to corona discharge sources often used in conjunction with both ion mobility spectrometers (IMS) and mass spectrometers for the analysis of nitrate-based explosives. The analysis of these compounds with conventional corona discharge sources is complicated by the production of high levels of NO_2^- and NO_3^- anions from atmospheric oxygen and nitrogen. These atmosphere-related anions interfere with the formation of analyte ions and mask the presence of ions characteristic of nitrate-based explosives [11]. These interferences are absent in DART because the electrical discharge is separated from the atmosphere.

An explosive mixture can be formed by the intimate mixture of any reducing agent and oxidizer. Other common IED ingredients include perchlorates, nitrates, and azides. These classes of compounds were all readily detected by DART operated in negative-ion mode and gave prominent ClO_4^-, NO_3^-, and N_3^- anions, respectively.

7.8   ARSON ACCELERANTS

The presence of arson accelerants on burned carpet samples can easily be detected using DART (Figure 7.16). One experiment involved spiking 50 μL of various solvents onto carpet patches, corresponding to a concentration of 35 mg/cm². Investigated solvents were isopropyl alcohol, xylene, paintbrush cleaner, methyl ethyl ketone, diesel fuel, and Coleman stove fuel. The patches were burned repeatedly and directly analyzed by DART.

As soon as the carpet remnant was held in front of the DART a clean mass spectrum was observed; exact mass measurements confirmed the identity of

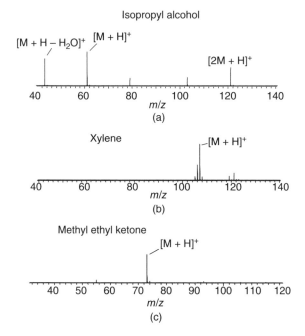

Figure 7.16 *Carpet burned with arson accelerants (a) isopropanol, (b) xylene, and (c) paint-brush cleaner (methyl ethyl ketone).*

the accelerant used. Interestingly, background peaks from the carpet material were insignificant compared to the peaks from the accelerant. In essence, the background from the carpet is "invisible" when compared to the solvent spectrum.

7.9 CHEMICAL WARFARE AGENTS

VX [(*O*-ethyl *S*-(2-diisopropylaminoethyl) methylphosphonothioate] is a highly toxic nerve agent. In its pure liquid form it has a vapor pressure of less than 0.0007 mm Hg at room temperature. When absorbed into a surface such as concrete, sand, or roadway asphalt, it presents an even lower vapor concentration. Nevertheless, DART was able to instantly detect the presence of VX on surfaces, with no sample preparation whatsoever (Figure 7.17). VX and other ultralow-volatility chemical agents were successfully detected on more than 40 different sample surfaces.

When VX is decontaminated by base hydrolysis, a pentacoordinate intermediate is formed that can decompose by two mechanisms. One of these produces a zwitterion salt known as EA2192, which is almost as toxic as VX. Detection of EA2192 is a significant analytical challenge because of its even lower vapor pressure. EA2192 is not amenable to gas chromatography/mass

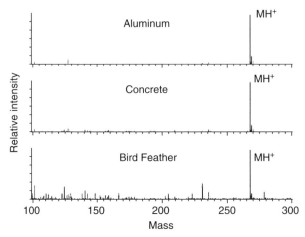

Figure 7.17   *The chemical agent VX detected on various surfaces: aluminum, concrete, and a bird feather.*

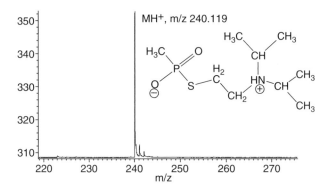

Figure 7.18   *Twenty picograms of EA2192 detected in a water matrix.*

spectrometry analysis. DART was able to detect 20 picograms of EA2192 at an excellent signal-to-noise ratio (Figure 7.18). Many other chemical agents were also tested. DART provided clean easily, interpretable mass spectra with the protonated neutral as the most abundant peak [12].

7.10   ELEVATED-TEMPERATURE DART FOR MATERIAL IDENTIFICATION

DART is useful for identification of unknown polymeric materials, such as fibers, plastics, and adhesives. It is helpful when examining these materials, to begin the analysis at a low gas temperature in order to detect proprietary additives, plasticizers, dyes, antioxidants, and so on. As the gas temperature is increased, characteristic polymer fragments are observed, indicative of specific polymer structures. DART spectra are much simpler than pyrolysis mass

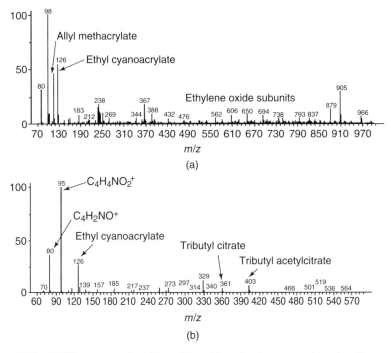

Figure 7.19 *DART mass spectra of two different brands of cyanoacrylate glue ("superglue").*

spectra. Fragment ions are seen at lower temperatures than those used in pyrolysis. Thus, heat alone might not be responsible for decomposing the polymer, and DART-specific mechanisms could play a role.

7.11 GLUES

Representative glues, resins, adhesives, and cements, such as epoxies, polyimide resins, PVC cement, and cyanoacrylates, were examined by DART. Glues from different manufacturers could be discerned within a compound class. Two different brands of cyanoacrylate glues ("superglues") were analyzed by DART, and not surprisingly both were found to contain ethyl cyanoacrylate. Differences in the brand formulation were apparent by comparison of their spectra (Figure 7.19).

7.12 PLASTICS

Low molecular weight polymers produce a distribution of protonated neutrals in the positive-ion mode (Figure 7.20). Larger polymers give spectra characterized by protonated oligomers. Polyethylene and polypropylene give a methylene series (Figure 7.20a), while nylon 6 (Figure 7.20b), polyethyleneimine (Figure 7.20c), and polystyrene (Figure 7.20d) give their characteristic spectra.

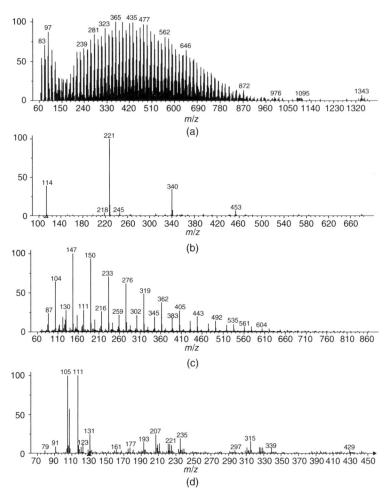

Figure 7.20 *DART mass spectra of polymers: (a) low-density polyethylene, (b) poly(caprolactam) or "nylon 6", (c) polyethyleneimine, and (d) polystyrene divinylbenzene beads (300–800 μm).*

Commercial plastics contain a variety of antioxidants, slip agents, and plasticizers, such as bis(ethylhexyl)phthalate, bis(ethylhexyl)adipate, and erucamide. These additives are easily detected by DART at temperatures that do not result in decomposition of the polymer. Specific combinations of additives could be useful in distinguishing plastics from different manufacturers.

7.13 FIBERS

Fibers of cellulose, cellulose acetate, and nitrocellulose all gave distinctive DART mass spectra characterized by low-mass fragments. Cellulose monomer and dimer were observed together with fragment peaks associated with water losses. Synthetic fibers gave mass spectra associated with the polymers used in their manufacture. For example, a fiber from a nylon glove showed peaks associated with nylon 6 (polycaprolactam), nylon 6/10 (poly(hexamethylene sebacamide), and ethoxylated nonylphenol (added to give the gloves a smooth, nonsticky texture).

7.14 IDENTIFICATION OF INKS

Ballpoint pen inks and marker inks comprise a wide variety of formulations. These range from dyes such as crystal violet (in black inks) and rhodamine (in red inks) to oils, lubricants, and anticlogging agents. DART mass spectra of ink lines on paper produce a characteristic mass spectral pattern for each formulation (Figure 7.21) [13]. These patterns differ not only by manufacturer, but also by different models from the same manufacturer.

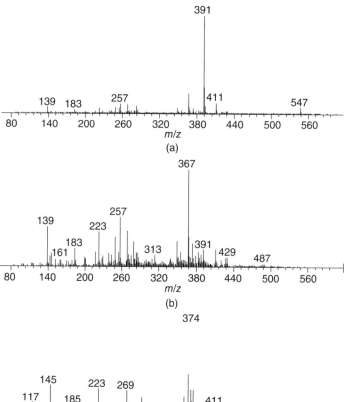

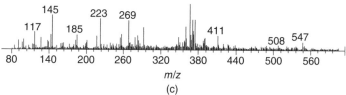

Figure 7.21 *Mass spectra of ink lines written on filter paper from three different black-ink ballpoint pens showing the characteristic pattern produced by the different ink formulations: (a) BIC Velocity, (b) BIC Atlantis, and (c) Cross Ballpen (refill cartridge).*

7.15 CONCLUSION

DART accepts the challenge of tackling real-world samples. It produces clean spectra that are free of artifacts. This makes DART mass spectra easier to interpret than those produced by other desorption methods. DART has already been adopted into several premier Department of Defense and Department of Justice forensic laboratories, demonstrating its utility and practical merits. A searchable library of DART spectra has been produced [12]. DART is fast, specific, and yet versatile enough to be the appliance of science.

ACKNOWLEDGMENTS

The authors gratefully acknowledge the facility support of the Research & Technology Directorate at the Edgewood Chemical Biological Center (ECBC), and fiscal support from the Agent Fate Program (DTO.42) by the Joint Science and Technology Office and the Department of Defense Chemical and Biological Defense Program.

REFERENCES

1. Cody RB, Laramée JA, Durst HD (2005). Versatile new ion source for the analysis of materials in open air under ambient conditions. *Analytical Chemistry* **77**(8):2297–2302. **(Accelerated Article)** DOI: 10.1021/ac050162j.
2. US Patent Numbers 6,949,741 7,112,785. Other patents pending.
3. Medway C, George S, Braithwaite R (1998). Opiate concentrations following the ingestion of poppy seed products—evidence for the poppy seed defense. *Forensic Science International* **96**:29–38.
4. Karas RP, Jagerdeo E, Deakin AL, LeBeau LA, Cody RB (2005). Quantitative analysis of gamma-hydroxybutyrate (GHB) in urine by direct analysis in real time (DART™) time-of-flight mass spectrometry. Poster presented at the Society of Forensic Toxicologists Meeting, Nashville, TN, Oct. 18–22.
5. Shen Z, Thomas JJ, Siuzdak G, Blackledge RD (2004). A case study on forensic polymer analysis by DIOS-MS: the suspect who gave us the SLIP. *Journal of Forensic Sciences* **49**(5):1028–1035.
6. Henderson JP (2005). Various techniques to visualize self-defense sprays. Master's Thesis, National University, San Diego, CA, USA.
7. Martz RM, Reutter DJ, Lasswell LD (1983). A comparison of ionization techniques for gas chromatography/mass spectroscopy analysis of dye and lachrymator residues from exploding bank security devices. *Journal of Forensic Sciences* **28**: 200–207.
8. Verweij AMA Lipman PJL (1993). *Journal of Chromatography A* **653**:359–362.
9. Henningsen DA, LeBeau M, Leibowitz JN, Schumacher LG (2005). Analysis of trace amount of bank dye and lachrymator from exploding bank devices by solid phase microextraction/mass spectrometry. Poster presented at the Pittsburgh Conference on Analytical Chemistry and Applied Spectroscopy, Orlando, FL.

10. Nielsen AT, Chafin AP, Christian SL, Moore DW, Nadler MP, Nissan RA, Vanderah DJ (1998). *Tetrahedron* **54**:11793–11812.
11. Hill CA, Thomas CLP (2003). *Analyst* **128**:55–60.
12. Restricted publications available from U.S. Army ECBC.
13. Jones RW, Cody RB, McClelland JF (2006). *Journal of Forensic Sciences.*

8

Forensic Analysis of Dyes in Fibers Via Mass Spectrometry

Linda A. Lewis

Senior Research Staff Member, Research Staff Member, Oak Ridge National Laboratory, Oak Ridge, Tennessee

Michael E. Sigman

Associate Professor of Chemistry and Assistant Director for Physical Evidence, National Center for Forensic Science, University of Central Florida, Orlando, Florida

8.1 INTRODUCTION

An important aspect of forensic fiber examinations involves the comparison of dyestuffs used to impart color on or in textile fibers. Information obtained from dyes used to color fibers can provide supporting evidence in forensic casework when comparing two fibers obtained from different locations. To determine that two fibers could have a common origin, it is necessary that they be shown to have the same dye components and that the ratio in which these components are present should be within a statistically acceptable range. Comparisons of absolute dye concentrations (i.e., nanograms dye per mm fiber) may not be necessary—or even advisable. Dye intensity may not be distributed uniformly along different fibers from the same coloring batch, or even along the length of a particular fiber. Thus, a forensic evaluation should

Forensic Analysis on the Cutting Edge: New Methods for Trace Evidence Analysis, Edited by Robert D. Blackledge.
Copyright © 2007 John Wiley & Sons, Inc.

comprise a qualitative evaluation of dye content (including processing additives and impurities) and a quantitative determination of the relative amounts in which each component is present.

A review of current textile dying techniques found that, in many cases, manufacturers use the same dye constituents in differing ratios to impart different colors to their products [1–3]. This practice facilitates computer-assisted production control. Textile dyers often use three dyes—a yellow, a red, and a blue—to produce the desired effect. Although there are numerous yellow, red, and blue dyes from which to choose, an individual textile manufacturer may use only a small selection to produce the myriad hues in the product line. Fortunately for the forensic process, commercial dyes often contain synthetic impurities that vary from batch to batch and the analysis of these compounds can be of considerable forensic value. Thus, the ability to determine dye-constituent ratios, as well as the actual dyes used in a coloring process, is virtually essential to definitively compare dyed fibers that appear similar using traditional screening methods, such as microspectrophotometry.

The predominant techniques currently employed in forensic fiber color examinations include microspectrophotometry [4, 5] and thin-layer chromatography (TLC) [6].

8.2 CONVENTIONAL FIBER COLOR COMPARISON METHODS EMPLOYED IN FORENSIC LABORATORIES

Microspectrophotometry is the most widely utilized color comparison technique in federal, state, and local forensic laboratories. To the forensic scientist, nondestructive analysis of evidence and application to extremely small sample sizes are the most attractive characteristics of this method. However, the lack of discriminatory power is an inherent limitation. Microspectrophotometry evaluates the spectral characteristics of the composite-dye mixture but says nothing about the individual dye components and the associated component ratios. Considerable diagnostic information is therefore left unexamined.

In contrast, TLC is a method in which dyes are extracted from the fiber, thus destroying the fiber, and subsequently separated and qualitatively compared. The method is relatively insensitive and may require more fiber for an analysis than the forensic scientist is willing to sacrifice. Furthermore, TLC is generally able to distinguish only the dyes present, but not the ratios in which they are present. When the dyes are extracted using the protocols set forth by the Federal Bureau of Investigation's Scientific Working Group on Materials Analysis, the extraction process facilitates classification of the dye (i.e., acidic, basic, reactive, azo, etc.) [7].

High-performance liquid chromatography (HPLC) [6] is a third analytical method that has been employed to a limited extent in the forensics laboratory for dye analysis. HPLC has been applied to acid [8], disperse [9, 10], and basic [11] forensic dye analysis. HPLC offers better separation than TLC and provides quantitative information. However, HPLC columns and gradient or iso-cratic elution systems are specific to only a limited group of similar compounds. Classification of the dyes on the basis of their extraction characteristics can be extremely helpful in identifying the appropriate column and chromatography conditions. The columns required for analysis are generally expensive, and large amounts of solvent may be needed compared to other separation methods. In addition, HPLC fiber length requirements are comparable to the lengths necessary for TLC analysis. Application of modern techniques of micro-HPLC may alleviate one or more of these shortcomings.

Each of the methods described previously is founded on either the composite- or individual-dye UV–visible (UV–Vis) absorbance characteristics. These methods provide a higher dye discrimination factor than dye-extraction properties alone. However, shortcomings are associated with these techniques in relation to trace-fiber color comparisons, as will be covered in the following section.

8.3 SHORTCOMINGS ASSOCIATED WITH UV–VIS BASED COMPARATIVE ANALYSIS FOR TRACE-FIBER COLOR EVALUATIONS

While not currently implemented, capillary electrophoresis (CE) is a method that has shown significant promise as a forensic "screening tool" for trace-fiber dye analysis [12, 13]. In a comprehensive study involving 12 carpet and 4 winding nylon 6 and nylon 6,6 samples with associated dyes, additives, and treatments, the ability to detect acid dyes extracted from nylon fibers between 1 and 3 mm, for dark and light colored fibers, respectively, was achieved using large volume stacking with polarity switching [12]. This technique was sensitive to dye constituents, as well as manufacturing additives and impurities yielding unique "fingerprint" electropherograms, as illustrated in Figure 8.1, which

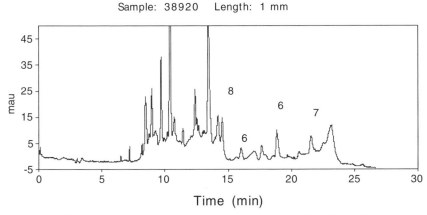

Figure 8.1 *CE electropherogram obtained from a 1 mm nylon 6 fiber. Peaks 6, 7, and 8 represent extracted commercial acid dyes Tectilon Blue 4RS, Tectilon Orange 3G, and Tectilon Red 2B, respectively. All other peaks resulted from additives and impurities.*

TABLE 8.1   Dyes and Colored Windings Used

Lab Code #	Commercial Name	CI Name
Dyes 1	Telon Yellow FRL01 200	Proprietary mix
2	Telon Red 2BN 200	Proprietary mix
3	Nylanthrene Red CRBS 200	Proprietary mix
4	Telon Red FRLS 175	Acid Red 337
5	Telon Blue BRL 200	Acid Blue 324
6	Tectilon Blue 4RS 200%	Unknown
7	Tectilon Orange 3G 200%	Acid Orange 156
8	Tectilon Red 2B 200%	Unknown
9	Dye-O = Lan Black RPL 150%	Acid Black 172

Manufacturers Code	Manufacturer	Dye Components	Manufacurers Color Description
Windings 9,900	Collins & Aikman	6, 7, 8	Brown
38,920	Collins & Aikman	6, 7, 8	Olive green
38,011	Collins & Aikman	6, 7, 9	Black

could be compared to a potential source material. The fiber illustrated in Figure 8.1 represents one winding sample in a set of three winding samples obtained from a commercial dyer in which the same three dye combinations were used to impart three very different colors (brown, olive green, and tan—Table 8.1) based solely on the ratios in which the same dyes were combined. The level of sensitivity required to distinguish slight variations in dye ratios was not found to be achievable using an on-column based detection system for UV–Vis analysis. Quantitative CE results indicated that the spectral-based detection system lacked the sensitivity required for precise comparative measurements.

As previously mentioned, dyers tend to use a three-color combination (blue, yellow, red) to yield a given color and shade. Additionally, the practice of using a combination of dye components of the same color (e.g., a combination of reds) to generate one of the three colors has been reported [12]. Figure 8.2 illustrates the CE electropherograms obtained from the three different red dyes listed below.

(a) Telon Red 2BN   Proprietary mix
(b) Nylanthrene Red CRBS   Proprietary mix
(c) Telon Red FRLS   Acid Red 337

As illustrated, the electropherograms in Figure 8.2a–c appear to have one to three major components, of which the UV–Vis spectrum for each component appears to be nearly identical. Mass analysis of this dye revealed that the red dyes (a) and (b) each contained the Acid Red 337 dye in (c), as well as one or more additional components. Thus, analysis of the UV–Vis property alone does not facilitate the discrimination power needed to distinguish between fibers colored with such complex mixtures.

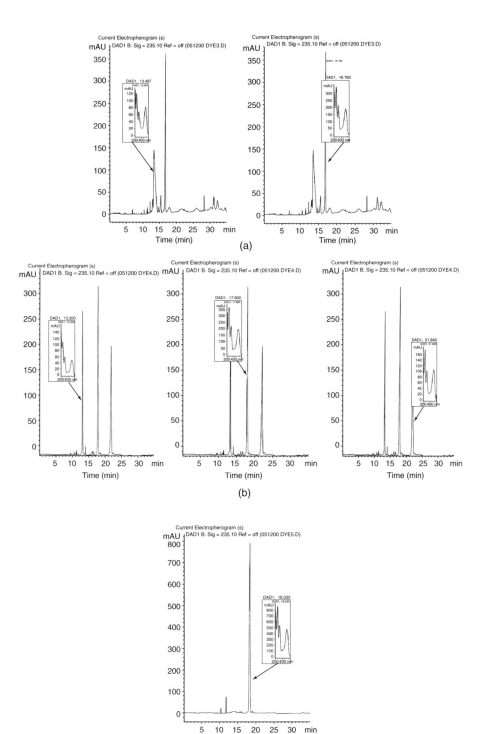

Figure 8.2 *Electropherogram and associated UV–Vis spectra for three different red dyes: (a) proprietary mix, (b) proprietary mix, and (c) Acid Red 337.*

An additional shortcoming in a UV–Vis based comparative analysis relates to the inherently broad Franck–Condon absorption bands exhibited by organic-based dyes. When dye components similar in base molecular structure that constitutes the chromophore are analyzed, the associated absorption bands may also be highly similar, making it difficult or impossible to discriminate between dye components by UV–Vis absorption spectroscopy alone. The difficulty in distinguishing between structurally similar dyes is illustrated in a recent article by Huang et al. [14]. During that study, seven pairs of common dyes with a common chromophore, yet different auxochromes, were examined both by UV–Vis spectroscopy and by liquid chromatography mass spectrometry (LC MS). The UV–Vis absorption profiles of these pairs were found to have very close maximum absorption wavelengths, as well as nearly indistinguishable absorption profiles. However, the LC MS results were able to differentiate between the dyes even in the case of Basic Red 9 and Basic Violet 14, where the associated chemical structures differ only by the replacement of a hydrogen atom with a methyl group. In this case the absorption maxima were 544 nm and 549 nm, respectively, and the absorption profiles were virtually indistinguishable. The fragmentation differences between these two dyes in an electrospray ionization (ESI) interface, positive ionization mode, are shown in Figure 8.3.

Thus, in order to discern between two trace fibers colored either with different ratios of the same dye constituents or with a dye similar but not identical in chemical structure, a higher level of sensitivity and discrimination power is required. Recent studies [14–16] have shown that mass-spectral-based detection methods, with or without chromatographic separation, provide for a more definitive comparison between two fibers based on the molecular ion mass (MS) and/or the parent-mass structure (MS/MS).

The remainder of this chapter provides an overview of the application of mass spectrometry to the qualitative and quantitative aspects of comparative

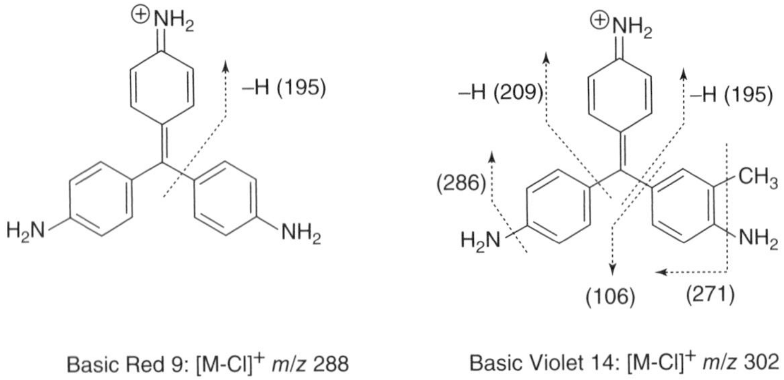

Basic Red 9: [M-Cl]$^+$ m/z 288   Basic Violet 14: [M-Cl]$^+$ m/z 302

Figure 8.3 *Fragmentation differences between Basic Red 9 and Basic Violet 14 in an electrospray ionization (ESI) interface, positive ionization mode.*

fiber-dye analysis. Qualitative identity is established primarily by comparison of the observed masses for each peak in the total ion chromatogram (TIC) of each fiber extract. In the case where the dye components are separated by chromatographic methods, a qualitative comparison of questioned and known samples involves direct comparison of the mass spectra of corresponding chromatographic peaks from each sample. The MS may be further verified by comparing the tandem mass spectra (MS/MS) for each of the dye peaks observed in the original mass spectrum, or by comparing the dye fragmentation patterns at different applied fragmentor voltages. Comparison of the relative intensities of the individual dye peaks during MS analysis provides a measure of the concentration of each dye in each of the fibers being compared.

8.4 GENERAL OVERVIEW OF MODERN DYE IONIZATION TECHNIQUES FOR MASS ANALYSIS

Of the mass spectral methods relating to trace-fiber color comparisons reported in recent literature, either a direct- or hyphenated-introduction technique into the mass spectrometer has been employed [14–16]. The hyphenated (separation prior to mass analysis) technique reported in these publications has been liquid chromatography (LC or HPLC). Mass spectrometry is a particularly powerful detector for separation techniques like LC for the reasons previously discussed. The challenge in coupling a mass spectrometer to a separation system is maintaining the vacuum level required in the mass analyzer while introducing gas-phase ions of nonvolatile analytes after emerging from a separation system. Interfaces employed for LC systems include electrospray ionization (ESI) and atmospheric pressure chemical ionization (APCI) techniques. In the case of ESI, a solvent is introduced at a higher flow rate than the sample, as the sample solution is sprayed from a needle held at high voltage creating a strong electric field (typically between 3 and 4 kV). Heat and a drying gas are usually employed to increase the rate of droplet evaporation as the sample solution is sprayed. As the droplet decreases in size, the electric-charge density on its surface increases. The mutual repulsion between like charges reaches a point that exceeds the forces of surface tension, resulting in the ejection of ions from the droplet (Taylor cone). Ions are then directed electrostatically into a mass analyzer. Singly or multiply charged ions may be generated by the electrospray process. The number of charges retained by an analyte can depend on such factors as solvent composition and pH, in addition to inherent chemical properties of the analyte. Typically, ESI is a "soft" ionization technique that generates molecular ions; however, the use of high fragmentor voltage within the ESI source has recently been reported as a means of analyzing the resulting fragmented dye constituents for comparative purposes [14].

In the case of APCI, the eluent passes through a heated pneumonic nebulizer. This process facilitates rapid droplet desolvation and vaporization. The vaporized sample molecules are directed to a corona discharge electrode held between 1.0 and 3.5 kV at atmospheric pressure. Ionization occurs via the corona discharge, creating reagent ions from the solvent vapor. Due to the high collision frequency at atmospheric pressure, chemical ionization of sample molecules is very efficient. Either positive or negative ions may be created within the interface. Proton transfer to the analyte occurs in the positive-ion mode, and electron transfer to the analyte or proton transfer to the solvent occurs in the negative-ion mode. Molecular ions are typically generated by this technique.

For both ESI and APCI interfaces, the sample is introduced into the vacuum region of the mass spectrometer as ions. With low solvent flow rates, direct introduction into the mass analyzer can be achieved using electrospray or chemical ionization. ESI and APCI are considered to be complementary ionization techniques, in that ESI ionizes polar and APCI ionizes nonpolar compounds more efficiently.

8.5 TRACE-FIBER COLOR DISCRIMINATION BY DIRECT ESI-MS ANALYSIS

The application of electrospray ionization mass spectrometry (ESI-MS) has been evaluated and found to be a promising method for both qualitative and quantitative comparative fiber-dye analysis with regard to acid dyes used to impart color on nylon fibers [15]. Qualitative identity is established primarily by comparison of the observed masses for each peak in the ESI-MS of each fiber extract. This may be verified further by comparing the tandem mass spectra (ESI-MS/MS) for each of the dye peaks observed in the ESI-MS. Comparison of the relative intensities of the individual dye peaks in the ESI-MS provides a measure of the concentration of each dye in each of the fibers being compared. Throughout the study, the negative-ion mode was utilized to evaluate the resulting acid-dye anions. Figure 8.4 presents the mass spectrum obtained for Acid Red 337 and is representative of the dye spectra analyzed in the study. (1) The cluster at m/z 410 corresponds to the expected anion for

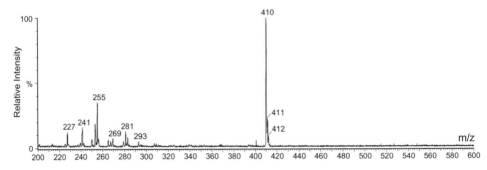

Figure 8.4 *ESI-MS of dye No. 4 (500 ppb in CH₃CN/H₂O 1:1).*

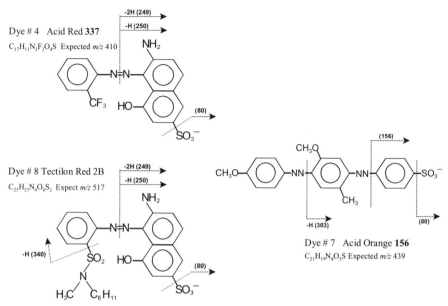

Figure 8.5 *Structures and observed CID fragmentations of those dyes for which structures are known.*

this dye (see Figure 8.5 for known dye structure of Acid Red 337) and the single-mass spacing of the ^{13}C isotope peaks (410, 411, 412) indicates a mono-anion. (2) Peaks between m/z 200 and 300 were impurities presumed to be present in the solvent. These peaks appeared in the spectra of all individual dyes, as well as "blank" spectra of solvent without the addition of dyes. These constituents did not interfere with dye analysis because the peaks were located outside the range of dye peaks encountered in the study (m/z 308–544). (3) There were no dye-fragment ions discernable in the spectrum under the analysis settings employed, indicative of the "gentle" nature of the ionization technique. For the type of aromatic azo dyes examined, fragmentation was not expected in the electrospray process, and multiple peaks found within a spectrum would most probably represent a mixture of analyte types rather than fragmentation.

Similar spectra were obtained for each of the dyes in Table 8.1, and the observed m/z values are listed in Table 8.2. Points of interest from the values listed in Table 8.2 include the following:

1. In cases where the chemical structure was known (dyes 4, 7, and 8), the observed mass values corresponded to expectation.
2. Several of the dyes (1, 2, 3, 6, and 9) showed multiple peaks indicative of mixtures. (Indeed dyes 1, 2, and 3 are stated by the manufacturer to be "proprietary mixes"; see Table 8.1.)

TABLE 8.2 Observed Relative ESI Anion Intensities and Charge States[a] for the Dyes Listed in Table 8.1

Dye #	Observed Anion m/z Values																			
	308	408	410	424	427	430	439	444/446	450	462.5	464	467	473.5	481.5	488	489/491	504	517	525	544
1					100^{-1}											$44/34^{-1}$	13^{-1}			
2			100^{-1}									90^{-1}								
3			100^{-1}					$56/22^{-1}$											53^{-1}	
4			100^{-1}																	
5									100^{-1}											
6											37^{-1}									100^{-1}
7							100^{-1}													
8																		100^{-1}		
9	100^{-3}	3^{-1}	3^{-1}	5^{-1}		6^{-2}				9^{-2}			16^{-1}	6^{-2}	2^{-1}					

[a]Observed charge state indicated by superscript; thus, XX^{-1} = singly charged cluster (^{13}C isotopes @ 1.0 Da); XX^{-2} = doubly charged cluster (^{13}C isotopes @ 0.5 Da); and XX^{-3} = triply charged cluster (^{13}C isotopes @ 0.3 Da) (requires "high resolution" to observe isotope peaks).

3. Dyes 2 and 3 appeared to have the same major component (dye 4), as well as one or two "additives." (The identical structure of the m/z 410 components of dyes 2, 3, and 4 was borne out by comparison of their CID spectra.

4. All dyes examined were singly charged (indicated by superscript −1 in Table 8.2) indicating monoacidic functionality *except* dye 9 (superscripts −2 and −3 for two and three charges, respectively).

5. Several of the peaks listed for dye 9 were closely related and apparently represent the same underlying triacidic structure. Thus, triply charged m/z 308.1 represents a trianion with mass 924.3 Da. m/z 462.7 corresponds to the protonated version of the same structure $[(924.3 + 1)/2 = 462.7]$. Likewise 473.7 and 481.7 correspond to the sodiated and potassiated structures, respectively. Thus, 308, 462.7, 473.7 and 481.7 do not represent different dyes in the mixture, but rather differently cationized versions of the same triacidic dye. However, the singly charged anion at m/z 424, and the doubly charged one at m/z 430 cannot be accommodated similarly and seem to be different dyes in the mixture. The peaks at 408, 410, and 488 were relatively too small to be used as markers in the ESI-MS fingerprint of dye 9. At somewhat lower signal-to-noise ratios (S/N), such low-intensity peaks could become indistinguishable from the background.

6. Each of the dyes examined had at least one peak that was distinct from every other dye in this particular set.

8.6 EXAMPLES OF NEGATIVE ION ESI-MS ANALYSIS OF COLORED NYLON WINDINGS

Green, brown, and black nylon "windings" (Table 8.1) colored with acid dyes were extracted using the FBI protocol for forensic fiber comparisons [7, 15]. Since the dyes were removed using a pyridine/water mixture, a solvent exchange step was necessary in order to prepare the samples for ESI-MS analysis. The pyridine solvent was evaporated from the fiber extract, and the residue was dissolved into 2-propanol/water 4:1 containing 0.1% ammonia. An extract equivalent to 1 mm for each winding was analyzed, and a representative spectrum from the brown winding is given in Figure 8.6. Upon inspection of the mass spectrum, several points of interest can be stated. The peaks expected for the component dyes of the brown winding (m/z 464 and 544 for dye 6, m/z 439 for dye 7, and m/z 517 for dye 8 given in Table 8.2) are clearly discernible, as are the "background" peaks (e.g., m/z 253, 255, 281) previously observed in the spectra of the individual dyes (cf. Figure 8.4). A set of peaks apparently representing a homologous series of analytes occurs at m/z 283, 297, . . . , 367. These do not obstruct that region of the spectrum that is most informative for dye analysis (m/z 400–560) and thus are not considered detrimental to the dye

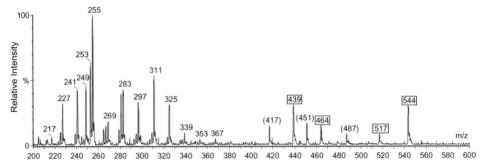

Figure 8.6 *ESI-MS of the extract from winding 1 ("brown") at the equivalent of 1 mm/60μL of 2-propanol/H₂O/0.1% NH₃. Dye peaks are marked with rectangles and extraneous peaks within the dye region are marked with parentheses.*

analysis. Peaks not attributable to the known dye components are also observed in the "dye region" of the spectrum: via m/z 417, 451, 487/489. They appear, albeit in varying relative abundances, in all of the winding extracts examined during this study and may represent degradation products of the nylon fiber generated during the extraction process or compounds that have been added during the dye-fixing process. The "extraneous" peaks that lie within the dye region of the spectrum have been marked with parentheses in Figure 8.6, and the recognized dye peaks have been marked with a rectangle. The average and standard deviations of the observed intensities for each dye component within the windings for repeated analysis were evaluated. The calculated standard deviations were found to be adequate to unambiguously differentiate between the brown and green windings that differ only by the proportions of dyes used to impart the color. The brown fiber was found to consist of dyes 6 (blue), 7 (orange), and 8 (red) in the proportion 10:10:3 based on signal intensity, whereas the proportion for the green fiber was found to be 16:10:3 for the same blue, orange, and red dyes. It is important to note that the differentiation of these winding extracts was possible through mass analysis, but not through UV–Vis based analysis.

8.7 EXAMPLES OF TANDEM MASS SPECTROMETRY (MS/MS) APPLICATIONS TO ELUCIDATE STRUCTURE

Molecular or quasimolecular ions observed in mass spectra only yield information regarding the mass of the dye, additive, or impurity. In the absence of fragmentation, structural information is not assessable. In order to obtain structural information about a precursor ion, collision induced dissociation (CID) spectra can aid by providing structural insight. In this example, direct injection ESI-MS on a triple quadrupole mass spectrometer was employed. Such a system can be deemed as "two dimensional" since separation of the ESI-MS ions do occur within the first mass filter of the triple quadrupole system. Even if CID fragmentation patterns do not provide all the information required to draw a definitive structure, the pattern of fragment peaks obtained from a particular precursor under defined dissociation conditions does provide a "fingerprint" of that precursor. It is thus particularly useful in establishing that the same analyte mass, observed in separate samples, represents the same structure. The appearance of a CID spectrum can vary significantly depending on the collision energy (CE) applied. For a "fingerprinting" experiment it is desirable to encompass as many product ions in a single CID spectrum as possible. Figure 8.7 shows the products of precursor m/z 410 (dye 4, Acid Red 337) acquired at four different CEs. At low CE (20 eV; Figure 8.7a), the precursor ion predominates, and m/z 80 is weak. At higher CE (35 eV; Figure 8.7d), m/z 80 is of moderate intensity, but m/z 410 has disappeared. In order to incorporate as much information as possible into a singe scan, the CE may be ramped as a function of m/z. Based on the results of Figures 8.7a–d, a CE ramp was applied to produce the spectrum displayed in Figure 8.7e. The ramp ran from 30 eV (at m/z 60) to 20 eV (at m/z 430). The "ideal" CE ramp for any given precursor ion will depend on its fragmentation characteristics. However, it can always be determined empirically. The same sequence of CID

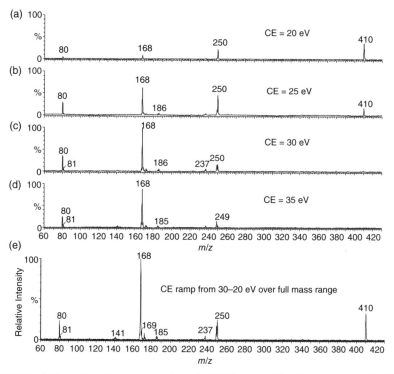

Figure 8.7 CID spectra of precursor anion m/z 410 from dye No. 4: (a–d) obtained with the collision energies indicated and (e) obtained with a collision energy ramp.

TABLE 8.3 Mass Range Scanned, Collision Energy (Ramp) Employed, and Fragments Observed for Each of the Precursor Ions Enumerated in Table 8.2

Precursor m/z	Dye # or Winding #	Scanned m/z Range	Collision Energy Ramp	Observed CID Fragments (m/z)
308	9	80–550	19	203 / 308 / 316 / 352 / 408 / 424 / 430
410	2, 3, 4	60–430	30–20	80 / 168 / 249 / 250 / 410
417	I, II, III	60–437	40–20	83 / 189
427	1	60–447	30	292 / 409 / 427
424	9			Too small for good CID @ 500 ppb of dye #9
430	9			Too small for good CID @ 500 ppb of dye #9
439	7	60–459	40–20	80 / 156 / 303 / 395 / 423 / 439
444/446	3	60–466	30–20	80 / 168 / 249 / 250 / (444/446)
450	5	60–470	35	344 / 386 / 450
451	I, II, III	60–471	35–30	83 / 125 / 207 / 225 / 333 / 433 / 451
462.5	9	60–482	30–20	142 / 172 / 267 / 281 / 352 / 424 / 463
464	6	60–484	35	344 / 400 / 464
473.5	9			Too small for good CID @ 500 ppb of dye #9
467	2	60–487	30–20	80 / 168 / 249 / 250 / 467
481.5	9			Too small for good CID @ 500 ppb of dye #9
487/489	I, II, III	60–510	25	283 / (487/489)
489/491	1	60–510	30–15	94 / 123 / 186
504	1			Too small for good CID @ 500 ppb of dye #1
517	8	60–537	30	80 / 168 / 249 / 250 / 340 / 517
525	3	60–535	30–20	168 / 185 / 249 / 250 / 264 / 340 / 525
544	6	60–564	35	356 / 420 / 480 / 544

experiments described previously for dye 4 was implemented for each of the precursor ions enumerated in Table 8.2, as well as those identified as "extraneous" in Figure 8.6. The results indicating the mass range to scan, the CE ramp to apply over that mass range, and the product ions observed from such a ramped experiment are listed in Table 8.3. Thus, each row in Table 8.3 describes a CID method and the resulting "fingerprint" for one of the dyes under investigation. Although it gives comfort to correlate the observed CID spectrum with the expected fragments of a known structure, it is not necessary to know the chemical structure of a compound in order for its CID spectrum to serve as a fingerprint. Observing two compounds (in this example dye anions) with identical ESI-MS peaks—which also produce identical CID spectra from those peaks—essentially ensures that the compounds are identical or at least very closely related in structure.

8.8 LC-MS ANALYSIS OF DYES EXTRACTED FROM TRACE FIBERS

The previous section focused on the direct ESI analysis of fiber-dye extracts (acid dyes and nylon fibers). This section covers the use of liquid chromatography to separate dye components as well as textile additives prior to mass analysis for a wider range of dye classes and fiber types. This section is organized based on the LC-MS interface used in the work discussed. While the research discussed here is not intended to be an exhaustive review of LC-MS analysis of dyes, the discussion should be useful in leading the interested reader to literature precedent that is relevant to specific needs.

Although the popularity of the particle beam and thermospray interfaces have declined with the increased popularity of the APCI and ESI interface, a significant number of LC-MS studies of dyes were performed with these older interfaces. In addition, the relatively high cost of LC-MS instruments will contribute to the persistence of these instruments in some laboratories for a period of time. It may therefore be useful to briefly discuss some of the previous dye analysis research utilizing particle beam and thermospray interfaces. The operational aspects of particle beam and thermospray interfaces have been discussed in much detail elsewhere [17]. In some cases, the dyes analyzed by these methods (and the newer APCI and ESI methods as well) came from sources other than forensic fiber samples. In fact, much of the development of LC-MS analysis techniques for dyes has been driven by environmental applications. The particle beam interface has been coupled to LC separation for the analysis of a series of triphenylmethane dyes [18]. Separation of the cationic triphenylmethane dyes was accomplished on a C18 column by eluting with 80% acetonitrile and 20% 0.1 M ammonium acetate. Electron ionization of the dyes resulted in fragmentation to produce phenyl and substituted phenyl radicals. Reduction of the triphenylmethane dye malachite green to form leucomalachite green in the ion source was observed. Voyksner and co-workers [19] have reductively cleaved a series of azo dyes and analyzed the resulting aromatic amines by LC-MS with a particle beam interface. In this case, the dyes were recovered from sludges and the aromatic amines resulting from tin chloride cleavage were separated on a C18 column and passed through a single-wavelength detector at 254 nm prior to entering the particle beam interface where they underwent electron ionization at 70 eV. A series of azo, diazo, and anthraquinone dyes were analyzed by Marbury and co-workers [20] for the purpose of comparing particle beam and thermospray interfaces. This study concluded that the particle beam interface had the advantage of providing for both electron ionization and chemical ionization, and that the electron ionization spectra offered the advantage of more extended fragmentation and therefore more structural information. However, the thermospray interface offered a higher sensitivity, while forming primarily $[M+H]^+$ ions and no structurally significant fragmentation. The need for increased structural information through analyte fragmentation has been an important theme in much of the thermospray ionization analysis of dyes. Similarly, Yinon and co-workers [21] examined a series of dyes by LC-MS employing C18 separation with a particle beam interface into a triple quadrupole instrument. Their findings were similar to those of Marbury et al. [20] in that the detection limits using the particle beam interface were two to three orders of magnitude worse than with thermospray ionization using a wire repeller. In a related comparison of

LC-MS interfaces, Straub and co-workers [22] analyzed a set of azo and diazo commercial dyes using thermospray, particle beam, and electrospray interfaces. These authors' findings were in general agreement with those discussed previously, and in addition they reported the benefits of using an electrospray interface in the negative ionization mode to analyze the sulfonated and multisulfonated azo dyes, which typically produced both singly and doubly charged anions.

The tendency to form nonfragmented ions in the thermospray interface through analyte adduct formation with charged species in solution has been viewed as a potential impediment to structure elucidation for some time. Tandem mass spectrometry coupled to a thermospray interface offers one method of obviating structure elucidation challenges associated with this soft ionization interface. Betowski and Ballard [23] reported the use of a thermospray interface and MS/MS techniques for the identification of dyes in 1984. Commercial sources of bromocresol purple, basic red 14, and basic orange 21 were chromatographed on a column packed with RP-2 sorbent and analyzed by a Finnigan MAT Triple Stage Quadrupole mass spectrometer. All three dyes were analyzed in the positive-ion mode and structural determination was possible based on the collision-activated dissociation spectra. Ballard and Betowski [23–25] later demonstrated the use of tandem MS/MS with thermospray ionization for the analysis of a set of 16 commercial dyes without prior chromatographic separation and dyes from industrial effluent using both LC separation and direct introduction techniques. Voyksner [26] demonstrated the use of thermospray ionization and use of a filament located at the tip of the thermospray vaporizer (operated at 200 eV, emission current of 0.5 A) to induce fragmentation of the $[M+H]^+$ ions formed from azo, diazo, and anthraquinone dyes. The dyes in this study were separated on reverse phase columns and produced primarily the $[M+H]^+$ ion without the added fragmentation induced by the filament on the thermospray ionizer. Filament-induced fragmentation was attributed to chemical ionization brought about through ionization of the eluting solvent with subsequent ionization of the analyte. This chemical ionization process was calculated to provide approximately 40 kcal/mol more energy than the thermospray ionization without the applied current, thus providing a less expensive alternative to tandem mass spectrometry. In a series of papers, Yinon and co-workers [27, 28] showed that a wire repeller introduced into the thermospray interface directly opposed to the ion-extraction funnel can facilitate higher sensitivity and increased fragmentation of dyes.

The particle beam and thermospray interfaces have become somewhat obsolete in the face of more recently developed APCI and ESI interfaces. As discussed previously, the APCI and ESI interfaces can be considered as complementary and both have been applied to the analysis of dyes. Sulfonated dyes have been separated by normal phase and ion exchange chromatographic methods with analysis by both ESI-MS and APCI-MS [29–33]. Smyth and co-workers [34] reported the use of C18 LC separation coupled with ESI-MS techniques on a quadrupole ion trap instrument to determine remazol black B and its degradation products. These authors reported that it was necessary to resort to ESI-MS^2 and ESI-MS^3 coupled with LC separation for unambiguous identification of these textile dyes. In related work, Baiocchi et al. [35] employed a C18-based LC separation of the sulfonated dye methyl orange and its catalytic decomposition products with ESI-MS, ESI-MS^2, and ESI-MS^3 analysis on a quadrupole ion trap instrument. It has generally been the authors' experience that singly sulfonated and disulfonated dyes may be separated and analyzed by reverse phase LC-MS methods; however, when dyes contain more than two sulfonate groups, good chromatographic separation becomes difficult on reverse phase columns.

The hyphenated technique of LC-MS has also been applied to the analysis of anthraquinone dyes in historical textiles [36]. Likewise, azo dyes have been examined by several research groups employing LC-MS [37–39]. Mayer and co-workers [37, 38] have analyzed disperse dyes from textile sources by LC-MS. In related work, ballpoint pen inks have also been analyzed by ESI-MS for forensic purposes [40]. In previous work by Yinon and Saar [41], thermospray ionization LC-MS was used to analyze known dyestuffs extracted from textile fibers. Several dyed fiber samples were investigated by extracting and analyzing the dye from a single 5–10 mm long piece of fiber with 5 μL of organic solvent. In a related nonforensic application, a combination of LC with diode-array detection and/or ESI-MS has been utilized for the determination of oxidative hair dyes used in commercial formulations [42].

In a recent report, acid, basic, disperse, and direct dyes were extracted and analyzed from nylon, acrylic, polyester/acetate, and cotton fibers, respectively [16]. When analyzing fiber-dye extracts from a range of dye classes and fiber types, several issues must be addressed including the adaptation of appropriate FBI standard extraction procedures (or equivalent) for subsequent LC-MS analysis, mechanisms to identify dye components (including impurities and additives), and the identification of an appropriate ion formation pathway (positive- or negative-ion mode). As a first step, fibers 5 mm in length were extracted with 20 μL 1:1 methanol in water in a sealed glass capillary with heat at 100 °C. The extracts were separated via HPLC using a programmed methanol/water gradient prior to mass analysis. Using an optimized protocol, some dyes (e.g., dyes containing sulfonated groups), especially in the presence of other easily ionized constituents, may not be readily detected in trace quantities when evaluating the total ion current (TIC) from the mass analysis; however, via referencing an in-line UV–Vis spectrum, the presence of light-absorbing constituents requiring subsequent mass analysis may be identified. For ion generation, the optimum fragmentor voltage was found to vary for specific dyes, but a compromise voltage was identified for unknown dye analysis. A voltage of 100 V in positive-ion mode and 40 V in negative-ion mode was selected. Under positive-ion mode, the predominant ions identified included $(M+H)^+$ and $(M+Na)^+$; and under negative-ion mode, numerous molecular ions including $(M–H)^-$, $(M\text{-}xNa + yH)^{-(x-y)}$ ($x = 1–4$, $y = 0–3$ for dye sodium salts), and $(M\text{-}xH + yNa)^{-(x-y)}$ ($x = 1$–4, $y = 0$–3 for acid dyes) were frequently observed. It was also noted that ionization efficiency was enhanced by the use of complex or ion-pair additives (acetic acid or triethylamine base) for analytes of low ionization potential, such as multiply sulfonated azo dyes. After obtaining information on the molecular ions formed from each dye component, the sample was analyzed at higher fragmentor voltages (e.g., 180 V) to induce dye fragmentation and aid in structure elucidation.

The use of triethylamine (TEA) was reported to facilitate the MS analysis of dyes containing three or more sulfate groups with the addition of TEA into the mobile phase at a 0.3% concentration. As previously mentioned, ESI and APCI are complementary ionization techniques for ionic and nonionic materials, respectively. As an alternative to electrospray ionization, the detection of monosulfonated and disulfonated azo dyes, in addition to a nonionic azo dye with a sulfonamide functional group, has been reported using APCI methods [29, 43]. Additionally, APCI has been reported as the method of choice for the ionization of indigotin and brominated indigotins (ESI failed to ionize the compounds, possibly due to internal hydrogen bonding) in the investigation of natural dyes from historical Coptic textiles; whereas positive and negative electrospray ionization was the method choice for flavanoid-, anthraquinone-, and indigo-based dyes.

8.9   PROPOSED PROTOCOLS TO COMPARE TRACE-FIBER EXTRACTS

As discussed previously, many forensic fiber analyses come down to a questioned and known comparison. In these cases, it is generally not necessary to positively identify each dye, impurity, and environmentally contributed component. In fact, this may prove impossible in many instances; however, it should always be possible to make a direct questioned and known comparison. It is important to point out that the methods discussed in this chapter are destructive and therefore are methods of last resort. The first step in any fiber comparison is a microscopic examination (stereo microscope followed by a polarized light microscope), which allows for the determination of general fiber shape, color, and the presence or absence of delustering agents in synthetic fibers. Subsequent analysis by FTIR spectroscopy and microspectrophotometry are also nondestructive tests that should be employed before the methods discussed in this chapter. When these nondestructive tests fail to discriminate between two fibers, dye extraction and analysis may be necessary.

8.9.1   Direct Infusion MS/MS Protocol

A protocol for the comparison of trace-fiber carpet evidence with a potential source material for the determination of origin has recently been proposed

[15]. However, the proposed comparison may also be applied fiber evidence in general. A more general comparative procedure for trace fibers is outlined below.

Proposition   Two fiber samples are believed to be identical in their dye content. Sample "A" is available in large quantity (50 mm or more). Sample "B" is available in small quantity only (1 mm can be spared for destructive analysis). They appear identical in color by microscopic examination.

Objective   Prove/disprove the proposition that the dye contents of the two fibers are identical using ESI-MS and CID.

Proposed Procedure

1. Extract a 5 mm long sample of "A" as well as a 1 mm samples of both "A" and "B" and dissolve the extracts in 500, 100, and 100 μL, respectively, of the appropriate solvent mixture. Using a 10 μL loop injector, sequentially analyze samples of the 1 mm extracts and obtain the mass spectra. If the spectra are not essentially identical, the proposition is refuted and the experiment is complete. If the spectra are essentially identical in *m/z* content and relative intensity, proceed to step 2.

2. If the mass spectra of dye ions from step 1 are represented in a "product ion" database (such as Table 8.3), confirm the assignments by running CID experiments under the previously established "ideal" conditions as indicated in the database.

3. If one or more of the major peaks in the mass spectra are not represented in a "product ion" database, use the 5 mm extract to establish the "ideal" conditions for obtaining a fingerprint CID of that (those) precursor ion(s) (as described in Section 8.7).

4. Using the CID conditions established in previous steps for each of the major peaks in the mass spectrum, do identical multifunction (one function per dye precursor mass) CID experiments on the 1 mm extracts of "A" and "B." If the CID spectra show identical fragmentation patterns for each pair (one from extract "A" and one from extract "B") of precursor ions, the two fibers can be deemed to be identical with regard to the composition of their dyes.

5. If the two fibers are deemed identical with regard to the composition of their dyes, evaluate the mass spectra signal intensities to ensure the dye proportions are similar.

6. Even if as many as eight precursor ions need to be examined by CID, with an analysis time of 1 min each, and at a flow rate of 5 μL/min, this will consume only 40 μL of solution. Together with the 10 μL consumed in step 1, this accounts for 50 μL of the original 100, leaving 50 μL for other methodologies, or for a repeat of this one as necessary.

The methodology just described addresses the problem of establishing that the dye content of different fibers is identical. For the examples reported, virgin samples obtained from manufacturers were extracted. It is highly probable that normal daily use of such fibers (e.g., washing, dry cleaning, prolonged exposure to sunlight, coffee spills) would perturb the analytical results to a greater or lesser degree. This would undoubtedly affect the results when comparing a used fiber with a pristine sample from the manufacturer. However, insofar as the history of two fibers is similar (e.g., comparing a carpet fiber found on a victim's clothing with a similar fiber taken from a suspect vehicle's floor mat) the degree of postmanufacturing alteration of the dye content would also be similar, and the forensic comparison should remain valid.

8.9.2   Generalized LC-MS and LC-MS/MS Protocol

Dye Extraction   Dye extraction protocols have been established by the FBI-SWGMAT and in most cases these extraction protocols are completely congruent with LC-MS analysis of fiber dyes. The SWGMAT protocols also offer the advantage of a preliminary classification of the dye type. It has been the authors' experience that a 1 cm single fiber will often yield sufficient dye when extracted into 20 µL of solvent to allow for four 5 µL aliquots for LC-MS analysis. Questioned and known samples should be extracted side-by-side on the same day and analyzed without extended delay.

Chromatographic Separation   Identifying optimal chromatographic separation conditions is admittedly the most problematic part of any LC-MS method, although a review of the literature shows that many separation schemes are fairly similar. Many dye types that can be readily extracted from textile fibers can also be separated chromatographically on reverse phase columns with methanol/water or acetonitrile/water mobile phases. Notable exceptions are acid dyes containing more than two sulfonate groups. These dyes are best separated on normal phase columns or by ion exchange methods. Given the large number of commercial dyes available and the variety of fiber types in the transferable fiber pool, prescription of a single separation method is not possible. Knowledge of the dye type based on the extraction protocol should be useful in identifying a separation scheme.

Mass Spectral Analysis   ESI and APCI ionization sources common on modern LC-MS instrumentation are well suited for the analysis of dyes. As discussed previously, these two techniques are complementary. For an ESI source, it is recommended that the 5 µL aliquots be run in both positive and negative ionization modes at low and high fragmentor voltages [16]. When an APCI source is used, more molecular fragmentation is generally observed and the analysis should be run in both positive and negative analysis modes. When chromatographically isolated components produce the same mass spectral profiles, MS/MS techniques may be required for further characterization of the dye components.

8.10   CONCLUSIONS

In order to definitively establish that two fiber samples are of common origin to a high degree of statistical certainty, it is necessary to demonstrate that their dye components are identical at the molecular level, and that those dyes are present in the same proportions in each fiber. The qualitative comparison is necessary because fiber manufacturers often use identical dyes in different proportions to create differently colored fibers. The combination of mass spectrometry and tandem mass spectrometry has been shown to provide both the qualitative and quantitative information required for such comparisons. The technique is sufficiently specific and sensitive to allow comparison of two fibers, one of which is available in lengths of as little as 1 mm.

ACKNOWLEDGMENTS

The authors thank the researchers who played an important role in assembling this chapter, including Dr. Albert Tuinman and Samuel Lewis. Also acknowledged are the agencies that supported this research, including the Federal Bureau of Investigation and the National Institute of Justice.

Disclaimer: Points of view in this Chapter are those of the authors and do not necessarily represent the official position of the U.S. Department of Justice.

REFERENCES

1. Mcgregor R, Arora MS, Jasper WJ (1977). Controlling nylon dyeing by dye and chemical metering. *Textile Research Journal* **67**:609–616.
2. Hiimeno K, Ohno S (inventors) (2004). Dystar Textilfabrben GmbH & Co., assignee. Non-azo disperse dye mixtures for dyeing synthetic textiles or fiber blends. European Patent Application EP 2004-11346.
3. McDonald R, Dornan R (inventors) (2001). J&P Coats, Limited, assignee. Making dye mixtures to produce a certain target color. GB Patent Application WO 2001-GB1644.
4. Erying MB (1994). Spectromicrography, colorimetry: sample, instrumental effects. *Analytica Chimica Acta* **288**:25–34.
5. Robertson J (1992). *Forensic Fiber Examination of Fibers*, Chapter 4. Ellis Horwood, New York.
6. Robertson J (1992). *Forensic Fiber Examination of Fibers*, Chapter 5. Ellis Horwood, New York.

7. Federal Bureau of Investigation (1999). Scientific Working Group on Materials Analysis. Forensic fiber examination guidelines. *Forensic Science Communications* **1**(1). Available at http://www.fbi.gov/hq/lab/fsc/backissu/april1999/houckapb.htm.

8. Laing DK, Gill R, Blacklaws C, Bickley HM (1988). Characterisation of acid dyes in forensic fibre analysis by high-performance liquid chromatography using narrow-bore columns, diode array detection. *Journal of Forensic Sciences* **442**:187–208.

9. West JC (1981). Extraction, analysis of disperse dyes on polyester textiles. *Journal of Chromatography* **208**:47–54.

10. Wheals BB, White PC, Patterson MD (1985). High-performance liquid chromatographic method utilising single or multi-wavelength detection for the comparison of disperse dyes extracted from polyester fibres. *Journal of Chromatography A* **350**:205–215.

11. Griffin R, Kee TG, Adams RW (1988). High-performance liquid chromatographic system for the separation of basic dyes. *Journal of Chromatography A* **445**:441–448.

12. Lewis L, Lewis S, DeVault G, Tuinma A (2002). Enhanced trace-evidence discrimination. Final Report to the FBI on DOE Project No. 2051-1119-Y1.

13. Xu X, Leijenhorst H, Van Den Horen P, Koeijar JD, Logtenberg H (2001). Analysis of single textile fibres by sample-induced isotachophoresis—micellar electrokinetic capillary chromatography. *Science & Justice* **41**:93–105.

14. Huang M, Russo R, Fookes BG, Sigman ME (2005). Analysis of fiber dyes by liquid chromatography mass spectrometry (LC-MS) with electrospray ionization: discriminating between dyes with indistinguishable UV–Visible absorption spectra. *Journal of Forensic Sciences* **50**(3):1–9.

15. Tuinman AA, Lewis LA, Lewis SA (2003). Trace-fiber color discrimination by electrospray ionization mass spectrometry: a tool for the analysis of dyes extracted from submillimeter nylon fibers. *Analytical Chemistry* **75**:2753–2760.

16. Huang M, Yinon J, Sigman ME (2004). Forensic identification of dyes extracted from textile fibers by liquid chromatography mass spectrometry (LC-MS). *Journal of Forensic Sciences* **49**(2):1–12.

17. Voyksner RD (1994). Atmospheric pressure ionization LC/MS: new solutions for environmental analysis. *Environmental Science Technology* **28**:118A–127A.

18. Turnipseed SB, Roybal JE, Rupp HS, Hurlbut JA, Long AR (1995). Particle beam liquid chromatography–mass spectrometry of triphenylmethane dyes: application to confirmation of malachite green in incurred catfish tissue. *Journal of Chromatography B* **670**:55–62.

19. Voyksner RD, Straub R, Keever JT, Freeman HS, Hsu WN (1993). Determination of aromatic amines originating from azodyes by chemical reduction combined with liquid chromatography/mass spectrometry. *Environmental Science Technology* **27**:1665–1672.

20. Marbury D, Tuschall J, Lynn B, Haney C, Voyksner R (1989). Comparison of particle-beam, thermospray HPLC/MS techniques or the analysis of dyes. *Advanced Mass Spectrometry* **11A**:206–207.

21. Yinon J, Jones TL, Betowski LD (1989). Particle beam liquid chromatography–electron impact mass spectrometry of dyes. *Journal of Chromatography* **482**: 75–85.

22. Straub R, Voyksner RD, Keever JT (1992). Thermospray, particle beam and electrospray liquid chromatography–mass spectrometry of azo dyes. *Journal of Chromatography* **627**:173–186.

23. Betowski LD, Ballard JM (1984). Identification of dyes by thermospray ionization and mass spectrometry/mass spectrometry. *Analytical Chemistry* **56**:2604–2607.

24. Ballard JM, Betowski LD (1986). Thermospray ionization and tandem mass spectrometry of dyes. *Organic Mass Spectrometry* **21**:575–588.

25. Betowski LD, Pyle SM, Ballard JM, Shaul GM (1987). Thermospray LC/MS/MS analysis of wastewater for disperse azo dyes. *Biomedical and Environmental Mass Spectrometry* **14**:343–354.

26. Voyksner RD (1985). Characterization of dyes in environmental samples by thermospray high-performance liquid chromatography/mass spectrometry. *Analytical Chemistry* **57**:2600–2605.

27. Yinon J, Jones TL, Betowski LD (1989). Enhanced sensitivity in liquid chromatography/thermospray mass spectrometry of dyes using a wire repeller. *Rapid Communications in Mass Spectrometry* **3**:38–41.

28. Yinon J, Jones TL, Betowski LD (1986). High-sensitivity thermospray ionization mass spectrometry of dyes. *Biomedical and Environmental Mass Spectrometry* **18**: 445–449.

29. Bruins AP, Weidolf LOG, Henion JD, Budde WL (1987). Determination of sulfonated azo dyes by liquid chromatography/atmospheric pressure ionization mass spectrometry. *Analytical Chemistry* **59**:2647–2652.

30. Ràfols C, Barceló D (1997). Determination of mono- and disulphonated azo dyes by liquid chromatography–atmospheric pressure ionization mass spectrometry. *Journal of Chromatography* **777**:177–192.

31. Holčapek M, Jandera P, Přikyl J (1999). Analysis of sulphonated dyes and intermediates by electrospray mass spectrometry. *Dyes and Pigments* **43**:127–137.

32. Holčapek M, Jandera P, Zderadička P (2001). High performance liquid chromatography–mass spectrometric analysis of sulphonated dyes and intermediates. *Journal of Chromatography A* **926**:175–186.

33. Socher G, Nussbaum R, Rissler K, Lankmayr E (2001). Analysis of sulfonated compounds by ion-exchange high-performance liquid chromatography–mass spectrometry. *Journal of Chromatography A* **912**:53–60.

34. Smyth WF, McClean S, O'Kane E, Banat I, McMullan G (1999). Application of electrospray mass spectrometry in the detection and determination of Remazol textile dyes. *Journal of Chromatography A* **854**:259–274.

35. Baiocchi C, Brussino MC, Pramauro E, Prevot AB, Palmisano L, Marci G (2002). Characterization of methyl orange and its photocatalytic degradation products by HPLC/UV-Vis diode array and atmospheric pressure ionization quadrupoleion trap mass spectrometry. *International Journal of Mass Spectrometry* **214**:257–256.

36. Novotná P, Pacáková V, Bosáková Z, Štulík K (1999). High-performance liquid chromatographic determination of some anthraquinone and naphthoquinone dyes occurring in historical textiles. *Journal of Chromatography A* **863**:235–241.

37. Mayer M, Kesners P, Mandel F (1998). Determination of carcinogenic azo-dyes by HPLC/MS analysis. *GIT Spezial Chromatography* **18**:20–23.

38. Mayer M, Frie A, Kesners P (1998). Determination of sensitizing disperse dyes by HPLC/MS analysis. *GIT Spezial Chromatography* **18**:65–66, 68–69.

39. Lemr K, Holčapek M, Jandera P, Lyčka A (2000). Analysis of metal complex azo dyes by high-performance liquid chromatography/electrospray ionization mass spectrometry and multistage mass spectrometry. *Rapid Communications of Mass Spectrometry* **14**:1881–1888.

40. Ng L-K, Lafontaine P, Brazeau L (2004). Ballpoint pen inks: characterization by positive and negative ion-electrospray ionization mass spectrometry for the forensic examination of writing inks. *Journal of Forensic Sciences* **47**:1238–1247.

41. Yinon J, Saar J (1991). Analysis of dyes extracted from textile fibers by thermospray high-performance liquid chromatography–mass spectrometry. *Journal of Chromatography* **586**:73–84.

42. Vincent U, Bordin G, Rodriguez AR (2002). Optimization and validation of an analytical procedure for the determination of oxidative hair dyes in commercial cosmetic formulations. *Journal of Cosmetic Science* **53**:101–119.

43. Andries P, Bruins L, Weidolf OG, Henion JD, Budde WL (1987). Determination of sulfonated azo dyes by liquid chromatography/atmospheric pressure ionization mass spectrometry. *Analytical Chemistry* **59**:2647—2652.

9

Characterization of Surface-Modified Fibers

Robert D. Blackledge

Retired, Former Senior Chemist, Naval Criminal Investigative Service Regional Forensic Laboratory, San Diego, California

Kurt Gaenzle

Senior Materials Engineer, Materials Engineering Laboratory, Naval Air Depot, North Island, San Diego, California

9.1 FIBERS AS ASSOCIATIVE EVIDENCE

Although fibers as associative evidence are hardly new, it was their importance in associating many of the victims in the Atlanta child murders [1] with the home of the suspect that established this class of evidence in the minds of the general public. In Chapter 1 on glitter, the properties of the "ideal contact trace" are enumerated in Section 1.2. Just as with glitter, fiber trace evidence in general meets these ideals: (1) individual fibers are nearly invisible and therefore if transfer or cross-transfer occurs between victim and assailant it is not highly likely that the assailant will notice this transfer and therefore make efforts to remove the transferred fibers; (2) individual fibers have a high probability of transfer and retention; (3) they are highly individualistic (although this varies considerably from fiber type to fiber type (white cotton fibers—no, specialty fibers and many carpet fibers—yes); (4) they can be quickly collected, separated, and concentrated; (5) traces (just a single fiber) may be character-

Forensic Analysis on the Cutting Edge: New Methods for Trace Evidence Analysis, Edited by Robert D. Blackledge.
Copyright © 2007 John Wiley & Sons, Inc.

ized; (6) properties may be measured and then compiled in a searchable database; and (7) most fiber types well survive most environmental insults.

9.2 SURFACE-MODIFIED FIBERS

Much has been written about the forensic characterization of fiber trace evidence. Michael Grieve, who first worked for the United States Army Criminal Investigation Laboratory located in Frankfurt, Germany (USACIL–Europe) during the Cold War and later at the Bundeskriminalamt (BKA) in Wiesbaden, Germany, no doubt did more than any other single individual to advance this field. Anyone new to this field would do well to begin with Mike's many books and articles [2].

Although fibers having a surface modification are not new, in recent years they have greatly proliferated both in terms of variety and in general use. Identifying that an otherwise relatively common fiber (e.g., a white cotton fiber) bears a surface modification can serve to place that fiber into a much smaller subclass of fiber trace evidence and therefore greatly increase its value as associative evidence. Today there are so many different types and manufacturers of surface-modified fibers that it would be impossible to adequately cover them all in a single chapter. Instead, this chapter will focus on just a few examples. Today, largely because of forensic laboratory accreditation pressures from ASCLD-LAB (American Society of Crime Laboratory Directors), many crime laboratory directors feel that they must have a "validated protocol" for every specific type of evidence their examiners might encounter. This is not only impossible; it's not even desirable. Such a philosophy produces "cookbook recipes" and examiners who follow a protocol by rote. (For more on this subject, readers are directed to the writings of Peter De Forest of the Forensic Science program at John Jay College of Criminal Justice, City University of New York.)

Our goal in this chapter will not be to provide readers with a "validated protocol" for the characterization of surface-modified fibers. Rather, we hope to provide a general approach or mindset. This general approach must be modified according to the specific type of surface-modified fiber encountered.

9.3 PRELIMINARY EXAMINATIONS

Many of the preliminary examinations/comparisons made between a questioned (Q) surface-modified fiber and a known (K) surface-modified fiber would be no different than the examinations already in routine use for fiber evidence (e.g., color comparison, presence and amount of delusterant, cross sectional shape). For the most part we will ignore these tests and readers are referred to the work of Grieve and others. However, some classical tests are worth mentioning because they may provide an examiner with a clue that he/she is in fact dealing with a surface-modified fiber. Although not always the case, typically in forensic science comparative exams, the analyst has a virtually unlimited supply of the K sample and only traces of the Q sample. If the K sample is a carpet, or perhaps a sweater or cap, from the information provided on a label it may be possible to determine what the nature of the K sample *should* be, whether it contains surface-modified fibers, and if so what their properties *should* be. (Of course, this must be confirmed by examinations performed on the K sample.) Let's consider some classical tests that might give an indication that one is dealing with a surface-modified fiber.

9.3.1   Infrared Spectra and Properties Measured by Polarized Light Microscopy

If the surface modification layer is sufficiently thick it may be observable under polarized light microscopy (PLM) either in a longitudinal view or in cross section. Although infrared microscopy in the transmittance mode (IR beam passes through the fiber and the spectrum indicates the absorption of certain wavelengths) would likely indicate the nature of the *core* of the fiber (e.g., nylon 6.6, rayon, polyester), attenuated total reflectance (ATR), a *surface* sampling method, might indicate there is a surface layer that is different from the core. Even ATR might not work if the surface layer is so thin that it is on the order of just a few molecules. However, if the measurements obtained with the Q sample by classical PLM (birefringence, etc.) are off compared to what they should be based on the FTIR identification of the fiber core, this could suggest there might be a surface modification.

9.3.2   Infrared Mapping with an FTIR Microscope

An FTIR microscope is used to map on a molecular level (rather than elemental) the distribution of different chemical species as the instrument scans a cross section of the sample. If one knows (or suspects) in advance what chemical species are present in the sample, then for each species a specific wavelength may be selected that is absorbed by that species but not by the others. For each wavelength selected a false color is chosen to represent the presence of the associated species. For example, let's suppose we know (or suspect) that a fiber has a core of polypropylene and a surface modification consisting of a silicone. We select a wavelength that is absorbed by the polypropylene (but not by the silicone) and assign it a false color—say, blue. We select a different wavelength that is absorbed by the silicone (but not by the polypropylene) and assign it a false color—say, red. Now with our FTIR microscope system we scan a cross section of the fiber. If we just perform a straight line scan across the diameter of the cross section (requires less memory) we will see on the system monitor a straight line that at its start and finish is red and in the middle is blue. Or, we could execute a two-dimensional scan (requires far more memory) and we would see on the monitor a representation of the entire cross section that would be a red ring around a blue circle (if the fiber had a circular cross section). Either way (linear or two dimensional) the color map shows both the presence and nature of the surface modification and also the nature of the fiber core. It also provides information about the thickness of the surface-modified layer. Although not a surface modification, this same FTIR mapping technique would be useful for characterizing bicomponent fibers.

9.3.3   Raman Mapping

Just about everything said in the previous section about FTIR mapping should also apply to Raman mapping. Preparing a fiber cross section might not even be necessary if the same confocal Raman *x-z* mapping procedure used for a single glitter particle was used (see Chapter 1, Section 1.3.9).

9.3.4   AATCC Test Method 118-2002

American Association of Textile Chemists and Colorists (AATCC) Test Method 118-2002, "Oil Repellency: Hydrocarbon Resistance Test," is used to detect the presence of a fluorochemical finish. A thin Teflon®-like coating can imbue to a fabric the qualities of water repellency and resistance to staining by oily materials. This test for a fluorochemical finish involves testing a swatch of material with a series of standard test liquids and observing at what point in this series the fabric is wetted (if the solvent does not wet the fabric you observe a clear well-rounded drop sitting on top of the fabric; if the solvent wets the fabric it will sink into the fabric and darken it). Although this test is designed for 20 cm × 20 cm swatches, it can be modified so as to be suitable for single fibers. Place a tiny drop of one of the Standard Test Liquids (the method gives information where these may be procured) onto a microscope slide. Carefully let a single fiber drop onto the top of the drop. Under the microscope observe the contact angle and whether the solvent wets the fiber. At the very least, this simple test could show whether the Q fiber is *excluded* or *included* as having a common origin with the source of K fibers. If the Q fiber is excluded, no further tests are necessary on this fiber. If the Q fiber is included, then one would proceed to more discriminating tests.

9.3.5   A Simple Example

At a website [3] (see www.nanotechproject.org/76) having to do with nano-technology, there is an absolutely delightful videoclip. In the videoclip an 11-year-old girl is acting like a reporter and is interviewing an expert on nanotechnology (her father, Andrew Maynard, Chief Science Advisor at the Woodrow Wilson Center's Project on Emerging Nanotechnologies). After several questions involving the basics, she asks if there are any practical examples of nanotechnology? Her father replies that the $70 silk tie he is wearing has been treated with a stain-resistant nanotech coating. The girl then asks if she can test its stain resistance. Her father takes it off and the girl and her younger brother apply ketchup, mustard, and coffee to the tie. After they wash the materials off and dry it, the father examines the tie and admits that it's as good as new. Although these tests were applied to a fabric, the same stain resistance tests could just as easily be applied to Q and K fibers.

9.4   DISTINGUISHING TESTS

A general rule for the selection and sequence of tests for the characterization of evidence samples is that one begins with tests that are not destructive of the sample, are quick and easy to perform, and do not require exotic or expensive instrumentation or chemicals. If at any stage in this test series a clear difference is obtained between the Q and K samples, no further tests are necessary. Saved for later are any tests that require expensive, sophisticated instrumentation and any tests that can only be performed if the evidence is sent to another laboratory. Absolutely last in any series of tests would be those that would be destructive of the sample. If destructive tests were necessary, if possible it would be best to partition the sample into halves (or thirds) so that there would be remaining sample to make available to a laboratory contracted by the defense, or to a referee laboratory. Additionally, it is always desirable to have remaining sample available for inspection by the jury.

9.4.1   Scanning Electron Microscopy/Energy Dispersive Spectroscopy

Examination of a fiber cross section by means of SEM/EDS (or related methods such as SEM/wavelength dispersive spectroscopy or transmission electron spectroscopy) may be able to show differences in the morphology between a surface-modified layer and the fiber core and/or a difference in elemental profiles. If differences in elemental profiles are seen between the surface layer and the core, elemental mapping may be performed. Elemental mapping is analogous to the infrared mapping discussed in Section 9.3.2 except that different elements are assigned false colors instead of different molecular types. Following are results obtained on different types of surface-modified fibers using SEM/EDS.

P2i Treated Samples   P2i (P2i Ltd., 2006) is a patented plasma process that was originally developed by the U.K. Defence Science and Technology Laboratory in collaboration with the University of Durham (www.p2ilabs.com/news/newsitem4.html). Its initial intended use was to protect armed forces and emergency services personnel against chemical attacks. In addition to medical textiles, the process has recently been expanded to include the filters used in the wells of DNA forensic microplates, pipette tips, and other devices. Because the surface coating is applied as a plasma, the surface layer is ultrathin, only a few molecules thick. It is so thin that the pore size of treated filters is not affected. One of the authors obtained several samples from Dr. Stephen Coulson, Technical Director, Porton Plasma Innovations (P2i) Ltd., Porton Down, U.K. The samples consisted of pairs that were identical in all respects except one of each pair had received the P2i plasma treatment and the other had not.

Scanning electron microscopy (SEM) images of the treated and untreated swatches revealed no noticeable differences (Figure 9.1). However, energy dispersive spectroscopy (EDS) showed a small fluorine peak with the

Figure 9.1 *Untreated swatch (left) and P2i treated swatch (right).*

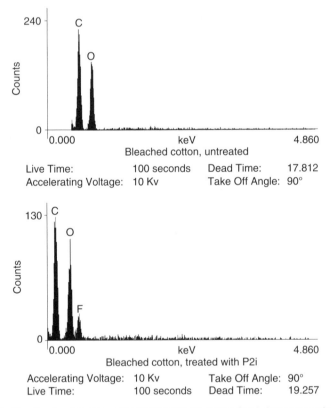

Figure 9.2 *Fluorine is absent in untreated bleached cotton (top), but present in P2i treated bleached cotton (bottom).*

P2i treated swatch of material that was absent with the untreated swatch (Figure 9.2).

Also obtained from P2i Limited were identical pairs of finished cotton except that one had received the P2i treatment while the other had not. Figure 9.3 shows the results of SEM/EDS with swatches from these samples.

Figures 9.2 and 9.3 clearly show that SEM/EDS is able to distinguish between P2i treated (some type of fluorocarbon) cloth swatches and identical untreated swatches, but for the purposes of forensic science we need to be able to distinguish between individual treated and untreated fibers. Although not as easy, Figure 9.4 shows the size of the area on a single fiber that was sampled and Figure 9.5 shows that treated and untreated fibers could be distinguished.

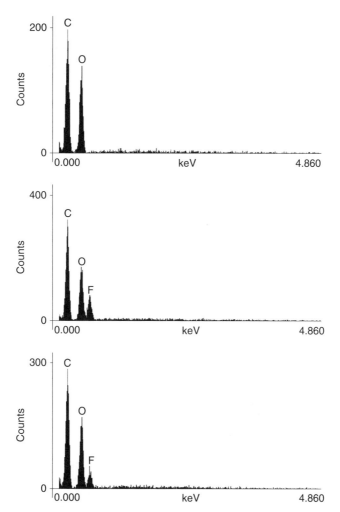

Figure 9.3 *Untreated finished cotton (top), P2i treated finished cotton (middle), and other side of P2i treated cotton (bottom). This demonstrates that the 2Pi plasma process produces a very thin repellant coating on all exposed surfaces.*

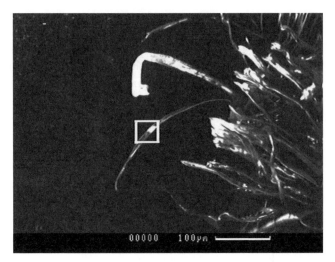

Figure 9.4 Bright area within box (center) shows the area sampled by SEM/EDS within a single fiber.

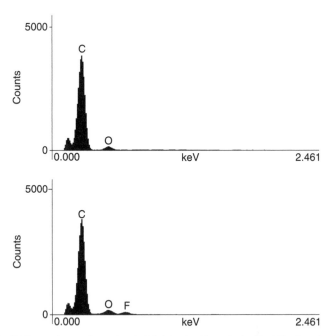

Figure 9.5 EDS spectrum of a untreated finished cotton single fiber (top) and a P2i treated single fiber (bottom). Note that the accelerating voltage was lowered to 3 Kv compared to the 10 kv used for the swatches. An added advantage of this lower accelerating voltage is that there is less penetration. Thus, more of the signal arises from the fluorocarbon layer at the surface. One could repeat the analysis using a higher accelerating voltage; the resulting decrease in the intensity of the fluorine peak would demonstrate that it is a property of a surface layer rather than being throughout the fiber.

3M Protective Finish Several different cloth samples were provided by Paul L. Johnson, Technical Service Specialist, 3M PMCS Division, St. Paul, Minnesota. Included in these samples were:

A1   100% cotton knit with 3M Protective Finish (a fluorocarbon)
A2   100% cotton knit with no protective treatment
B1   100% cotton woven with 3M Protective Finish
B2   100% cotton woven with no protective treatment
B3   100% cotton woven with non-3M Protective Finish (brand name not provided)

Although SEM showed no differences between A1 and A2, with EDS they are clearly distinguished by the fluorine peak seen in A1 (Figures 9.6 and 9.7).

EDS results with the cotton woven samples were even more interesting. All three samples showed a peak for silicon that was not seen in the A1 and A2 samples. Both B1 and B3 showed a peak for fluorine, and B2 did not (Figures 9.8, 9.9, and 9.10).

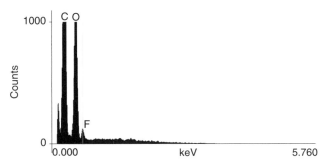

Figure 9.6 *EDS spectrum of A1, 100% cotton knit with 3M Protective Finish (a fluorocarbon).*

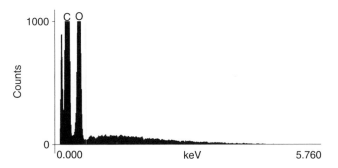

Figure 9.7 *EDS spectrum of A2, 100% cotton knit with no protective treatment.*

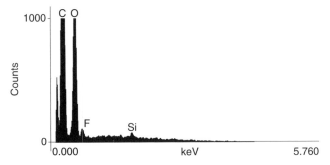

Figure 9.8   *B1, 100% cotton woven with 3M Protective Finish.*

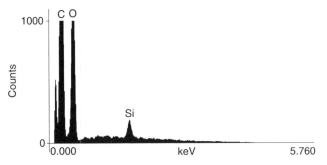

Figure 9.9   *B2, 100% cotton woven with no protective treatment*

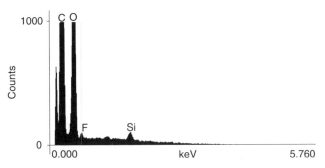

Figure 9.10   *B3, 100% cotton woven with non-3M Protective Finish (brand name not provided).*

Use Low Voltage to Detect Qualitative Differences in Surface Chemistry.   Generally, SEM voltage is set around 20 Kv for routine EDS analysis. Figures 9.2 through 9.10 have shown that lowering the SEM voltage will increase the signal for those elements at or near the sample surface. A rule of thumb is to lower the SEM voltage to approximately three times the x-ray line voltage of interest [4]. Using two different SEM voltages in the forensic examination of a questioned sample can indicate whether the presence of an element is primarily restricted to the surface or near surface (a

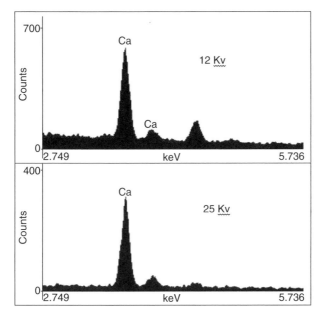

Figure 9.11 *Active® Glass by Pilkington Glass. EDS of titanium dioxide rich exterior facing surface. Decrease in Ti peak at higher voltage shows the glass is enriched at or near the surface rather than being uniformly distributed. [Other conditions are constant: Take Off Angel: 90°; Live Time: 60 seconds; Dead Time: ~55 seconds]*

surface modification or laminate), or is instead evenly distributed throughout the sample. Because it illustrates this principle so graphically, a relatively new type of window glass is used for illustration. Active® Glass was introduced by Pilkington Glass in 2001 [5, 6]. Intended for use only on exterior windows, its exterior-facing surface has self-cleaning properties due to it being rich in titanium dioxide that together with ultraviolet rays from the sun produce a catalytic effect that breaks down surface dirt to carbon dioxide and water. Because the exterior surface is hydrophilic, rainwater (or water from a hose) tends to sheet down the glass rather than form droplets and upon drying there are no water spots. Using SEM voltages of 12 and 25 Kv to examine the exterior-facing surface of Active Glass (Figure 9.11) not only detects titanium, but also shows it to be more abundant at or near the surface rather than being uniformly distributed (like the calcium) throughout the glass.

9.4.2   Gas Chromatography/Mass Spectrometry

Gas chromatography/mass spectrometry (GC/MS) may be of value in those cases where a fiber's surface treatment can be removed with appropriate solvents. Of course, one must insure that that the resulting extract will not cause damage to the column's liquid phase.

An example where GC/MS can distinguish treated from untreated fibers can be found in garments designed for outdoor wear that have been treated with an insecticide. Residues may be present on clothing items if one has applied insecticides using an aerosol spray, but there are also clothing items available commercially that have been treated with an insecticide that is claimed to be semi-permanent. BUZZ OFF™ (BUZZ OFF Insect Shield LLC, Greensboro, NC—see www.buzzoff.com) is one example. At its website the maker claims that BUZZ OFF treated garments have no unpleasant insecticide odor and yet may be laundered up to 25 times without removing the insecticide. However, the maker cautions that dry cleaning solvents will remove it. An approximately $1\,cm^2$ swatch was cut from a BUZZ OFF treated scarf and extracted with a few drops of chloroform. The insecticide bonded to the fabric in BUZZ OFF treated garments is permethrin. Figure 9.11 also shows the total ion chromatogram (TIC) obtained when approximately $1\,\mu L$ of the extract was injected into a GC/MS. The molecular formula for permethrin is shown in Figure 9.15.

The two succeeding peaks in Figures 9.12 and 9.13 are both permethrin. The mass spectrum of the peak at 13.75 minutes is shown in Figure 9.14. As can be seen from its formula (Figure 9.15), permethrin can exist as *cis* and *trans*

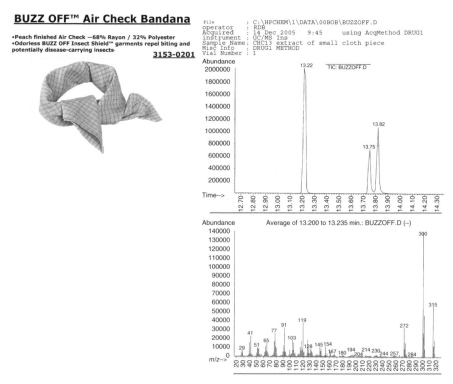

Figure 9.12 BUZZ OFF™ treated bandana (top left) purchased from ExOfficio (www.exofficio. com). GC/MS total ion chromatogram (top right) from $CHCl_3$ extract of a $1\,cm^2$ swatch. Electron ionization mass spectrum of first peak (bottom right), which is bumetrizole (Figure 9.12), a UV absorber that helps to prevent fading in sunlight.

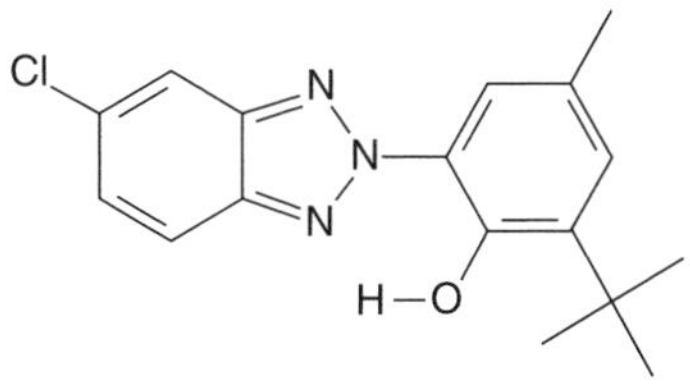

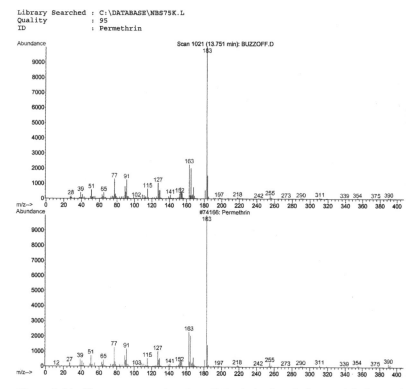

Figure 9.13 *Molecular formula of bumetrizole, $C_{17}H_{18}ClN_3O$; molecular weight = 315.8.*

Figure 9.14 *Mass spectrum of peak at 13.7 min* (top) *and of permethrin* (bottom).

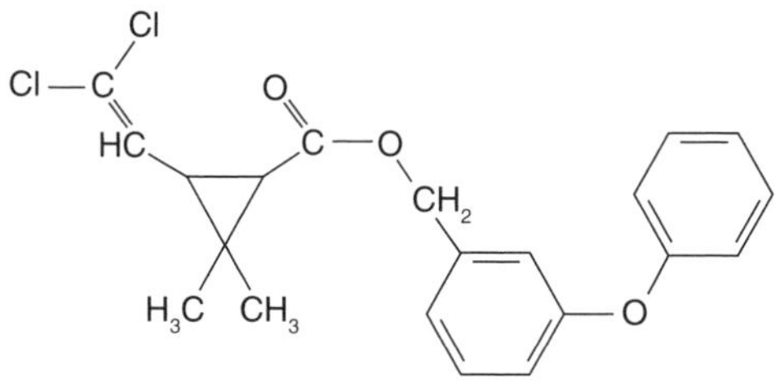

Figure 9.15 *Molecular formula of permethrin, $C_{21}H_{20}Cl_2O_3$; molecular weight = 391.3.*

isomers. An approximate 40:60 *cis/trans* ratio is used in many commercial products.

Insecticide treated clothing may also be used by the U.S. military. "Military personnel sometimes must be rapidly deployed to areas where life-threatening, insect-borne diseases are prevalent. This places such personnel at an increased risk of contracting diseases such as malaria, scrub typhus, leishmaniasis, and Lyme disease. The suddenness of deployments and movement after deployment often precludes the use of protection or control measures. To protect against specific disease risks from insect bites, the U.S. Army has formulated a clothing impregnant containing permethrin, a pyrethroid insecticide that is effective against disease vectors such as mosquitoes, ticks, and other arthropods. The Army proposes to use permethrin-impregnated fabric to manufacture battle-dress uniforms (BDUs). BDUs, made from either 100% cotton fabric or 50% nylon and 50% cotton fabric, are used to camouflage soldiers" (Health Effects of Permethrin-impregnated Army Battle-Dress Uniforms, Subcommittee to Review Permethrin Toxicity from Military Uniforms, National Academy Press, Washington, DC, 1994).

Antimalaria mosquito nets (Olyset®) impregnated with permethrin are being produced by Sumitomo Chemical [7]. These last 3–5 years and currently cost less than $5.

A hypothetical scenario. A hiker in Alaska wearing BUZZ OFF treated clothing is attacked and killed by a grizzly bear. Searchers locate a bear in the area and shoot it with a tranquilizing dart. Is this the bear that attacked the hiker? As the great beast momentarily lies comatose, tape lifts are made of its muzzle and forepaws. Along with any blood and tissue recovered with the tape lifts, as well as bite marks and tracks, a subsequent examination at a crime lab of any fibers recovered from the tape lifts and their comparison with fibers from the hiker's clothing can help to include/exclude this particular bear.

9.4.3 Pyrolysis Gas Chromatography/Mass Spectrometry

Because it destroys that part of the sample examined, pyrolysis gas chromatography/mass spectrometry (Py GC/MS) if used at all would always be the last examination in any analysis scheme. Additionally, from the point of view of the criminalist, Py GC/MS has several disadvantages. (1) At trial it is always desirable to have actual physical evidence to show the jury. (2) If the Q sample has been entirely consumed in analysis, opposing council will request that any testimony related to it be ruled inadmissible since defense council would not have had an opportunity to have the evidence examined by their experts. Although this ploy seldom succeeds, it no doubt has a psychological impact on the jury. (3) In the hands of inexperienced practitioners, Py GC/MS often has poor reproducibility.

More than anything else, criminalists want to get it *right!* Everything that they do in their analysis is discoverable. Subsequent to running any test on the evidence and examining the data, an analyst may come to the conclusion that the test as run has insufficient reproducibility and is not trustworthy. Therefore, the data from that test should not be used in trying to answer the question: "Could the Q sample and the K sample have originated from a common source?" In industry (quality control/quality assurance) or in scientific research this would be perfectly reasonable and acceptable. But just let a forensic scientist do this and they will be accused of attempting a cover up and selectively using only those analysis results that support the prosecution!

Sadly, for the above reasons, criminalists may decide that it is better not to make any attempt to compare the Q and K samples using Py GC/MS. And yet, for selected sample types if run correctly, Py GC/MS can be highly discriminating and show excellent reproducibility [5, 6]. Pyrolysis methods should be considered when the sample under consideration is polymeric in nature. Some examples are plastics, rubbers, PVC tapes, photocopy toners, paints, and fibers. In pyrolysis air (or oxygen) is not present, so combustion does not occur. Instead the bonds in the polymeric materials are broken by heat and smaller fragments are produced. These smaller fragments are characteristic of the larger molecules (polymers) from which they originated, and they are now sufficiently small (volatile) that they will pass through a GC column and can then be detected and identified by the type of mass spectrometers typically found in crime laboratories. (Those readers unfamiliar with the forensic application of Py GC and/or Py GC/MS are directed to reviews by Saferstein [7], Blackledge [8, 9] and Challinor [10].)

Py GC/MS can readily identify artificial fibers made of nylon, rayon, polyester, and so on. However, there are nondestructive methods (polarized light microscopy, infrared microscopy, Raman microscopy, etc.) that can do this just as well. So what possible role could Py GC/MS have in the identification/comparison of fibers having a surface modification?

As used in the past, pyrolysis methods would entirely consume the sample and all that would remain would be a residue containing nonvolatile inorganic materials. If a fiber bearing a thin surface modification were to be pyrolized under the conditions used in the past, the vast majority of the pyrolysis products would have originated from the fiber's polymeric core. If a mass spectrometer were to be used as a detector, any traces originating from the surface modification would likely be lost under this huge signal.

However, suppose if instead of pyrolizing the entire fiber sample we just gave the sample a very brief burst of heat (flash pyrolysis) that would be sufficient to vaporize and fragment the surface modification layer but would largely leave the core of the fiber intact? In such a case, although the overall signal to a detector (mass spectrometer) would be smaller, a much higher percentage of it would be due to the surface modification layer. Several of the cloth samples provided by Paul L. Johnson (Technical Service Specialist, 3M PMCS Division, St. Paul, Minnesota), that were examined by SEM/EDS, were also examined by a method of pyrolysis GC/MS. Included in these samples were 100% cotton woven samples that were identical except:

B1   100% cotton woven with 3M Protective Finish (a fluorocarbon)

B2   100% cotton woven with no protective treatment

B3   100% cotton woven with a non-3M Protective Finish (brand name not provided)

Although testing of these samples is still incomplete, following are some of the preliminary results obtained by Toshio Tsuchiya, Chief of Laboratory, Japan Analytical Industry Co., Ltd., Tokyo, Japan.

Analytical Purpose   To determine the ability of short-time pyrolysis (1 second) to identify the existence of a surface treatment on the cloths, to discriminate between samples, and to determine the reproducibility of the pyrograms.

Analytical Conditions

Pyrolyzer, JPS-700 and GC/MS. Curie point pyrolyzer, JPS-700 coupled onto a GC/MS (Shimadzu QP-6202, EI, 70 eV). Pyrolysis temperature: 423 °C for 1 second.

GC Condition. Column: DB-5ms, i.d. 0.25 mm × 30 m length, $t = 0.25 \mu m$, 40–280 °C, 10 °C/min. Carrier gas (He) flow rate: 1.0 mL/min; split ratio: 1/100.

Sample Treatment. Samples were cut 3 mm × 10 mm with scissors (~5 mg).

Results   Pyrograms obtained with the B1, B2, and B3 samples are shown in Figure 9.16. All three samples could be distinguished, and the presence of isopentil trifluoro acetate in B1 could possibly be used as a marker for the fluorocarbon surface treatment. It is interesting that no indications of a fluorocarbon were seen in B3 even though EDS of the same sample showed a weak fluorine peak. Perhaps a higher pyrolysis temperature is needed even though it would also increase the background from the fiber core.

Of course, the samples were small strips of cloth (3 mm × 10 mm) rather than single fibers. Nevertheless, as preliminary results they are encouraging. Several things could be tried in an effort to bring the sensitivity of the method down to the single fiber level needed for forensic science. With a split ratio of 1:100, 99% of the pyrolysate vapors are being lost to vent. Splitless injection with cryogenic trapping [14, 15] could possibly improve the amount of sample reaching the detector.

Mass spectrometer detector response could also be improved by using selected ion monitoring (SIM) rather than a general scan. In trace-evidence analysis one typically has a virtually unlimited amount of the K sample and very limited amounts of the Q sample. Using the K sample and larger sample amounts, the retention times of the peaks characteristic of the surface treatment and major fragment ions in those peaks could be determined. This would form the basis for the development of an SIM method. Matching retention

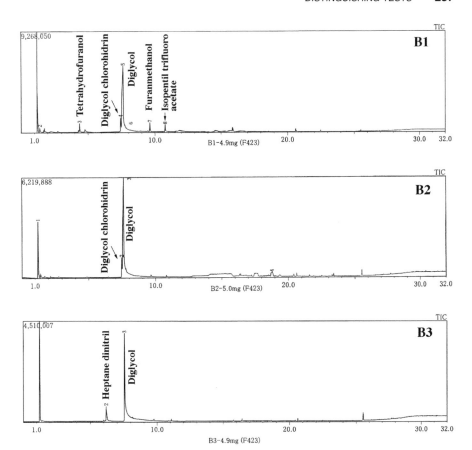

Figure 9.16 *Pyrograms obtained from the B-series samples. B1 = 100% cotton woven with 3M Protective Finish (a fluorocarbon); B2 = 100% cotton woven with no protective finish; and B3 = 100% cotton woven with a non-3M Protective Finish (brand name not provided).*

times and relative abundances of the selected SIM ions in the K and Q samples would be a strong argument for their having a possible common origin.

For any mass spectrometer to work the sample must be ionized. There are a number of ways of producing sample ions for introduction into a mass spectrometer. Among these methods, EI (the method most commonly used in the mass spectrometers found in forensic laboratories) is a relatively harsh method. EI typically produces much fragmentation of the analyte molecules, and in some cases a molecular ion is either not seen or is in very low abundance. Such extensive fragmentation reduces the signal seen for any particular ion. Among the other softer (less fragmentation) methods of producing ions, chemical ionization is the most likely to be available as an option on the mass spectrometers found in today's forensic laboratories. If the pyrolysate vapors were introduced into the mass spectrometer under conditions of either positive or negative chemical ionization, the signal from the molecular ions might be greatly increased compared to the EI method. Combining cryogenic focusing as the pyrolysate vapors enter the GC with chemical ionization as the separated pyrolysis products enter the mass spectrometer might produce even greater sensitivity.

There is at least one last approach that is worth mentioning, even though its utility might be limited to those fibers having a fluorocarbon (or some other halogen) type surface treatment. Capillary column GC with an electron capture detector (ECD) is an extremely sensitive method for the detection of halogen-containing molecules. For selected species its sensitivity can even exceed various methods of mass spectrometry. Although capillary column GC with ECD detection has been used extensively for the detection of pesticide residues by environmental laboratories, it has received comparatively little attention in most forensic laboratories. ECD detectors utilize a radioactive β-electron emission source. Surely the licensing requirements attendant to the use of ECD detectors have been a contributing factor to their lack of use in forensic laboratories. However, a recent article [13] described a forensic science application for a pulsed-discharge electron capture detector (PDECD). PDECDs do not use a radioactive source and yet have comparable detection limits. Combined with splitless injection and cryogenic trapping of the pyrolysate vapors, using an PDECD on the GC might well bring the sensitivity into the single-fiber sample range.

ACKNOWLEDGMENTS

We thank Dr. Stephen Coulson, Technical Director, Porton Plasma Innovations (P2i) Ltd., Porton Down, U.K., for providing matching pairs of samples that had and had not received their P2i surface modification treatment. We also thank Paul L. Johnson, Technical Service Specialist, 3M PMCS Division, St. Paul, Minnesota, USA, for providing matching fabric samples that had received no surface treatment, had received the 3M surface treatment, or had received a competitor's surface treatment. And for the pyrolysis gas chromatography/mass spectrometry analysis we thank Japan Analytical Industry Co., Ltd., Tokyo, Japan, and especially Toshio Tsuchiya, Chief of Laboratory.

REFERENCES

1. Deadman HA (1984). Fiber evidence and the Wayne Williams trial. *FBI Law Enforcement Bulletin* **53**(3 & 5), March (part 1):12–20; May (part 2):10–19.
2. Grieve MC (1994). Fibres and forensic science—new ideas, developments, and techniques. *Forensic Science Review* **6**(1):50–80.
3. Project on Emerging Nanotechnologies of the Woodrow Wilson International Center for Scholars. www.nanotechproject.org/76. Accessed 27 Oct. 2006.
4. *Scanning Electron Microscopy and X-ray Microanalysis* (2003). Goldstein G, Newbury DE, Joy DC, Lyman CE, Echlin P, Lifshin E, Sawyer LC, and Michael JR, authors, 3rd Ed., Kluwer/Springer Publishers.

5. Blackledge RD, Springer F (2003). Self-cleaning window glass: a new subclass of glass trace evidence. Oral presentation B134, 2003 Annual Meeting of the American Academy of Forensic Sciences, Chicago, Illinois.

6. Maggay CD, Blackledge RD, Springer F (2004). Self-cleaning window glass: breakage transfer process validation and subclass/brand characterization. Oral presentation B45, 2004 Annual Meeting of the American Academy of Forensic Sciences, New Orleans, Louisiana.

7. Preventing malaria. *Chemical & Engineering News* **May 29**:19 (2006).

8. Walker JQ (1977). Pyrolysis gas chromatographic correlation trials of the American Society for Testing and Materials. *Journal of Chromatographic Science* **15 July**:267–274.

9. Wampler TP, Levy EJ (1987). Reproducibility in pyrolysis: recent developments. American Laboratory **12**:75–82.

10. Saferstein R (1985). Forensic aspects of analytical pyrolysis. In *Pyrolysis and GC in Polymer Analysis*, Liebman SA, Levy EJ (eds.). Marcel Deckker, New York, pp. 339–371.

11. Blackledge R (1992). Applications of pyrolysis gas chromatography in forensic science. *Forensic Science Review* **4**(1):1–16.

12. Blackledge R (2000). Pyrolysis gas chromatography in forensic Science. In *Encyclopedia of Analytical Chemistry*, Meyers RA (ed.). John Wiley & Sons, Hoboken, NJ.

13. Challinor J (1995). Examination of forensic evidence. In *Analytical Pyrolysis Handbook*, Wampler T (ed.). Marcel Dekker, New York, pp. 207–241.

14. Wampler TP, Bowe WA, Higgins J, Levy EJ (1985). Systems approach to automatic cryofocusing in purge and trap, headspace, and pyrolytic analysis. *American Laboratory* **Aug**:82–87.

15. Wampler TP, Levy EJ (1985). Cryogenic focusing of pyrolysis products for direct (splitless) capillary gas chromatography. *Journal of Analytical and Applied Pyrolysis* **8**:65–71.

16. Collin OL, Niegel C, DeRhodes KE, McCord BR, Jackson GP (2006). Fast gas chromatography of explosive compounds using a pulsed-discharge electron capture detector. *Journal of Forensic Sciences* **51**(4):815–818.

10

Characterization of Smokeless Powders

Wayne Moorehead

Senior Forensic Scientist, Orange County Sheriff–Coroner, Forensic Science Services Division, Santa Ana, California

10.1 INTRODUCTION

Although there are many types of explosive compounds, nearly all explosives fall into two major categories: low and high. A low explosive deflagrates, burning at a rate of less than 1000 meters per second, and requires confinement to explode [1]. A high explosive detonates, burning at a rate exceeding 1000 meters per second, and does not have to be confined to explode.

Smokeless powder, used in nearly all firearms from the beginning of the twentieth century, typically falls into the low explosive category, requiring confinement in such items as a pipe or ammunition cartridge case to explode. Smokeless powders are used in the commercial and civilian reloading and recreational markets primarily by sport shooters and hunters, as well as military munitions.

While having a useful and legitimate purpose, smokeless powders used for reloading, called canister powders (Figure 10.1), represent a readily available explosive for making improvised explosive devices (IED) by the criminal or terrorist elements. Pipe bombs in the United States are the most common IED with the majority containing smokeless powder [2, 3]. These smokeless powder

Forensic Analysis on the Cutting Edge: New Methods for Trace Evidence Analysis, Edited by Robert D. Blackledge.

Figure 10.1 Three examples of canisters of smokeless powder.

particles can be recovered and submitted to a forensic laboratory for analysis. After a thorough examination and laboratory characterization the forensic examiner can identify the particles and attempt the determination of smokeless powder brand.

10.2 PURPOSE OF ANALYSIS

Two purposes exist for the analysis of smokeless powder in forensic science: the identification of particles as smokeless powder and determination of brand origin.

10.2.1 Identification of Smokeless Powder

The identification of a particle as smokeless powder assists in the reconstruction of criminal incidents, such as resolving distance determinations in a shooting. This is a simple method that can be accomplished by examining the morphology of the powder using a stereo light microscope (SLM), performing a solubility test using acetone, and analyzing the acetone extract with either transmitted or attenuated total reflectance (ATR) Fourier transform infrared (FTIR) spectroscopy. A gas chromatograph (GC) or liquid chromatograph (LC) with a mass spectrometer (MS) detector may be used to confirm the particle identity. These methods are also useful in determining the brand of a questioned powder.

10.2.2 Determining Brand

This work focuses primarily on the examination of pre blast or intact postblast powders. Smokeless powder may be recovered from an IED under several different situations. (1) The device may function (explode), resulting in a small number of the smokeless powder kernels remaining intact and scattered around the scene. (2) The device may malfunction, resulting in ample unconsumed original explosive remaining. (3) The device may not function at all, resulting in the bomb squad rendering the IED safe, allowing the recovery of the majority of the intact powder from the device. Regardless of the means for recovering the unexploded powder, evidence submission to the crime laboratory is required for determining the brand. The identification of the smokeless powder brand can be important for investigative and adjudicative reasons.

Examination of unexploded, canister smokeless powder by morphology, micrometry (or dimensional measurements), mass, FTIR spectroscopy, GC/MS, LC/MS, or capillary electrophoresis (CE) to determine the brand or to generate a short list of possible powder brands represents the goal of smokeless powder characterization. Smokeless powder brands listed by morphology are presented later in this chapter for illustrative purposes only and are not meant to imply a complete listing of all brands of powder. Only continuous monitoring of the literature and analysis of different brands and lot numbers of smokeless powder can provide adequate information for brand identification.

10.3 BRIEF HISTORY OF SMOKELESS POWDER

Black gunpowder, or "black powder," anteceded smokeless powder as a propellant. While the date of discovery of black powder is unknown, some authorities cite India, Arabia, and China as the source. Knowledge of its use spans many centuries. Generally, Europe was ignorant of explosive mixtures until a scholar and friar, Roger Bacon, in England, first documented the use of black powder in his work, *Opus Majus* in 1267 [4, 5]. All reviewers of this work agree that Roger Bacon made no mention of the use of black powder as a propellant for a firearm. Indeed, the formulation in his written work lacks the correct proportions to be useful for more than pyrotechnic adventure. A monk named "der schwarzer Berthold" is believed to be the first European inventor to use black powder as a propellant in guns around AD 1313 [4, 7]. Canons and mortars were manufactured for war shortly after the death of Friar Bacon, but before AD 1350 [8].

Hand-held guns were not invented until the formula for black powder changed and the strength of barrels improved. During the 1400s the firearm developed into a common battlefield weapon with the black powder formulation essentially the same as it is today, a mixture of potassium nitrate, sulfur, and carbon in the approximate ratio $75:10:15$.

In the middle of the nineteenth century, chemists were experimenting with the use of strong acids, either alone or in combination, on different materials. Braconnot, in 1833, prepared nitrocellulose by exposing materials of vegetable origin, such as sawdust, paper, and starch, to nitric acid [9]. The product from his method of preparation, called "xyloidine," contained no more than 5% or 6% nitrogen [10, Volume 2], but when exposed to flame, burned to completion with no residue. Other scientists continued work along the same lines, until 1845, when Schonbein stood out from the others because he was the first to pay "special attention to the properties of the product obtained from cotton which he has named guncotton" [10]. In 1846, another researcher, Pelouze, investigating the nitration of cellulose materials, made a distinction between guncotton, "pyroxylin," and xyloidine. Xyloidine, made from starch, was a completely solubilized substance, whereas pyroxylin, made from cotton, retained its original fibrous, cellulosic form. In August 1846, Schonbein went to England where he provided the manufacturing procedure to John Taylor, who received an English patent in October of the same year. The Schonbein process became well known due to the patent and involved mixing one volume of nitric acid to three volumes of sulfuric acid, then immersing the cotton in the acid mixture for 1 hour at 50–60 °C, followed by a water wash until acid free. The material was then pressed to remove water, dunked in a weak potassium carbonate solution, pressed again, and then rinsed with a dilute solution of potassium nitrate. This product was then dried at 150 °F. Gladstone determined that pyroxylin contained 12.75% nitrogen and proved that the pyroxylin contained more nitrogen than xyloidine [10]. In 1847, Crum nitrated cotton to its fullest and computed the nitrogen to be 13.69%. Unfortunately, explosions of manufacturing plants for nitrated cellulose began to occur across Europe and England due to the instability of the product. Various methods were unsuccessfully tried until 1862, when pulping of guncotton was undertaken successfully by Abel [4]. His process removed the unstable impurities in the cotton. Abel further pressed the material into different forms hoping to reduce the violence of the explosion in guns. He did discover that guncotton was safer to handle when wet and that a small amount of dry guncotton could detonate the wet guncotton when using a newly invented fulminate of mercury detonator.

Compression of the nitrocellulose (NC) and addition of "phlegmatizing" materials (a material that reduces sensitivity) to slow the burning rate were tried. Captain Schultze from Austria invented a powder that burned too rapidly for rifles but could be used in shotguns by keeping the cellulose nitrate in fibrous form. Applications of the fibrous type of powder remain today.

The French had an expressed need to develop an alternative to black gunpowder. On the battlefield, black gunpowder gave away concealed positions and shortened visibility to only a few yards because of the volume of thick white smoke produced when burning. In 1876, the French Army incorporated smokeless gunpowder and shortly afterward, it was in the hands of civilians [12]. In 1884, French physicist Paul Vieille systematically experimented with NC, finding that it burned in parallel layers and was dependent on the smallest dimension of the kernel for the total burning time [10, Volume 3]. Vieille discovered that two types of guncotton could be made: a colloidal powder soluble in diethyl ether/ethanol solution and a material that remained insoluble in the solution. The former material consisted of high nitrogen containing cellulose and constituted a doughy mass, which could be formed into various shapes. This became known as Poudre B, the "first" of the modern smokeless gunpowder propellants. Poudre B and all powders made afterward with only NC as the primary ingredient became known as "single base" smokeless gunpowder [4]. In 1884, Alfred Nobel found that nitroglycerin (NG) dissolved NC in about equal proportions [13]. The usage of NC and NG together in a powder became known generically as a "double base" smokeless powder. Alfred Nobel made the new type of smokeless powder and called it ballistite. The importance of NC solubility in NG was the removal of a volatile solvent, a hazardous process that was no longer required. The nonvolatile solvent, NG, became part of the explosive. Nobel received patents in France (1887) and England (1889) for inventing ballistite. At this time, Nobel also suggested the use of diphenylamine as a stabilizer for nitrocellulose. Germany introduced it into their powders shortly afterward but the United States and France hesitated until around 1910 [10, Volume 3; 11]. England manufactured a product called "cordite," a mixture of guncotton and NG. This was made colloidal by the addition of acetone as a solvent and mineral jelly, which was then made into round to oval cross-sectional cords by being forced through dies.

Reduction of the amount of NG and volatile solvents in NC powders led to further experimentation to find a less troublesome method. Use of nonvolatile solvents involving gelatinizing agents reduced the amount of NG required for solubilizing the NC. Because of the reduced time in manufacturing, this kind of powder was used from 1912 onward. World War I saw the need to replace the hard-to-manufacture NG. The replacement provided enhanced explosive power to the NC, with aromatic nitro high explosives, such as dinitrotoluene (DNT). The introduction of this product benefited the user by prolonging the life of the weapon through the reduction of the heat of explosion in the barrel. Further work showed that diethylene glycol dinitrate (DGDN or nitroglycol) solubilized NC better than NG producing a "more uniform gelatinized mass" [10, Volume 3] that improved manufacturing. Between the two world wars experimentation continued to find improvements in smokeless powder. Inorganic chemicals, such as potassium sulfate, used to inhibit the flash of a gun, were incorporated into the powder during manufacture. However, since World War II, few significant events in the development of smokeless gunpowder have occurred.

10.4 CHARACTERIZATION TOWARD SMOKELESS POWDER IDENTIFICATION

Occasionally in a shooting crime, determining muzzle-to-target distance is required. Particles present around the wound must be analyzed to determine if they are smokeless powder. The identification of particles as smokeless powder involves a microscopic examination using a stereo light microscope, solubility in acetone, and at least one instrumental method. Common instrumental methods include transmitted FTIR spectroscopy, ATR-FTIR spectroscopy, GC/MS, LC/MS, or CE to confirm the identity of the particles as smokeless powder.

Postburn powders are frequently not in their original condition. The morphology may not be consistent with the expected morphology of known powders because they have been altered by burning. During the training for smokeless powder analysis, the analyst should examine unburned powder with a SLM to become familiar with the myriad of morphologies, colors, textures, coating styles, and approximate sizes. Using the unburned powders, the analyst should burn small portions of each powder and reexamine the results with the SLM to gain knowledge of their postburn morphology and characteristics. One of the instrumental methods mentioned previously can additionally be utilized to confirm the identification of a particle as smokeless powder.

10.5 CHARACTERIZATION TOWARD BRAND IDENTIFICATION

Powders require a combination of tools in an analytical scheme to narrow the number of brands to one or a few. Each step of the analytical scheme should narrow the possibilities. No single method is yet capable of assisting the analyst in brand identification, except the finding of colored smokeless powder kernels. While the rest of the powder may be black to light green, these colored kernels along with the morphology may lead to a single brand. There may be some powders that cannot be individualized to a particular brand and must be included in a short list of like powders. Some powders are sold by different manufacturers with the same number designation (e.g., Hodgdon H4227 and IMR 4227). No assumptions about the chemical or physical characteristics of the powder should be made. The analytical scheme begins with a categorization by morphology and proceeds to other types of physical and chemical characteristics.

10.5.1 Characterization by Morphology

Visual examination of smokeless powder reveals a variety of different characteristics. On the macroscopic level, the visual examination includes general morphology, luster, color, and the presence of colored kernels (colored dots). Some smokeless powder manufacturers intentionally add a particular colored kernel to their powders for identification. For some powders the morphology and the finding of a colored kernel can lead to an identification of a particular brand of smokeless gunpowder [14]. Examples of several brands with general morphologies and colored kernels can be found in Table 10.1. Except for the finding of unique colored kernels, macroscopic examination does not provide brand discrimination among similar morphologies. The microscopic level morphology combined with other techniques (micrometry, mass, FTIR spectroscopy, GC/MS, LC/MS, CE) proves to be valuable in brand identification.

TABLE 10.1   Colored Kernels Leading to Brand Identification

Dot Color(s)	Shape	Comment	Manufacturer/Brand
Blue	Disc		Alliant Blue Dot
Blue	Disc		Hercules Blue Dot
Green	Disc		Alliant Green Dot
Green	Disc		Hercules Green Dot
Pink-red	Disc		Alliant Red Dot
Pink-red	Disc		Hercules Red Dot
Red, green, and yellow-green	Tube		Alliant Reloder 7
Red, yellow, and blue-green	Tube		Hercules Reloder 7
White	Lamella		Alcan 5
Yellow	Disc	Bright	DuPont Hi-Skor 700-X
Yellow	Disc	Pastel, some perforated	DuPont Hi-Skor 800-X
Yellow	Disc	Perforation is common	Hodgdon Clays
Yellow	Lamella	Rare to see	Alcan 7

10.5.2   Micromorphology

The Association of Firearm and Toolmark Examiners [15] and Wallace and Midkiff [2] have each published terms describing the morphology of smokeless powder. The terms used to describe the powders are lamella, ball, tubular, disc, flattened ball, agglomerated ball, cracked ball, flake, irregular flake, and perforated disc. These terms provide a basis for describing micromorphological features. Not only does magnified visual examination permit classifying a powder into one of the major categories, it assists in distinguishing and properly categorizing certain powders (e.g., flattened ball powders that macroscopically appear to be disc shaped). The microscope also permits subcategorization of morphology of the kernels observed to help segregate brands.

A stereo microscope with an appropriate ring light, fiberoptic light cables, or oblique illumination provides excellent resolution for morphological examination. The major category and subcategory morphologies used to describe smokeless powders are supplied in this chapter with photomicrographs of some kernel types. For classification into a subcategory of morphology, the dimensional measurements (micrometry) of the smokeless kernels may be required.

Based on a collection of over 100 powders, the manufacturer and brands of a particular morphological category are presented. This listing is incomplete and forensic scientists are encouraged to further develop and complete the list on their own. A brand of powder listed with one manufacturer may also have been sold by a different manufacturer because one company purchased another or was sold off by a parent company to become an independent corporation. Examples of the same brands being sold by different companies include IMR powders once manufactured by DuPont and Alliant powders once manufactured by the Hercules Powder Company. While a brand may be sold by different manufacturers, no assumptions about the powder brand being the same or different should be made. Only by confirming the physical and chemical properties can a correct brand categorization be made. While new brands are being introduced to the marketplace, some longstanding brands are being terminated. Brands no longer being manufactured may remain available on the shelf for many years. The list of brands at the end of each morphological group represents only the powders examined in the collection at the time of the study and does not represent a complete list of all brands.

Lamella Lamella shaped powders are manufactured in strips, then cut across the length. The lamella kernels have a tablet or plate dimension whose general morphology is rhomb, diamond, and square shaped, but the smokeless powder industry sometimes uses the term "trapezoid" [16] for the entire group (Figure 10.2). Occasionally trapezoid kernels can be found in these powders, but the majority are rhomb to square in shape. Some kernels have relatively small acute angles with large obtuse angles (diamonds) but some can approximate 90 degrees (rectangular to nearly square). Colored dots, if present, are of the same general morphology as the powder and are helpful in the brand identification.

Examples of lamella brands include:

Alcan: AL-5, AL-7.

Ball Only true "ball" powders fall into this category containing primarily ball shaped kernels with some ovoid kernels present (Figure 10.3). No or few flattened ball or flattened oval kernels are observed. If there is a significant presence of flattened ball or flattened oval kernels, then the powder is categorized

Figure 10.2 Lamella powder—Alcan AL-7 at 10×.

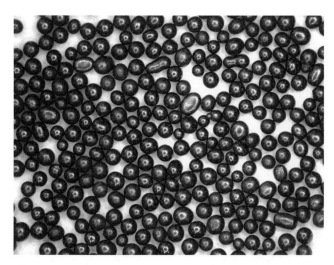

Figure 10.3 *True ball powder—Hodgdon H380, at 10×. Also note the oval kernels present in the powder.*

as flattened ball. The Olin Company has registered the term BALL® and Hodgdon Powder Co., Inc. has registered the term Spherical® to describe their three-dimensional round powders [17]. Since Primex Technologies is the only manufacturer of sphere shaped powders in the United States [18], the generic term "ball" will be used to describe the morphological shape of a smokeless gunpowder kernel that has the characteristic sphere or near sphere shape, regardless of the company selling the product. No infringement of copyright or registered name is intended.

Examples of ball brands include:

Accurate Arms Company, Inc.: 9, 2700, 4100.

Hodgdon Powder Co., Inc.: H380, H870.

Tubular   Tube powders are extruded powders whose length approximates or significantly exceeds the diameter. Tube powders are further categorized into long and short tube powders. Typically, when viewed on end, tubular kernels have visible perforations. Some kernels may have a thick graphite coating, which covers or obscures the perforation. The perforation can be revealed by cutting the center of the kernel perpendicular to the length, exposing the internal cross section. The powder industry does not intentionally seal the ends concealing the perforation [2]. Some analysts report these kernels as rods, with no perforation, but in all of the American and several European-made powders examined, the cylindrical particles have always had a perforation, though sometimes covered with graphite. The graphite is added to reduce static charge, slow the burning rate, and allow easier pouring for reloading cartridges. The

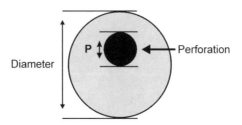

Figure 10.4 *End view of short tubular powder exhibiting a perforation.*

Figure 10.5 *Hercules Reloder 12, a short tube powder, shows some kernels with scrap, at 10×.*

distance from the edge of the perforation to the margin of the diameter of the kernel is called the web. Due to the positioning and size of the perforation, frequently off-center, the web distance on each side of the perforation can be different. In order to equilibrate the web, the perforation is subtracted from the total diameter and this number is divided by 2 (Figure 10.4). This method effectively presents an ideal web measurement for each kernel.

Occasionally, defective cutting distorts the ends of the kernels resulting in a short excess of material called scrap (Figure 10.5) or leaves longer irregularities called tails. These defects may distort the end diameter(s) of the kernels. The scrap or tail should not be incorporated into measurements of length or diameter. Other irregularities of kernel morphology are rare to find.

Long Tube Powders Long tube powders have a length dimension that exceeds 1.7 mm, a dimension significantly exceeding its diameter (Figure 10.6). These dimensions cause the kernels to lie with the long direction parallel to a

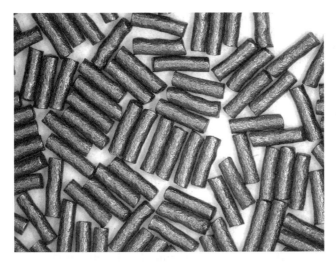

Figure 10.6   *Long tubular powder IMR 4198, at 10×.*

Figure 10.7   *Norma 203 exhibiting bias cut, at 10×.*

flat surface much like a pencil on a table. In this orientation, both the diameter and length can be measured. Because the cutting process creates scrap and distortions at the ends of the kernels, diameter measurements must be made across the center of the long direction of the kernel.

Kernels that have been cut at an angle other than normal, or perpendicular to the length direction, are known as bias cut (Figure 10.7). Typically this suggests a foreign manufacturer because many tubular powders of non-U.S. origin are bias cut [15]. Note that companies in the United States may purchase smokeless powders from foreign manufacturers. No examined long tube powders contain colored dots.

Examples of long tube brands include:

Accurate Arms Company, Inc.: 3100, 4350, MAGPRO, XMR-2495, XMR-3100.

DuPont: IMR 4064, IMR 4198, IMR 7828.

Hercules Powder Company: Hi-Vel 2.

Hodgdon Powder Co., Inc.: H4198, H4350, H4831.

IMR Powder Company: IMR 3031, IMR 4064, IMR 4198, IMR 4350, IMR 4831, IMR 7828.

Vihtavuori Oy: N170.

Short Tube Powders Short tube powders have a length approximating the diameter or exceed it by a small amount. The length for short tube powders ranges from approximately 0.35 mm to less than 1.70 mm [19]. When poured onto a flat surface, many of these powders have kernels that lie with their cross-sectional diameter in a vertical position, while other kernels lie with their long dimension parallel to the surface (Figure 10.8). Short tube powders may also exhibit a bias cut. The bias cut causes many of the short tube kernels to fall over, lying on their long dimension. Additionally, some kernels have scrap or tails that may be observed. Unlike the long tubes, the short tubes have morphology, size, color, and presence of colored dots that can immediately distinguish some of the powders from one another. Nearly all short tube brands have a perforation in the kernel.

Figure 10.8 *Short tubular powder including visible perforations, scrap, and tailing—IMR 4227, at 10×.*

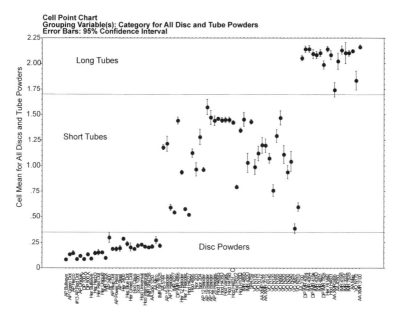

Figure 10.9 *Graph of length measurement displaying tubular and disc powders separated.*

Determining whether the powder is a long or short tube immediately reduces the number of possible brand sources for the powder. The list of potential candidates may become an important investigative lead, particularly as the number of possible powders diminishes. Displayed in Figure 10.9 are powder measurements of the length or thickness dimension of tubular and disc powders, respectively. The length or thickness dimension clearly places them into their respective subcategories.

Examples of short tube brands include:

Accurate Arms Company, Inc.: 5744, XMR-2015, XMP-5744.

Alliant Powder: Reloder 7, Reloder 12, Reloder 15, Reloder 19, Reloder 22, Reloder 25.

DuPont: IMR 4895.

Hercules Powder Company: Reloder 7, Reloder 12.

Hodgdon Powder Co., Inc.: 4895, H322, H1000, H4198, H4227, H4350, H4895, H4831SC, Varget.

IMR Powder Company: IMR 4227, IMR 4320, IMR 4895, SR4759.

Norma Precision AB: 203, 204.

Vihtavuori Oy: 3N37, N110, N120, N135, N140, N165, N310, N320, N330, N340, N560.

Disc Disc shaped powders are extruded tubes or circular strips whose diameter usually exceeds its length, better termed its thickness. The preferred orientation of disc powders is to lie with their diameter easily measurable. The majority of kernels in a disc powder can be smooth flat discs or rough irregular fragments depending on the brand. Disc shaped powders can be further sub-categorized (Table 10.2) as thick or thin (<0.18 mm) discs as well as large (>1.25 mm) or small diameter based on the visual or measured dimensions of typical kernels in a brand of powder. Finding a few thin disc kernels in a thick disc powder (or the opposite) are expected, but the powder is placed into a subcategory based on its primary visual features and its overall mean thickness and diameter. Additional features may be observed with disc powders. Thick disc kernels may stack like pancakes (Figure 10.10) while thin disc kernels may fold into taco shell or potato chip shapes (Figure 10.11).

As with tubular powders, some brands of disc powder are more susceptible to scrap and tailing than others.

Examples of thin disc small diameter brands include:

Alliant Powder: Bullseye.
Hercules Powder Company: Bullseye.

Examples of thin disc large diameter brands include:

Alliant Powder: Green Dot, Herco, Red Dot, Unique.
DuPont: Hi-Skor 700-X, Hi-Skor 800-X.

TABLE 10.2 Categorization of Disc Powders

Thin Disc—Small Diameter	Thin Disc—Large Diameter
Alliant Powder Bullseye	Alliant Powder Green Dot
Hercules Bullseye	Alliant Powder Herco
	Alliant Powder Red Dot
	Alliant Powder Unique
	DuPont Hi-Skor 700-X
	DuPont Hi-Skor 800-X
	Hercules Herco
	Hercules Red Dot
	Hercules Unique
	IMR Hi-Skor 700-X
Thick Disc—Small Diameter	**Thick Disc—Large Diameter**
Alliant Powder 2400	Accurate Arms Solo 1250
Alliant Powder Power Pistol	Accurate Arms Nitro 100
Hercules 2400	Alliant Powder Blue Dot
Hodgdon Universal	Alliant Powder Steel
IMR SR4756	Hercules Blue Dot
IMR SR7625	Hercules Green Dot
	Hodgdon International
	Hodgdon Clays
	Scot Powder 1000

Figure 10.10 *Stacked thick disc powder—Alliant Powder Power Pistol, at 10×.*

Figure 10.11 *Thin disc powder showing potato chip and taco shapes—Hercules Red Dot, at 10×.*

Hercules Powder Company: Green Dot, Herco, Red Dot, Unique.
IMR Powder Company: Hi-Skor 700-X.

Examples of thick disc small diameter brands include:

Alliant Powder: 2400, Power Pistol.
Hercules Powder Company: 2400.

Hodgdon Powder Co., Inc: Universal.
IMR Powder Company: SR4756, SR 7625.

Examples of thick disc powders large diameter brands include:

Accurate Arms Company, Inc.: Nitro 100, Solo 1250.
Alliant Powder: Blue Dot, Steel.
Hercules Powder Company: Blue Dot, Green Dot.
Hodgdon Powder Co., Inc.: Clays, International.

Flattened Ball This category contains the widest variety of morphology of the smokeless powders (Figure 10.12). In addition to the classic flattened ball shape (i.e., a ball or sphere that has two flat edges on opposite sides), the following additional shapes can be found: agglomerated ball, ball, cracked ball, oval/oblong (stretched lengthwise) flattened ball with two flat surfaces on opposite sides, angular flat ball, and broken/irregular flattened ball. The flattened ball characteristics are described next; ball powders were previously described.

The "classic" flattened ball powder consists of a sphere that has been "smashed": that is, two of the kernel's surfaces are flat and nearly parallel as a result of passing through rollers [18]. Some kernels may retain a nearly round form while others may become oval. Cracked ball kernels are ball powders (round or oval) that have been pressed or rolled to the point where the edge has split apart or cracked from one to several times [15]. Agglomerated ball

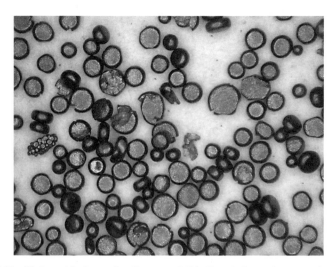

Figure 10.12 *Flattened ball powder. Represented in the photograph are agglomerated ball, angular flattened ball, ball, cracked flattened ball, cracked oval flattened ball, and irregular kernels from Accurate Arms 7, at 10×.*

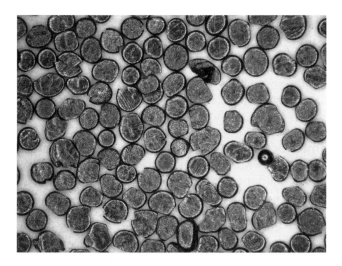

Figure 10.13 *Flattened ball powder showing different groups of morphology of Winchester Western 571, at 10×. Note the single ball powder kernel present among the flattened ball kernels.*

kernels appear much smaller than the typical kernels in the powder; adhering to each other they sometimes appear similar to a bunch of grapes. Angular flattened ball kernels appear to have acute angles, some with rounded portions, possibly indicating their heritage as a flattened ball kernel. Depending on the brand, broken-irregular kernels are present in powders as a rare to common occurrence. In some powders, broken-irregular and flattened angular kernels constitute a considerable portion of the kernels compared to the flattened ball and oval kernels (Figure 10.13).

Examples of flattened ball brands include:

Accurate Arms Company, Inc.: 2, 2 Improved, 5, 7, 1680, 2230, 2460, 2520, 8700,

Hodgdon Powder Co., Inc.: BL-C(2), H110, H335, H414, H870, HP-38, HS-6, HS-7, TITEGROUP, TITEWAD, Trap 100.

Olin & Winchester: 231, 296, 452AA, 540, 571, 680, 748, 760, WAAP, WAP, WMR, WSF, WSL, WST.

10.5.3 Other Characteristics

After examination of the powder for morphology, several other physical characteristics may be used for differentiation: color, luster, and colored dots.

Color On the microscopic level color is a marginally good discriminator. Most of the powders are gray to black due to the graphite coating. The interiors may be amber, yellow, green, or black. Some brands may exhibit variations of a color, especially a variety of green (e.g., dark gray-green, light gray-green, brown-green, yellow-green), which provide better discrimination.

Luster   Luster represents the shininess of the kernel in reflected light. The graphite coating may present either a shiny or dull surface. Without the coating, the appearance is typically dull.

Colored Kernels or Dots   The finding of a colored kernel, such as red, white, blue, green, and/or yellow, might point to a specific brand or limited brands of powder. The colored dot appears vastly different from other black, gray, or green variation colored kernels in a powder, and it represents a small fraction of the total powder. The identification of a brand may take place at this point. A red or pink disc shaped kernel can identify Red Dot brand smokeless gunpowder [14]. Careful observation is required to fully describe the colored kernel (e.g., is the kernel a bright yellow or pastel yellow and is a perforation present). Each characteristic potentially identifies one brand of powder over another.

After the powder has been categorized by morphology, physical and chemical methods can be utilized to further characterize the powder to attempt brand identification. One method is measuring the kernel's dimensions and statistically evaluating them against a database of known brands.

10.6   MICROMETRY

Measuring objects by use of a microscope is called micrometry, which can accurately measure small linear distances [20]. As early as 1923, Taylor and Rinkenbach [21], referring to analysis of smokeless powder, suggested that "measurements of grain size are made on 20 grains selected at random, and the results are averaged . . . the diameter is measured on the stage of a microscope equipped for the purpose." Micrometry of a sufficient quantity of kernels, typically a minimum of 25, provides statistically significant numbers that may be utilized toward the individualization of smokeless powders. The combination of a stereo microscope equipped with a digital camera linked to a computer with image measuring software provides adequate measurement of the kernels.

Measurements of length, width, area, kernel and perforation diameter, thickness, and occasionally diagonal dimensions should be given in millimeter units (Table 10.3). The web would be calculated as described earlier. Because of varying morphologies, powders require different dimensions to be measured (e.g., a lamella powder requires different dimensional measurements than a tube powder). Irregularities such as scrap or tails should not be included in the measurement.

TABLE 10.3 Linear and Area Measurement for Morphology

Kernel Shape	Length	Width	Thickness	Diameter	Area	Perforation	Web
Lamella	X	X	X		X		
Tube	X			X		X	X
Ball				X			
Disc			X	X			
Flattened ball			X	X			
Oval	X	X	X				
Irregular, broken, and cracked	X	X	X		X		

The following represents a useful example of equipment for creating a micrometry database of smokeless powders. A stereo light microscope with a zoom magnification of 6.3× to 40× using a 1× planachromat objective with the right eyepiece containing a micrometer to calibrate the magnification settings (10×, 15×, 20×, and 30×) of the microscope. The magnification settings are marked on the magnification changing wheel of the microscope at the time of calibration to permit returning to a calibrated magnification, as close as possible, for calibrated measurement or photomicrography. An illuminator using a daylight color correction filter and a ring light attached to the microscope objective provides illumination. Use an image capturing and measuring software package, such as Easy Measure or Easy Doc Imaging Software by Mideo Systems, Inc. (Huntington Beach, CA, USA) or an equivalent software package, which can be calibrated using a 1.00 mm stage micrometer at each magnification. The measurement software has linear, diameter, and area measurement features. The diameter feature can determine the diameter from a partly rounded kernel whose complete diameter was absent, such as an angular flattened ball kernel, or can determine the area by outlining the region. Of course, any combination of equivalent equipment is acceptable.

The collection of data from an unknown powder requires statistical evaluation against the database for brand elimination. While the Student's unpaired *t*-test is appropriate for one-to-one or one-to-two comparisons, the Bonferroni adjustment to the *t*-test is required when multiple sample tests are run against the unknown [22, 23]. Not all statistical software packages have this statistical feature.

After dimensional measurements are completed, the next physical characteristic for brand identification is the mass of a set of kernels.

10.7 MASS

The mass of the smokeless powders represents another parameter to obtain and compare against a database. Only unburned and unreacted kernels may be used for mass determination. Because of the impracticality of measuring each kernel, one successful method includes counting three sets of 50 kernels and one set of 150 kernels. The 150 kernel set is divided by 3 and the three sets of 50 can be used to build the database. The powders are counted on a rough filter paper (or equivalent) to reduce kernel movement while using a stereo light microscope with proper illumination. A truly representative random sampling is important as the kernels may separate based on size, thereby skewing results. The mass is useful for statistically individualizing the powders after morphology and micrometry parameter determination, but mass is not significant as a single parameter of characterization. After the physical methods have been used to characterize a powder, some brands may remain indistinguishable; therefore, instrumental methods such as FTIR spectroscopy can be employed.

10.8 FTIR SPECTROSCOPY

Infrared spectroscopy has been used for propellant analysis since at least 1953 [24]. In this early article, Pristera provides spectra of diethyl ether soluble components of smokeless powders. He was attempting to identify components present in propellants by comparing an unknown spectrum, typically containing more than one compound, with pure library standards. The identification of all components present in the IR spectrum may not be possible and are typically not required for the determination of smokeless powder for distance determinations. The goal for brand identification is to use the spectrum to find a particular brand of smokeless powder or a narrowed list of possible smokeless powders. Infrared spectroscopy of smokeless powders is a destructive technique that should be used last in the analytical scheme. Once a kernel is prepared for infrared spectroscopy, the original morphology no longer exists. The midregion of infrared energy (2.5–15 µm or 4000 to 600 cm^{-1}) is the most useful for infrared spectroscopy of smokeless powder.

Two different methods exist for infrared spectroscopy analysis of smokeless powders: transmitted micro-FTIR [25] and attenuated total reflection FTIR (ATR-FTIR) [26].

10.8.1 Transmission Micro-FTIR

With the transmission micro-FTIR method, place 3–6 kernels in a spot well depending on the kernel size, add 2–5 drops of analytical reagent (AR) grade acetone to dissolve the kernels, and expose the interior components. Allow the acetone to evaporate and repeat adding solvent and evaporating. Add 1–2 drops of AR acetone, swirl the liquid in the spot well, and use a capillary pipette to transfer the liquid to a boron carbide pellet mortar (SPEX Industries, Edison, NJ, USA). For parameters used in transmission micro-FTIR see Table 10.4. Add infrared grade potassium bromide (KBr) and mix. Using a micropellet press, a KBr pellet is made. If the sample proves too concentrated,

TABLE 10.4   Parameters Used in Transmission FTIR

Detector	MCT	Number of sample scans	160
Beamsplitter	KBr	Number of background scans	160
Source	IR	Resolution	$2\,cm^{-1}$
Sample gain	Automatic	Aperture	25

TABLE 10.5   Parameters Used in ATR-FTIR

Objective	ATR zinc selenide	Angle of incidence	$45°$
Detector	MCT	Number of sample scans	32
Beamsplitter	KBr	Number of background scans	32
Source	IR	Resolution	$4\,cm^{-1}$
Aperture	$100\,\mu m$		

dilute with KBr and repress until an acceptable spectrum can be obtained. To visualize the minor peaks, the samples may need to be more concentrated than a "textbook" spectrum.

10.8.2   ATR-FTIR

With the ATR-FTIR method, place 0.05 g of smokeless powder kernels in 500 μL AR acetone and allow to stand for 5 minutes to permit dissolution. Transfer the solvent to a reflective slide and allow the solvent to evaporate leaving a thin film for use with the 10× ATR objective. The ATR detector measures in the mid-IR region from 4000 to $650\,cm^{-1}$. For parameters used in transmission micro-FTIR see Table 10.5.

Both FTIR methods result in similar spectra for respective brands of smokeless powder with the transmission method having slightly more discrimination. Using acetone as the solvent, the primary ingredient detected is nitrocellulose. FTIR spectroscopy alone is insufficient for brand identification but can be useful after a morphological examination. Micrometry greatly assists in reducing the number of possible powders for consideration when combined with infrared spectroscopy.

As an example of the usefulness of FTIR spectroscopy in brand identification, two long tube powders that are indistinguishable by morphology, micrometry, and mass are examined by using transmission FTIR spectroscopy (Figures 10.14 and 10.15). The spectra show slight differences in composition between the two powders, allowing a questioned powder to be brand identified. These slight differences in the infrared spectra may be difficult to discern and a method that separates and identifies the components would be beneficial. Chromatography with mass spectrometry is a commonly used method in many areas of forensic science to accomplish separation and identification.

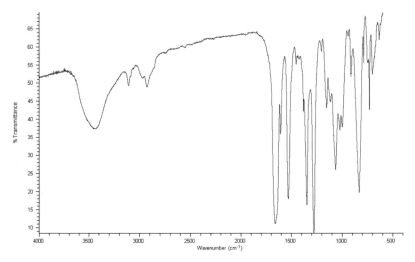

Figure 10.14 *Accurate Arms 3100 FTIR spectrum.*

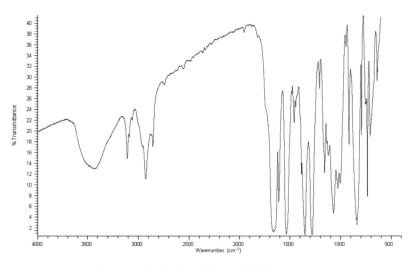

Figure 10.15 *DuPont IMR 4350 FTIR spectrum.*

10.9 CHROMATOGRAPHY WITH MASS SPECTROMETRY

Two primary methods of instrumental chromatography used in the crime laboratory have been gas and liquid chromatography. The concept of mass spectrometry (MS) has been considered since the late 1800s and put into actual use in the early part of the twentieth century [27]. Gas chromatographs with a mass spectrometer detector (GC/MS) have been in existence since the late 1950s [28]. Liquid chromatography with a mass spectrometer detector

(LC/MS) has been used for explosives analysis since the 1970s [29–31]. The forensic scientist should decide which of these primary chromatography systems to use based on the availability of the instrument in his/her laboratory, familiarity with the technique, and the thermal stability of the compound(s). GC and LC are used to separate the extracted smokeless powder components for detection with MS.

10.9.1 Gas Chromatography

In a test that included federal, state, international, military, city, and academic laboratories for the determination of smokeless powder, 13 of the 19 laboratories used GC or GC/MS [32]. With GC, the sample must meet four requirements: the sample must (1) be soluble in a compatible solvent for the GC, (2) be sufficiently thermally stable, (3) be volatile enough to pass through the column at permitted oven temperatures, and (4) have some affinity for the column stationary phase. GC/MS tends to be more versatile, particularly when using a capillary column, allowing a large variety of compounds to be analyzed without changing the column, flow rates, or other significant parameters. Using a temperature programmable injection port or an on-column injection method that begins close to room temperature may reduce issues with thermally unstable compounds in the GC or GC/MS.

As few as three kernels and up to 100 mg of smokeless powder are exposed to methanol, methylene chloride, or chloroform or a combination of solvents for extraction of the nonnitrocellulose components. The sample is then vortexed (from 15 seconds to 10 minutes) or allowed to stand overnight, then injected into the splitless injector [33] or on-column injector in the GC [34]. The typical flow rates of the GC carrier gas, helium in this study, approximate 0.5–1.0 mL/min at the initial oven temperature. A nonpolar column such as a methyl silicone (equivalent to a DB-1) or a phenyl silicone (equivalent to a SE-54 or DB-5), 15–20 meters, 0.25 mm diameter with a 0.25 μm film thickness, will permit adequate separations. Suggested GC method conditions are included in Table 10.6 for splitless injection and Table 10.7 for on-column injection.

If a temperature programmable or an on-column injector is not available, establish the splitless injector temperature at 175 °C, a compromise temperature between the heat sensitive compounds and the higher temperature

TABLE 10.6 GC Method Conditions for Spitless Injectors

Injector temperature	175 °C	
Initial oven temperature	90 °C	Hold 2 minutes
Oven temperature program	20 °C/min	
Final oven temperature	300 °C	Hold 1 minute
Injection volume	1 μL	
Transfer line temperature	300 °C	
Detector temperature	300 °C	

**TABLE 10.7 GC Method Conditions for
On-column Injection**

Injector temperature	175°C
Initial oven temperature	90°C
Oven temperature program	20°C/min
Final oven temperature	265°C
Injection volume	0.1–0.2 μL
Transfer line temperature	250°C
Detector temperature	350°C

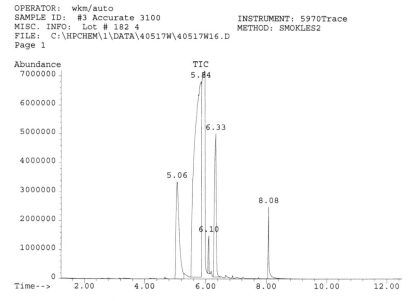

```
OPERATOR:  wkm/auto
SAMPLE ID:  #3 Accurate 3100            INSTRUMENT: 5970Trace
MISC. INFO:  Lot # 182 4               METHOD: SMOKLES2
FILE:  C:\HPCHEM\1\DATA\40517W\40517W16.D
Page 1
```

Figure 10.16 *Accurate Arms 3100 GC/MS.*

compounds. Some thermally sensitive compounds may not survive at this temperature. If an on-column or temperature programmable injector is available, then set the initial temperature at 30° or close to room temperature. The splitless mode allows the examination of minor components present in the smokeless powder that are often missing in the typical split injector ratios.

Combined with morphology and micrometry, GC/MS permits a better brand identification than either infrared spectroscopy method. The same two long tube powders in the example in infrared spectroscopy that are indistinguishable by morphology, micrometry, and mass were examined by GC/MS (Figures 10.16 and 10.17). These chromatograms clearly show a difference between the two powders. Because other powders with different morphology may have the same profile, immediately diluting-and-shooting a questioned powder on the GC/MS does not provide a brand or limited list of brands. Only a multimethod approach allows a brand or limited list of brands to be determined.

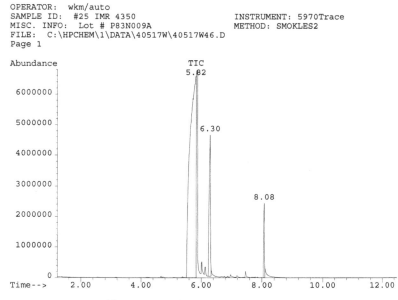

```
OPERATOR:  wkm/auto
SAMPLE ID:   #25 IMR 4350                    INSTRUMENT: 5970Trace
MISC. INFO:  Lot # P83N009A                  METHOD: SMOKLES2
FILE:   C:\HPCHEM\1\DATA\40517W\40517W46.D
Page 1
```

Figure 10.17 *DuPont IMR 4350 GC/MS.*

10.9.2 Liquid Chromatography

Liquid chromatography (LC) allows nonvolatile and thermally sensitive compounds to be analyzed [31]. LC permits the separation of thermally sensitive compounds that can be found among the nonnitrocellulose portions of the powder [35, 36]. Requirements for analysis include a compound's solubility in the mobile phase solvent(s), their affinity for the stationary phase, and discovery with an appropriate detector. LC tends to be dedicated to particular analyses such as dicyandiamide and sodium benzoate [37], or an area of analysis such as smokeless powders [36].

Smokeless powder is eluted with methanol or acetonitrile solvent, depending on the LC solvent system, and filtered to remove particulates [32, 38]. Several different methods for LC analysis use a reversed phase system employing a C18 (or equivalent) column typically 250 mm long with a diameter of 4.6 or 5.7 mm [35]. The common solvent systems used are methanol/water or acetonitrile/water at approximately 50:50 to 65:35 ratios [35, 38–40]. When the MS is not used as the detection mechanism, an ultraviolet spectrometer diode array detector reading absorbance at 254 nm is typically employed.

Both chromatography techniques, GC and LC, isolate the individual nonnitrocellulose components by separation and then introduce the sample into the mass spectrometer for identification. The mass range at the lower end begins between 30 and 45 atomic mass units (amu) and at the upper end from 400 to 550 amu [33, 36]. Electron impact (EI) and both positive and negative chemical ionization (CI) [34, 40, 41] are used in the mass spectrometer for obtaining identifying masses of the extracted smokeless powder components.

Determining the identification of each of the components in the smokeless powder and their relative concentrations in the powder allows better characterization of the powder for branding. By examining ratios of specific peaks, better characterization may be accomplished resulting in more specificity in the brand identification. Complexing morphology with either the GC/MS or LC/MS can provide excellent brand information for the investigator or court. Use of the LC/MS has the advantage of identifying heat sensitive compounds.

10.10 CONCLUSION

When a case containing smokeless powder is submitted for analysis, useful investigative information can be provided in a short period of time by the micromorphology of the kernels. Micromorphology, color, and luster typically produce a list of possible brands from the large number of reloading smokeless powders available. When colored dots are present in a powder, a particular brand may be identified at this point. If no colored dots are found, performing micrometry on the kernels can be accomplished rapidly, providing the brand or a short list of brands during the early stages of an investigation. In many situations, the addition of micrometry to the micromorphology can identify a single brand, but for some brands it simply narrows the number of possibilities. After the morphological categorization and micrometric analysis, mass may provide some additional discrimination, but instrumental methods provide better discrimination. GC/MS and LC/MS afford more useful criteria for individualization than either transmission FTIR or ATR-FTIR spectroscopy. Nearly all powder brands will be identified at this point. Any unresolved brands would be on a short list of only two or three smokeless powders.

By using the methods described in this chapter to establish a database for smokeless powders, useful brand information may be derived from any questioned smokeless powder submitted. The physical methods (e.g., micromorphology and micrometry) may be applied in the field by forensic scientists to give bomb technicians and detectives useful brand information early in the scene investigation. Because a large library of data must be available for comparison, this chapter provides information for the beginning of building a smokeless powder library needed for brand identification.

REFERENCES

1. Newhouser C (1984). *Introduction to Explosives*. FBI Bomb Data Center, U.S. Department of Justice, Federal Bureau of Investigation, Washington, DC.

2. Wallace CL, Midkiff C (1993). Smokeless powder characterization, an investigative tool in pipe bombings. In *Advances in Analysis and Detection of Explosives,* Yinon J (ed.). Kluwer Academic Publishers, The Netherlands.

3. Primedia (2005). Professional security training network presents defending and responding to the threat of bombing. Available at http://www.pwpl.com/security/downloads/summaries/9000265summ.pdf.

4. Davis TL (1943). *The Chemistry of Powder & Explosives.* Angriff Press, Hollywood, CA.

5. Brock A (1949). *A history of Fireworks.* George G. Harrap & Co. Ltd, London.

6. Coxe WH (1933). *Smokeless Shotgun Powders: Their Development, Composition, and Ballistic Characteristics.* E.I. du Pont de Nemours & Company, Wilmington, DE.

7. Federoff B (ed.) (1962). *Encyclopedia of Explosives and Related Items*, Volume 2. Picatinny Arsenal, Dover, NJ.

8. Hatcher JS (1966). *Hatcher's Notebook.* 3rd ed. Stackpole Company, Harrisburg, PA.

9. Quinchon J, Tranchant J (1989). *Nitrocelluloses—The Materials and Their Applications in Propellants, Explosives and Other Industries.* Ellis Horwwod Limited, West Sussex, England.

10. Urbanski T (1984). *Chemistry and Technology of Explosives*, Volumes 2 and 3. Permagon Press, Oxford, UK.

11. Departments of the Army and Air Force (1967). *Military Explosives.* TM 9-1300-214, Washington, DC.

12. Rowe WF (1988). Firearms identification. In *Forensic Science Handbook*, Volume II, Saferstein R (ed.). Prentice Hall, Englewood Cliffs, NJ, Chapter 8.

13. Fant K (translated by Ruuth M) (1993). *Alfred Nobel—a Biography.* Arcade Publishing, New York.

14. Matunas E (1983). Hercules and its powders. *Handloader* **103**:22–27 (May–June).

15. AFTE Standardization Committee (1994). *Glossary of the Association of Firearms and Toolmark Examiners*, 3rd ed. AFTE, Chicago, IL.

16. Wootters J (1991). Alcan AL-8. In *Propellant Profiles*, Wolfe D (ed.). Wolfe Publishing Co., Prescott, AZ, p. 7.

17. Hodgdon (1997). Hodgdon profiles—B. E. Hodgdon, Chairman Emeritus. Retrieved 29 June 1998 from the World Wide Web: http://www.hodgdon.com/profiles/beh.htm.

18. Simpson L (2000). *Hodgdon spherical powders. Shooting Times*, 80–82.

19. Moorehead W (2005). The characterization of reloading smokeless powders toward brand identification. In: *Proceedings of the American Academy of Forensic Sciences, Annual Meeting*, New Orleans, Publication Printers Corp., Denver, CO.

20. McCrone W (1978). *Polarized Light Microscopy.* Ann Arbor Science Publishers, Ann Arbor, MI.

21. Taylor C, Rinkenbach W (1923). *Explosives: Their Materials, Constitution, and Analysis.* Bulletin 219. U.S. Government Printing Office, Washington, DC.

22. SAS Institute (1999). *Statview Reference Manual*, SAS Institute Inc., Cary, NC.

23. Simon S (2005). Bonferonni correction. Available at http://www.cmh.edu/stats/ask/bonferroni.asp.

24. Pristera F (1953). Analysis of propellants by infrared spectroscopy. *Analytical Chemistry*, **25**(6):844–856.

25. Moorehead W (2000). The characterization of reloading smokeless powders using morphology, micrometry, and infrared spectroscopy for brand identification. Thesis, California State University at Los Angeles, CA.

26. Dowell B (2005). Differentiation between manufacturers and brands of unburned smokeless gunpowder particles using ATR-FTIR microspectroscopy and individual morphology Thesis, National University, San Diego, CA.

27. Borman S (1998). A brief history of mass spectrometry instrumentation. *Chemical & Engineering News* **Jan**:39–75. See also http://masspec.scripps.edu/information/history/perspectives/borman.html.

28. McLafferty FW, Turecek F (1993). *Interpretation of Mass Spectra*, University Science Books, Sausalito, CA.

29. McFadden WH (1973). *Techniques of Combined Gas Chromatography/Mass Spectrometry*. John Wiley & Sons, Hoboken, NJ.

30. McFadden WH (1979). Interfacing chromatography and mass spectrometry. *Journal of Chromatographic Science* **17**:2–16.

31. Yinon J (1983). *Analysis of explosives by LC/MS. In: Proceedings of the International Symposium on the Analysis and Detection of Explosives.* U.S. Government Printing Office, Washington, DC.

32. MacCrehan WA, Reardon MR (2002). A quantitative comparison of smokeless powder measurements. *Journal of Forensic Sciences* **47**(5):996–1001.

33. Moorehead W (2005). The characterization of reloading smokeless powders toward brand identification—Part II. Unpublished research.

34. Martz RM, Lasswell LD (1983). Smokeless powder identification. In: *Proceedings of the International Symposium on the Analysis and Detection of Explosives*. FBI Academy, Quantico, VA. U.S. Government Printing Office, Washington, DC, pp. 245–254.

35. MacCrehan WA, Reardon MR, Duewer DL (2002). A quantitative comparison of smokeless powder measurements. *Journal of Forensic Sciences* **47**(6):1283–1287.

36. Bender EC (1983). Analysis of smokeless powders by HPLC. In: *Proceedings of the International Symposium on the Analysis and Detection of Explosives*. U.S. Government Printing Office, Washington, DC, pp. 309–320.

37. Bender EC (1989). The analysis of dicyandiamide and sodium benzoate in pyrodex by HPLC. *Crime Laboratory Digest* **10**:76–77.

38. Stine GY (1991). An investigation into propellant stability. *Analytical Chemistry* **63**(8):475A–478A.

39. Parker CE et al. (1982). Analysis of explosives by liquid chromatography–negative ion chemical ionization mass spectrometry. *Journal of Forensic Sciences* **27**(3):495–505.

40. Yinon J (1991). Forensic identification of explosives by mass spectrometry and allied techniques. *Forensic Science Review* **3**(17):18–27.

41. Berberich DW, Yost RA, Fetterolf DD (1988). Analysis of explosives by liquid chromatography/thermospray/mass spectrometry. *Journal of Forensic Sciences* **33**(4):946–959.

11

Glass Cuts

Helen R. Griffin

Forensic Scientist III, Ventura County Sheriff's Department Forensic Sciences Laboratory, Ventura, California

A woman says she was raped and shows the police the clothing that was cut with the assailant's knife. A man says he was stabbed at a party, but the accused says the man fell onto a broken beer bottle. These are the types of scenarios in which examination of the textile damage can answer probative questions. Was the bra cut with a knife or with scissors? Did a knife or a shard of broken glass make the cut in the shirt? In order to answer these questions the forensic scientist must be aware of what to look for when different objects are used to damage fabric. Also, what makes a cut look worn versus recent? What distinguishes cutting from tearing?

Numerous studies have been done to determine the answers to these questions. Chapter 4 of *Forensic Examination of Fibers, Second Edition* [1] lists thirty-four references covering 1954 through 1998 and additional sources are listed at the end of this chapter. The studies involve recreations of scenarios. A number of different weapons have been tested on a variety of fabrics. The results have provided forensic scientists with a starting point for examining fabric damage. The end point is usually to do a recreation of the specific case scenario under examination.

Determinations of the identifying characteristics of cuts made by sharp objects penetrating textiles (stab cuts) include a comparison of stab cuts made by knives and glass [2, 3]. In addition, Monahan and Harding [2] noted that

Forensic Analysis on the Cutting Edge: New Methods for Trace Evidence Analysis, Edited by Robert D. Blackledge.

slash cuts start and finish with a V shape in which the cut yarns are sometimes interlaced with uncut yarns. The presence of uncut yarns can also occur in the middle of a slash cut. Taupin [3] noted that rolling a sand bag covered with cloth in broken glass resulted in discontinuous cuts and that the number of cuts depended on the number of shards or points on a broken bottle.

11.1 A HOMICIDE

December 1994, a young mother arrived at work with her face bruised and swollen after a fight with her boyfriend. When her supervisor called the police, she was fearful. "My boyfriend will kill me if you report this." She was right.

Her boyfriend was incarcerated pending his March trial date, but was mistakenly released in late February. Five days later, while her daughter was visiting her grandparent's, the woman's body was found raped and strangled in her home.

A broken window in the back of the apartment was examined. When the glass from the window was pieced back together the evidence showed that it was broken from the outside.

Glass fragments similar to glass from the broken window were found on the boyfriend's jacket. During the course of examining the jacket for glass fragments, it was noted that the lower portions of the right sleeve had multiple small cuts. These cuts were examined up to 40× magnification with a stereo microscope. The jacket was a tight-weave cloth fabric and the cuts did not appear typical of knife or scissors cuts. Where cutting occurred, the fibers were cleanly cut. However, cutting occurred *only* on the surface yarns of the weave. Also, some yarns were skipped over and did not show any cutting. Intermixed with the cut fibers were pulled fibers. Furthermore, two of the cuts were parallel and spaced approximately 4 millimeters apart along their lengths. A simulation of these cuts is shown in Figure 11.1.

A number of these characteristics had been observed in cuts on clothing submitted from burglaries involving broken windows. Simulation cuts were made in an undamaged portion of the jacket. A sharp knife blade, a dull knife blade, and a freshly broken piece of glass were used to make the cuts. The glass made cuts similar to those observed in the damaged area of the jacket's right sleeve. The sharp knife cut cleanly through the entire weave even with very light pressure. The dull knife resulted in more tearing than observed in the damage to the right sleeve when light pressure was used. With heavy pressure the dull knife cut through all layers of yarns in the weave. This data was combined with the presence of glass fragments on the jacket similar to glass from the broken window. The evidence supported the hypothesis that the boyfriend broke the window, reached through to unlock it, and gained entry through it into the home.

Figure 11.2 shows cuts made by a knife and by a piece of broken glass in a tight-weave fabric.

Figure 11.1 *Simulation cuts made using two pieces of glass on a tight-weave jacket. The bottom image is a higher magnification of the top image.*

11.2 A ROBBERY

There is a sense of justice when the perpetrator of a crime becomes the victim. This was the case in April 2000 when two men decided to rob a jewelry store. One of the employees shot and wounded one of the would-be thieves. Cornered in the back of the store, the men broke the glass from a door and crawled through. Some time later and a mile away the police found a man with a gunshot wound. Unluckily for the suspect, he had worn a nylon jogging suit that day.

The pants were shredded in the right front shin (Figure 11.3) and left front knee areas. There was so much damage to the right leg fabric that a "pocket" had been formed between the outer fabric and the inner lining. This "pocket"

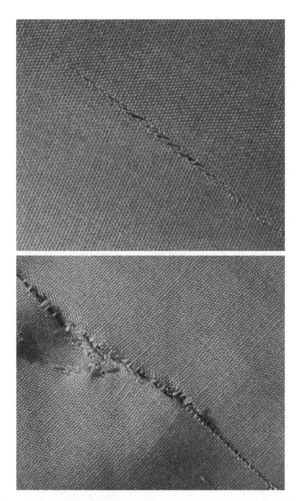

Figure 11.2 Comparison of a knife cut (top) and a glass cut (bottom) in a tight-weave jacket. Both of these cuts go through all yarn layers.

Figure 11.3 Shredding in the right front shin area of a pair of jogging pants.

contained fragments of glass that were similar to the glass in the door at the jewelry store (Figure 11.4).

Most of the damage to the pants consisted of cuts rather than tears. Most of the cuts in the outer fabric went through the weave, but some cut only through the surface yarns (Figure 11.5). The cuts randomly changed direction,

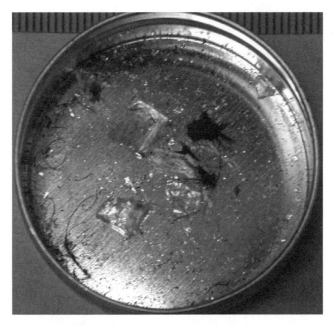

Figure 11.4 *A metal vial containing tempered glass fragments and cloth fragments collected from between the outer fabric and inner lining of the jogging pants.*

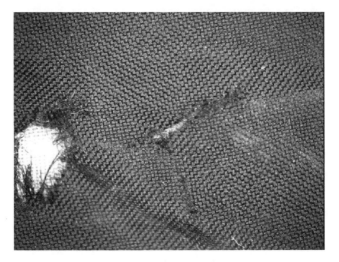

Figure 11.5 *A close-up of some of the damage in the jogging pants. There are cuts that go through the fabric and cuts that go through only the surface yarns.*

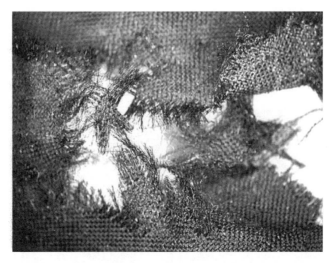

Figure 11.6 *A close-up of damage to the jogging pants where the cutting occurred in multiple directions.*

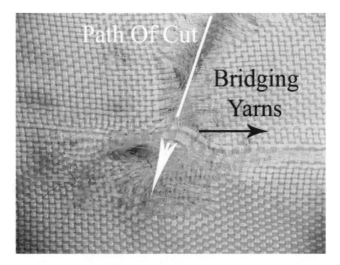

Figure 11.7 *Bridging yarns interrupt the continuity of this cut.*

with some small pieces of fabric cut completely away from the bulk (Figures 11.4 and 11.6). A cut in the lining had yarns that bridged across the two sides of the cut (Figure11.7).

11.3 A HIT AND RUN

A young man was walking to the store when a car struck him. Instead of stopping to assist, the driver took off.

Figure 11.8 *A shard of glass is evident in a small cut near the top of the picture. A close-up of a cut that is evident below the glass shard.*

One of the items of evidence submitted for examination was the victim's jacket. The objective of the examination was to compare glass from the jacket to glass from the windshield of the car.

Examination of the jacket revealed a large shard of glass (approximately 6 millimeters in length) embedded in the back of the jacket's hood (Figure 11.8). Cuts in the hood showed cleanly cut, unfrayed edges with uniform length of the fibers both within and between yarns characteristic of cuts made with a sharp object (Figure 11.9). Even though a sharp object made the cuts, some cuts were only through the surface yarns rather than through the full thickness of the fabric.

The shard of glass had properties similar to glass from the windshield of the car.

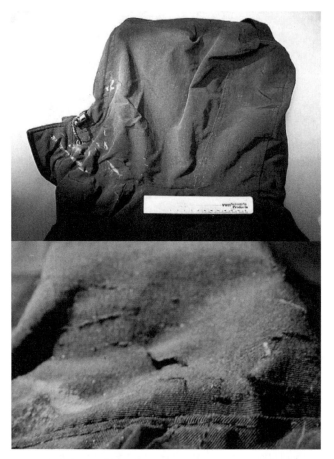

Figure 11.9 *The jacket hood is shown in the top photograph, with a closer view of some of the cuts in the bottom photograph. Cuts made by a sharp object are present.*

11.4   CUTTING VERSUS TEARING

The response of a fabric to a cutting or tearing action is dependent on the type of tool, fabric construction (weave, knit, or nonwoven), fiber type (especially natural versus man-made), fabric resistance, and the dynamics of the interaction (direction of cutting, folds in the fabric, etc.).

The first step in examining fabric damage is to look at the direction of the damage with respect to the structure of the fabric. Fabrics have a path of least resistance for tearing. The path will be dependent on the fabric but, for woven and knit fabrics, can readily be determined by testing. Figure 11.10 shows tears made in a tight-weave cloth (65% polyester/35% cotton), a jogging suit (100% nylon), a denim jacket (100% cotton), a knit shirt (100% cotton), a leather jacket, and a sweatshirt (50% polyester/50% cotton).

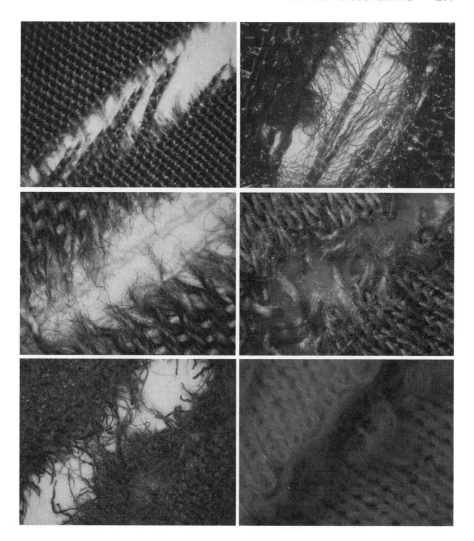

Figure 11.10 *Tears made in (clockwise from the* top left*) a polyester/cotton weave, a nylon weave, cotton denim, a cotton knit, leather, and a polyester/cotton knit.*

If the path of the damage is against the direction of tearing, cutting is suspected. Examination of the damaged edges with and without low power magnification will reveal whether the yarn ends follow a neat regular path associated with cutting or have random lengths associated with tearing. This characteristic extends into the yarn ends. Uniform, neat termination of the fiber ends indicates cutting, while ragged fiber ends indicate tearing. The contrast between cutting and tearing is well demonstrated in the jogging suit fabric shown in Figure 11.11. Figure 11.12 shows knife cuts made in a tight-weave cloth, a jogging suit, a denim jacket, a knit shirt, a leather jacket, and a sweatshirt.

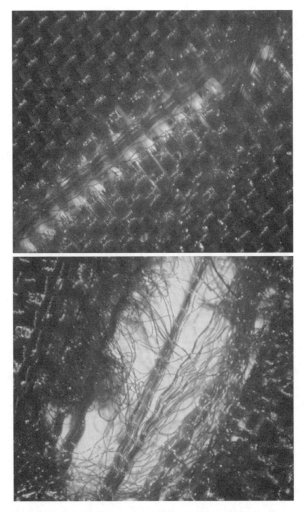

Figure 11.11 *A magnified view of the yarn and fiber ends contrasting a cut* (top) *and a tear* (bottom) *in a nylon weave.*

For knit fabrics, loops cut through on both sides will form short yarn lengths that are only loosely attached (Figure 11.13).

Stretching of the fabric out of knit or weave is more typical in tearing than cutting.

Scissors make cuts similar to knife cuts for a portion of the blade length, but must then be opened to cut again. This can result in a cut with numerous direction changes. The point at which the cut changes direction can also have associated small strips of fabric cut on two sides. Figure 11.14 shows scissors cuts made in a tight-weave cloth, a jogging suit, a denim jacket, a knit shirt, a leather jacket, and a sweatshirt.

All of these characteristics can be masked by blood [2].

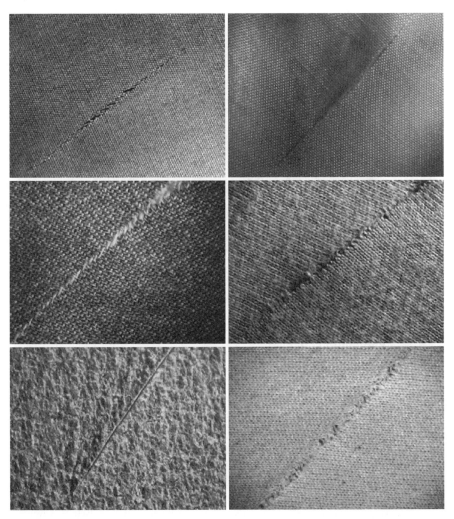

Figure 11.12 *Knife cuts made in (clockwise from the* top left*) a polyester/cotton weave, a nylon weave, cotton denim, a cotton knit, leather, and a polyester/cotton knit.*

11.5 SLASH CUTS MADE BY GLASS

A study was undertaken to determine the appearance of cuts in fabric making forced contact with broken glass. A frame was built that could be clamped around pieces of glass to simulate a window frame (Figure 11.15). Articles of clothing were purchased from a thrift store. The fabrics were selected to cover the range of woven, knit, and nonwoven fabrics routinely seen in casework.

Figure 11.13 *Knit fabrics cut through both sides of a loop form short yarn lengths that are only loosely attached.*

11.5.1 Associated Glass

The first step in a glass examination is to look for glass fragments using low power magnification and good illumination. This is done prior to a brush down or tape lift in order to document the quantity and location of glass fragments. The glass is then collected in order to compare it with glass from a known source. This is done even if a source glass has not yet been determined. The low tack adhesive on Post-it® notes is ideal for lifting glass from a fabric surface, as the adhesive does not readily contaminate the glass and most glass fragments are easily removed from the adhesive (Post-it® notes are specified as they have been found superior to other brands for this application). If no glass is observed, it may still be present in the debris from a brush down or on tape lifts. Glass fragments that are retained by clothing are commonly 0.2 millimeter or less in size.

11.5.2 Fabric Type

The type of fabric is determined in order to assess the cuts. Woven fabrics, knit fabrics, and nonwoven fabrics all cut differently, as do different types of weaves and knits. The fiber types are also important in assessing the damage.

11.5.3 Blade Characteristics

If you look at a piece of broken glass under low magnification, you can see that there are both thin, sharp blades and broad blades along the edges

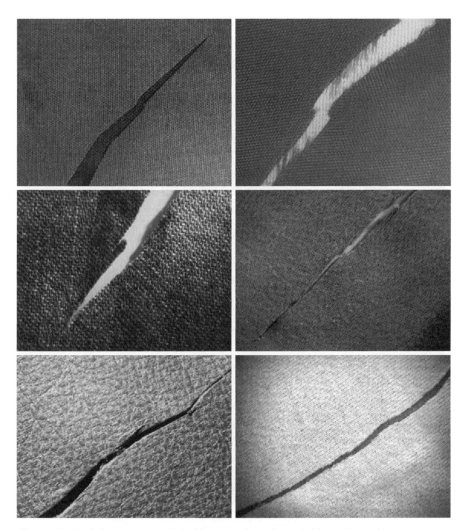

Figure 11.14 *Scissors cuts made in (clockwise from the* top left) *a polyester/cotton weave, a nylon weave, a cotton denim, a cotton knit, leather, and a polyester/cotton knit.*

(Figure 11.16). There are usually a number of these blades present on a broken glass surface. Also, in a broken window, there are often a number of broken shards around the center of the break (Figure 11.15).

Sharp Cuts and Tearing   As noted in previous studies [4], the ease of penetration of a blade is directly proportional to the cross-sectional area of the blade's tip. In the case of slash cuts, glass cuts show a variety of blade characteristics, including sharp, shallow blades with very small cross-sectional area on the edge.

Figure 11.17 shows examples of glass-caused tearing.

Figure 11.15 *The frame used to hold the glass in the original experiments.*

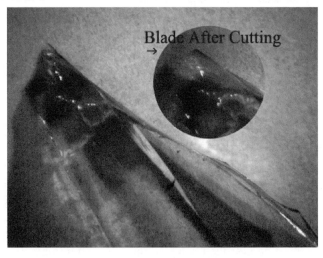

Figure 11.16 *A magnified image of a shard of glass. After cutting, the blade loses the thin sharp appearance.*

Discontinuous Cuts   Monahan and Harding [2] observed that slash cuts could have cut yarns interlaced with uncut yarns. Taupin cut yarns [3] noted that rolling a sand bag covered with cloth in broken glass resulted in discontinuous cuts. Figure 11.18 shows examples of glass cuts with these characteristics. In fact, it

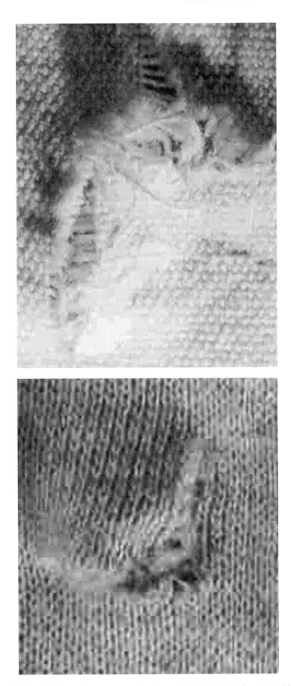

Figure 11.17   Tearing in a weave (top) and a knit (bottom) caused by glass.

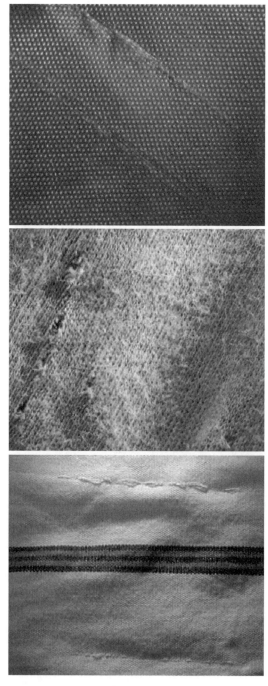

Figure 11.18   _Discontinuous cuts caused by glass._

is not uncommon for the glass blade to slice cleanly through the surface yarns, while leaving the underlying yarns undamaged as shown in Figure 11.19. When all of the yarns are cut except a few, the remaining yarns bridge the gap (Figure 11.20).

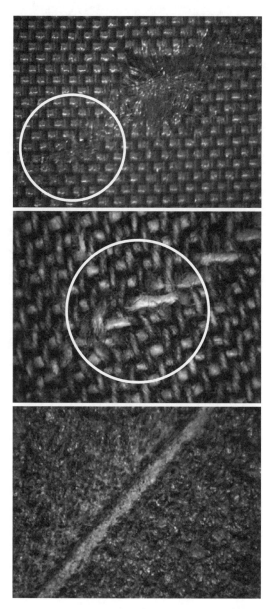

Figure 11.19 *Cuts in a nylon weave* (top), *cotton denim* (middle), *and leather* (bottom), *where the surface yarns/fibers are cut, but the underneath yarns/fibers are undisturbed.*

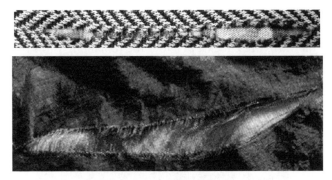

Figure 11.20 Bridging yarns in a knit (top) and weave (bottom).

Figure 11.21 Parallel cuts formed by glass cutting leather.

Parallel Cuts There are often multiple blades formed on the broken glass surface and it is common to see cuts running parallel to each other when a fabric is cut by glass. This is very evident in leather (Figure 11.21) and is also shown in Figure 11.22 in a woven fabric.

Direction Changes The dynamics of a cut always need to be considered when assessing fabric damage. The blade, target, or both can be moving while the cut is being formed. Rapid changes in the direction of the cutting have been observed in slash cuts made by glass. In the pants associated with the robbery, this extended to small pieces of fabric being cut from the pants (Figures 11.4 and 11.6).

Recent or Worn Damage As with any other type of cut, wearing will fray the fiber ends within each yarn. The yarn ends become tangled and mixed with

Figure 11.22 *Parallel cuts formed by glass cutting a woven fabric.*

foreign fibers during washing. Figure 11.23 shows the same cuts as Figure 11.12 after washing.

An Added Feature in Leather Figure 11.24 shows how a small tag of leather was left at the end of a glass cut.

11.6 CONCLUSION

When examining physical evidence, there are tests that give definitive answers and tests that are more like pieces of a puzzle. If you have enough puzzle pieces a picture starts to form. Combining information on clothing damage with the presence of glass that could be from a point of entry or exit from a crime adds valuable assistance in solving the puzzle.

Knives and glass can form similar cuts. The tears that glass forms in clothing look similar to the tears formed when clothing catches on something like a protruding nail. Although each damaged area taken individually could be made by something other than glass, the combined properties of all of the damage present can sometimes be very convincingly linked to broken glass.

ACKNOWLEDGMENTS

The author would like to thank the Washington State Patrol and Ventura County Sheriff's Department for providing the facilities and time to do this study. Thanks also to Val Copeland for assistance with the initial portion of this study presented at the American Academy of Forensic Sciences meeting in Seattle, Washington, February 22, 2001 and the California Association of Criminalists meeting in Ventura, California, October 27, 2004.

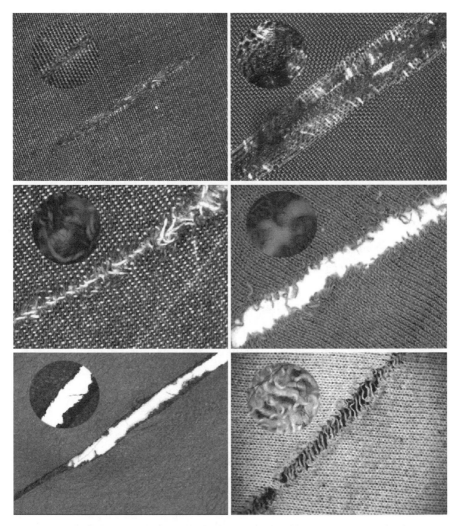

Figure 11.23 *The same cuts as shown in Figure 11.12 are shown here after washing. Insets show close-ups of the yarn ends.*

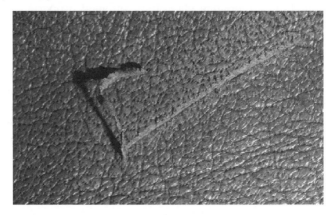

Figure 11.24 *A tag of leather at the end of a glass cut.*

REFERENCES

1. Taupin JM, Adolf FP, Robertson J (1999). Examination of damage to textiles. In: *Forensic Examination of Fibers*, 2nd ed., Robertson J, Grieve M (eds.). Taylor & Francis, London.
2. Monahan DL, Harding HWJ (1990). Damage to clothing—cuts and tears. *Journal of Forensic Sciences* **35**(4):901–912.
3. Taupin JM (1998). Testing conflicting scenarios—a role for simulation experiments in damage analysis of clothing. *Journal of Forensic Sciences* **43**(4):891–896.
4. Knight B (1975). The dynamics of stab wounds. *Journal of Forensic Sciences* **6**(3):249–255.

ADDITIONAL SOURCES

Alakija P, Bowling GP, Gunn B (1998). Stellate clothing defects with different firearms, projectiles, ranges, and fabrics. *Journal of Forensic Sciences* **43**(6):1148–1152.

Booth RF, Lott PF (1998). Distinguishing between new and slightly worn underwear: a case study. *Journal of Forensic Science* **43**(1):203–204.

Causin V, Marega C, Schiavone S (2004). Cuts and tears on a paper towel: a case report on an unusual examination of damage. *Forensic Science International* **148**:157–162.

Costello PA, Lawton ME (1990). Do stab-cuts reflect the weapon which made them? *Journal of the Forensic Science Society* **30**:89–95.

Green MA (1978). Stab wound dynamics—a recording technique for use in medico-legal investigations. *Journal of the Forensic Science Society* **18**:161–163.

Hearle JWS, Lomas B, Cooke WD (1998). *Atlas of Fibre Fracture and Damage to Textiles*, 2nd ed. Woodhead Publishing, Cambridge, UK.

McHugh S, Daeid NN (2003). An investigation into the correlation of knife damage on clothing and wounds in skin. *Forensic Science International (Forensic Medicine and Pathology Abstracts)* **136** (Supplement 1): 238 (MED-FP-08).

Nute HD (1988). Effect of fabric structure on damage to cloth from a contact shot. *Southern Association of Forensic Scientists* **16**(1):15–18.

Sitiene R, Varnaite J, Zakaras A (2004). Complex investigation of body and clothing injuries during the identification of the assault instrument. *Forensic Science International* **146**(S):S59–S60.

Taupin JM (1998). Damage to a wire security screen: adapting the principles of clothing damage analysis. *Journal of Forensic Sciences* **43**(4):897–900.

Taupin JM (1999). Comparing the alleged weapon with damage to clothing—the value of multiple layers and fabrics, *Journal of Forensic Sciences* **44**(1):205–207.

Taupin JM (2000). Clothing damage analysis and the phenomenon of the false sexual assault. *Journal of Forensic Sciences* **45**(3):568–572.

12

Forensic Examination of Pressure Sensitive Tape

Jenny M. Smith

*Criminalist, Missouri State Highway Patrol Crime Laboratory,
Jefferson City, Missouri*

12.1 INTRODUCTION

A roll of tape is common to most households, workshops, and places of business. It is everywhere and readily available. Tape is handy for packaging, sealing, masking, and repairs but has also found usefulness in the world of crime. It is used as ligatures, restraints, and blindfolds and to construct improvised explosive devises (IEDs). Some dealers in the marijuana trade wrap their bundles in plastic wrap and duct tape in an attempt to thwart the detection by the drug-detecting dog. It has thus become increasing common for tape products to be submitted to the crime lab for analysis. Typically, the request will be to compare a questioned piece of tape to a known source. Alternately, the lab may be asked to identify a possible source of questioned tape as to the manufacturer and distributor.

   While there are many different classes of tape, only a few are commonly encountered in the forensic lab. The emphasis of the descriptions and methods in this chapter will deal mainly with duct tape, black electrical tape, and packaging tape, with mention of filament and masking tapes. However, the methods for one type of tape can easily be applied to most other classes of tape. It

Forensic Analysis on the Cutting Edge: New Methods for Trace Evidence Analysis, Edited by
Robert D. Blackledge.

should be emphasized that the tape market is ever changing and that there are no hard and fast rules as to the characteristics of each class of tape. There will always be exceptions especially with changing market trends.

12.2 PRODUCT VARIABILITY

As with any commercial product, whether it is lipstick, glitter, shoe polish, or tire rubber, the value of the product as class evidence depends on the variability of the product. Unscented white candle wax or white cotton fibers are comparatively generic. It would be difficult to differentiate these items from different sources. The value of tape products as evidence is due to their variability that has been widely reported [1–11]. There are many different classes of tape products; duct, masking, electrical, packaging, office, and so on, and generally the more complex the product the more variable it will be. For example, duct tape as a class of tape is more complex and varies more than matte office tape. This consideration should always engage the analyst, as probability questions are likely to arise when one is asked to testify. Probabilities cannot be assigned to these commercial products because they constantly change. But the analyst should have at least a general idea of the product's variability if only to assure one's self that it is worth even comparing.

The design and construction of the tape product will vary depending on its commercial end use. Thus, there are different general classifications of tape. Within a given class of tape, macroscopic similarities will be evident. However, often upon closer analysis of the chemical and physical characteristics of different rolls of tape within a given class, differences can be found. In duct tape, for example, one can expect to find differences between rolls of tape manufactured by different companies. However, differences might also be found between batches of duct tape manufactured at the same plant (Figure 12.1).

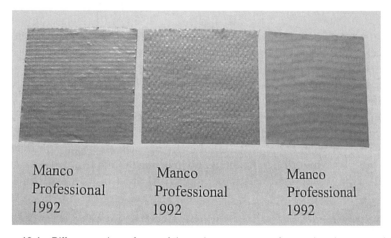

Figure 12.1 *Differences in surfaces of three duct tapes manufactured at the same plant.*

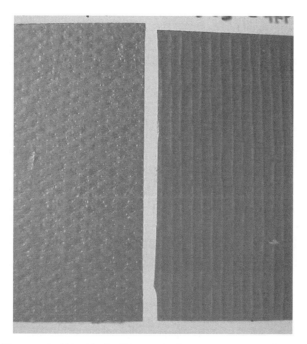

Figure 12.2 *The stated width of both of these rolls from the same manufacturer was 50.8 mm. However, they are not the same width.*

This can be due to equipment upgrades or changes in the various components due to market fluctuations. But differences might even be found between rolls from the same batch of duct tape, for example, slit with different slitting machines giving slightly different tape widths or that show a different offset of yarns at the edges (Figures 12.2 and 12.3).

While studies show that tape products have sufficient variability to be useful as an investigative tool, within-roll studies have shown consistency in construction throughout one end of a roll to the other [3, 7].

12.3 TAPE CONSTRUCTION

Pressure sensitive tapes consist of a flexible backing and an adhesive that when applied to a surface forms a bond with slight pressure. The bond with the surface is reversible and can be broken usually without damage to the surface. From this simple design of a backing and adhesive, tape can become increasingly complex depending on its end use.

Figure 12.4 shows the layers of a typical tape product. Not all of these layers will necessarily be present in every product.

Release Coat   The release coat is a layer on top of the backing that reduces unwind tension of tape from the roll. Common release coat materials

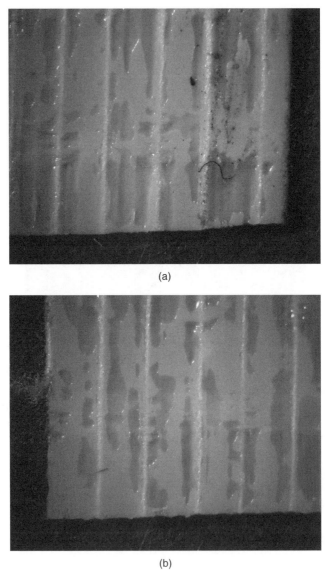

(a)

(b)

Figure 12.3   (a) Looking at the adhesive side of a piece of duct tape, note the yarn offset from edge. (b) This piece of duct tape shows a different offset than that in (a).

include siloxanes, stearates, carbamates, acrylates, polyethylene, and poly (vinyl acetate). This layer is difficult to isolate for analysis.

Backsize   The backsize serves to smooth a surface that is rough or porous (not shown on diagram of Figure 12.4). It is applied to the nonadhesive side. Commonly, poly (vinyl chloride), poly (vinyl acetate), or acrylics are used.

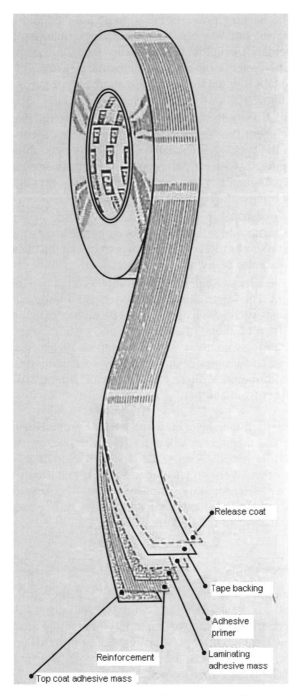

Figure 12.4   *The layers found in pressure sensitive tapes.*

Backing   The backing provides the support for the adhesive and can be made of fabric (medical tape), paper or crepe (masking tape), polyethylene (duct tape), cellulose acetate (office tape), polypropylene (packaging tape), polyester (strapping and filament tape), or poly (vinyl chloride) (electrical tape). When the backing is a polymer, it may also be referred to as the film.

Primer Coat   If the adhesive does not bond well to the backing material, a bonding agent (primer coat) may be used. This agent will bond well with both the backing and adhesive. Nitrile rubbers, chlorinated hydrocarbons, polyethylene, or polypropylene might be used and could be a blend of components or copolymers. The primer coat might also be used to bond fabric reinforcement to the backing.

Fabric Reinforcement   In duct tape the fabric used as reinforcement is known as scrim and is a loosely woven gauze-like fabric. Filament or strapping tape uses rows of fiberglass yarns along the machine direction, but these can also be nylon or polyester yarns.

Adhesive   The adhesive can be either a single component or multicomponent system. The single component system is typically acrylic based, but other systems such as silicones exist. The multicomponent system basically consists of a blend of elastomers and resin tackifiers. This can be made increasingly complex with the addition of fillers, stabilizers, and extenders. The formulas of any company's adhesives are typically proprietary information. Adhesive analysis is a good way to distinguish one manufacturer from another.

Of the above listed tape layers it is the backing, adhesive, and fabric reinforcement that lend themselves well to analysis using the commonly available tools within the forensic lab. A more detailed discussion of these components will follow. The other layers—release coat, backsize and primer coat—are thin and difficult to separate. Separation of these from the backing for analysis may not be practical but if present and to an appreciable degree they may contribute to the bulk analysis of the backing material.

12.3.1   Tape Backings

The tape backing (sometimes called the "film") of pressure sensitive tapes provides support for the adhesive and reinforcement fibers (if reinforcement is present). There are many different materials that may be used for a backing. Of the common commercial tapes that are found in hardware stores, large retail outlets, and supermarkets, the list is smaller. Polyethylene, polypropylene, poly (vinyl chloride), paper, polyester, cloth, cellulose acetate, cellophane, and poly (vinyl acetate) are all used as backings for common commercial tapes and this includes duct tape, packaging tape, electrical tape, office tapes, and hospital tapes, and so find their way into forensic labs as evidence.

When a backing is colored or opaque, there is certainly something added to that backing to make it so. There may be colorants, fillers, cross-linkers, plasticizers, stabilizers, and fire retardants added to the polymer. Common colorants include aluminum powder, carbon black, titanium dioxide, iron oxides, and pigments. Phenolics or isocyanates may be present as cross-linkers, zinc oxide or zinc dibutyl dithiocarbamate may support those cross-linkers, lead carbonate is a UV blocker, and antimony trioxide is a fire retardant. Fillers such as calcium carbonate, talc, and dolomite may also be found in certain types of tape backings. Even tape backings that appear essentially clear and colorless may have substances added that may only be detected by viewing under plane polarized or cross-polarized light.

Some plastics have very desirable properties for a particular end use, such as its heat resistance or low conductivity, but the plastic may be too brittle for a flexible tape backing. PVC is the best example of this. A plasticizer is a low molecular weight substance, usually oil that is mixed with a hard plastic to make it more flexible. It does this by creating spaces between the polymer chains adding "free volume," thus allowing for greater chain flexibility. The most common plasticizers are phthalates (dioctyl phthalate or DOP), aromatic oils, sebecates or adapates (hexyl, octyl, and nonyl), and aliphatic oils. Other materials that may be used include aryl phosphates (cresol), stearates, mineral oil, castor oil, and rosin oil.

So far the discussion of backings has emphasized the chemical composition but other physical attributes may aid the examiner in comparison of known and questioned pieces of tape. The backing thickness can vary from tape to tape and typically ranges from 1.5 to 4.0 mils. This can be assessed with calipers or by cross section. The cross sectioning is recommended and not only allows for side-by-side comparison of tape thickness but it is the easiest way to determine if the tape backing is multilayered [12]. Cross sections should include the whole tape without separation of the adhesive. These multilayered tape backings have been found in some duct tapes where several layers of polyethylene have been noted with only one layer containing the aluminum colorant. A tie layer may also be found on the adhesive side of duct tape backings. These are typically a low melting polymer such as ethylene vinyl acetate (EVA) or ethylene methacrylic acid (EMA).

The equipment used to manufacture tape may leave marks or striations on the polymer surface. These marks may be transitory; that is, they may show up in one part of the tape but not another within the same batch or even the same roll or may be continuous throughout the length (Figure 12.5).

12.3.2 Adhesive Formulations

The simple formulation of PSAs is that of an elastomer combined with a resin tackifier. The most common elastomer is derived from natural and synthetic rubbers found in calendered/laminate tapes. Natural rubber is essentially *cis*-polyisoprene with added proteins. Random copolymers such as styrene

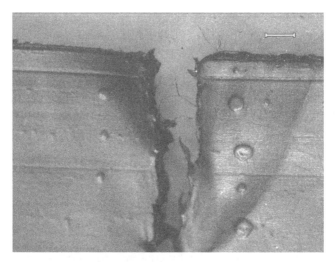

Figure 12.5   *An abduction case with a questioned and known for comparison. A 10× stereo-scopic view shows striation marks on the tape surface.*

Styrene
End-Block

Rubber
Mid-Block

Styrene
End-Block

Two phase morphology gives strength at the end block and
elasticity at the mid block.

Figure 12.6   *Block copolymer model showing the plastic end blocks and rubber midblocks.*

butadiene (SBR) and block copolymers such as styrene isoprene styrene (SIS) and styrene butadiene styrene (SBS) may also be used either alone as in hot melt/extruded adhesives or in combination with natural rubber. The styrene-based copolymers have come into wide use in the tape industry with the styrene serving as a plastic end block and the isoprene or butadiene serving as a rubbery/flexible midblock. This two-phase morphology produces both strength and elasticity (Figure 12.6). Other synthetic rubbers that can be found in tape adhesives are butyl rubber and polybutadiene.

The addition of a tackifier makes the elastomer "sticky" by lowering the glass transition temperature and viscosity and thus increasing the flow and wettability. Aliphatic C5 hydrocarbons, polyterpenes, and natural rosin esters commonly serve as tackifiers.

The acrylate adhesives (ethyl and butyl) can have "stickiness" without the addition of a tackifier. Silicone adhesives are another family of PSAs that do not need a tackifier. Silicone-based adhesives are less commonly encountered in forensic casework. These are found in high-end specialty tapes designed to resist weathering and temperature extremes.

Beyond the elastomer and tackifier, the "black art" of adhesive formulations lies in the complex addition of fillers, colorants, extenders, and cross-linkers. It is these additives that often allow a tape to be traced to a certain manufacturer. Fillers serve to reinforce, add bulk, and lower the cost without adversely altering the adhesive properties. Calcite (calcium carbonate), kaolin clay (aluminum silicates), titanium dioxide (usually rutile), dolomite (calcium magnesium carbonate), and talc (magnesium silicates) are common fillers. In addition, zinc oxide may serve as a filler, reinforcer, cure accelerator, colorant, or cross-linker. A given formulation may have one or more of these additives and in different ratios. Add these to the different elastomers and tackifiers and the possibilities are almost endless.

An adhesive with opacity and color will contain fillers. Clear adhesives such as acrylates and silicones may or may not contain fillers.

12.3.3 Common Reinforcement Fabrics

A reinforcement fabric is found in several classes of tape products and provide added strength. They are usually embedded in the adhesive layer but may be found embedded in the backing as in gaffer's tape. Duct tape and strapping and filament tapes are the most commonly encountered reinforced tapes. Less common are hospital tapes and gaffer's tape.

Typically, the fabric reinforcement can be seen through the adhesive. Cotton, polyester, glass filament, and nylon are common fibers found in reinforcement fabrics. They may be in a loose weave such as that found in duct tapes, a tight cloth, as in surgical tape, or in straight bundles, as in filament tape. The bundles of fibers may be loose or spun. Sometimes the loose bundles will be "texturized" or crimped to give more bulk. The spun bundles may have a "Z" or "S" twist. The reinforcement fibers may have varying degrees of fluorescence (Figure 12.7).

12.4 DUCT TAPE

Within the industry, duct tape is referred to as a polycoated cloth tape. The common name, "duct tape," would suggest that this product is used in the heating and air conditioning construction and repair industry, but the typical duct tape product would not be suited for such applications. In fact, it is used for many things other than duct work. It was first utilized for its water repellency by the military in WWII to seal ammunition boxes. Its most basic construction is a polyethylene film backing, laminated to a loosely woven gauze called scrim as a reinforcement fabric, and coated with a thick layer of natural rubber-based adhesive. It is marketed in widths of 48 mm and 50.8 mm but other widths are possible.

The backing of duct tape is low density polyethylene (LDPE). Aluminum powder is added to give the typical silver/gray color, although it can be found

Figure 12.7 *(See color insert.) Fluorescence of scrim fibers under long wave UV.*

in most any color and also with screen printing such as camouflage designs. Usually the color is dispersed in one layer of film, but the film may be multiple layers with the color in only one of these layers. Talc has been found in duct tape backings also, presumably as a filler. Other fillers may be possible. The LDPE used in duct tape backing is flexible enough without the need for a plasticizer.

Duct tape adhesive typically has a natural rubber base that may be blended with a synthetic rubber, such as an SIS or SBS copolymer. It is the formulation of the adhesive that is most apt to identify a particular duct tape to a particular manufacturer. As might be expected, these formulations are proprietary. Fillers are added to reinforce, cut cost, and add bulk. One or more fillers may be found and include calcite, dolomite, talc, and kaolinite. Adhesives that have a cream or yellowish color often contain titanium dioxide and possibly zinc oxide. Zinc oxide may also serve as a curing agent and cross-linker (Figure 12.8).

The reinforcement fabric of duct tape is a loosely woven fabric called scrim. The scrim can sometimes be seen through the adhesive. A "scrim" count or "yarn" count is quoted as the number of warp yarns (machine direction) per inch and the number of fill yarns (cross direction) per inch (Figure 12.9). A

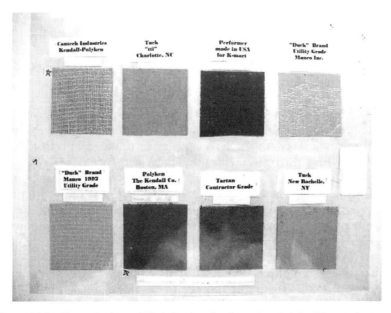

Figure 12.8 *(See color insert.) Variation in adhesive color of eight different duct tapes.*

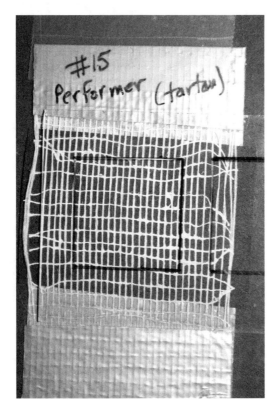

Figure 12.9 *A glass slide is marked with a 1 inch square box laid on top of the weave. The "scrim" or "yarn" count is the number of yarns along the warp (machine direction) and fill yarns (across) per inch. This scrim has a count of 18 × 10.*

typical scrim count of a low-end utility grade duct tape would be 18×9 or 18 yarn bundles in the warp direction and 9 in the fill direction. High-end tapes may have a scrim count up to 45×25. The design of the scrim is to allow for easy tearing across the tape. Weave patterns may be a basket weave with an over and under weave or a weft insertion weave where the fill yarns insert into a warp that can be a chain stitch (Figure 12.10). The basket weave scrim will allow easy tearing across the tape if the warp yarns are spun cotton bundles. However, they may also be blended cotton/polyester (65:35) in the warp direction in spun bundles. The fill yarns are typically polyester but may also be blended with cotton. They may be in loose bundles of texturized (crimped) fibers or straight fibers. The fibers of the weft insertion weave are polyester in both the warp and fill directions. The design of the knit weave allows easy tearing across the tape.

The scrim fibers may exhibit fluorescence in warp, fill, or both. In addition, they may have different cross-sectional shapes and varying degrees of delusterant and diameters (Figure 12.11). The fiber bundles might be "Z" or "S"

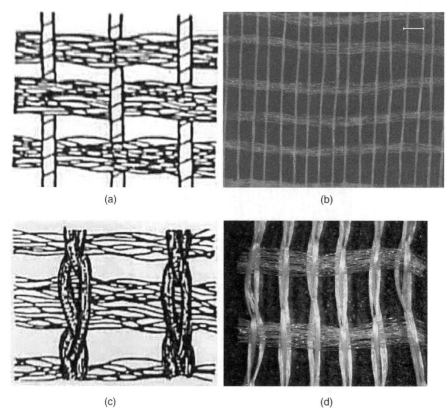

(a)

(b)

(c)

(d)

Figure 12.10 *(a, b) Basket weave scrim pattern from duct tape reinforcement. (c, d) Weft insertion scrim pattern from duct tape reinforcement.*

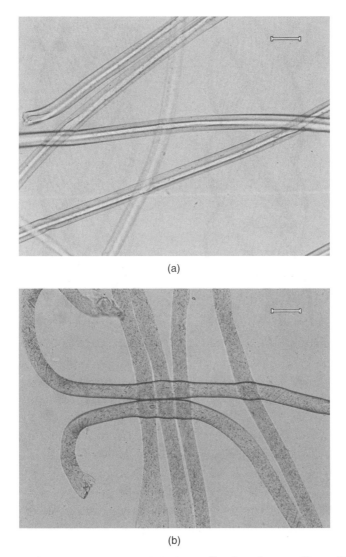

(a)

(b)

Figure 12.11 *(a) A lightly delustered trilobal polyester fiber from duct tape fill yarn. (b) A heavily delustered round polyester fiber from duct tape fill yarn.*

twist or may have no twist (Figure 12.12). In addition there are several differ-ent basket weave patterns used in scrim fabric lending yet even more vari-ability to the mix. As the tape market evolves many more variations on these common features could emerge. Figures 12.13 and 12.14 show some fibers and cross shapes that may be found in scrim fabrics.

Rayon scrim fibers can be found in some duct tapes produced outside the United States. There are now some tapes on the market sold as "clear" or

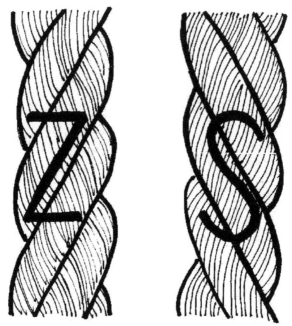

Figure 12.12   *The spun fibers in duct tape scrim may be an S or Z twist.*

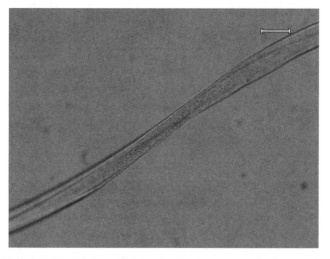

Figure 12.13   *A polygonal polyester fiber common to duct tape scrim.*

"transparent" duct tapes. They defy many of the common features described previously of the typical gray duct tape. They may have polypropylene instead of polyethylene backings. The adhesive may be an acrylate without fillers, or a UV protected SIS system, and the reinforcement fabric may not be a typical

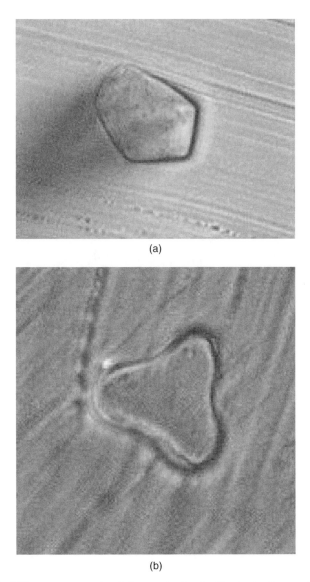

(a)

(b)

Figure 12.14 *Different cross-sectional shapes of polyester fiber in duct tape scrim: (a) polygonal, (b) trilobal, (c) round, and (d) triangular.*

gauze or knit weave. It should be emphasized that any of the components of duct tape can change with the market trends.

The analysis of duct tape should include elemental analysis of the backing and adhesive. Since the film backing may have multiple layers, both sides should be analyzed. FTIR spectroscopy is very useful for evaluating both the organic and many of the inorganic components of the adhesive and the backing.

(c)

(d)

Figure 12.14   *Continued*

Polarized light microscopy (PLM) offers much information about the inorganic fillers in the adhesive and rounds out the information gained by the FTIR and elemental analyses. PLM is also very useful for characterizing the fiber reinforcement. In fact, it offers more information on the fibers than any of the other instrumental tests. A check of the scrim fabric for fluorescent properties is recommended. There have not been any published studies at the time of this writing on the added benefit of pyrolysis gas chromatography/mass spectrometry (GC/MS) of duct tape components.

12.5 ELECTRICAL TAPE

Electrical tape is designed for use on electrical components and thus must have electrical insulating properties and be corrosion resistant. There are many grades of electrical tape but the familiar $\frac{3}{4}$ inch wide black vinyl tape available in most large retail outlets is what is most commonly encountered in the forensic lab. This discussion will focus primarily on this class of electrical tape. It may be referred to as black electrical, vinyl, or plastic tape. This tape might often be associated with improvised explosive devices.

The PVC backing of electrical tape would be almost inflexible except for the addition of plasticizer in up to a 2:1 ratio of PVC to plasticizer. As expected, the plasticizer figures prominently in the analysis of these tape backings, the most common plasticizer being dioctyl phthlate (DOP), an aromatic oil. Aliphatic sebecates and adipates may also be encountered. The black color comes from the addition of carbon black. Other compounds are added to these backings in part to add bulk but also to add desirable properties. Fillers such as calcite, rutile titanium dioxide, kaolinite, and talc may also cut costs, if they can be bought cheaper than the PVC. Antimony oxide, phosphorus, or chlorinated hydrocarbons may serve as fire retardants. Lead carbonate, lead sulfate, Ca, Pb, Cd, and Ba stearates, dibutyl tin, and diphenyl urea represent a partial list of stabilizing compounds. The elemental analysis of the backing alone offers a great deal of discrimination between electrical tapes [4, 5]. It should be noted that not all of these additives are present all the time and, even if they were, they may not be within the detection limits of the instrumentation available to most crime labs.

Electrical tape backings are variable not only in the above additives but also in their macroscopic appearance. A physical examination of the tape surface under oblique lighting shows variation in the surface textures from one manufacturer to another. These are from the film processing equipment (Figure 12.15).

In the examination of electrical tapes, the predominance of plasticizer in the backing can mask the detection of the other components (Figures 12.16 and 12.17). Also, the carbon black in the PVC raises the refractive index and thus causes dispersion effects in Fourier transform infrared (FTIR) analysis, giving sloping baselines [13]. (Figure 12.18). The inorganic additives may be separated from the backing by dissolving the PVC in tetrahydrofuran (THF). The spun solid precipitate can then be separated and analyzed. The plasticizer can also be isolated from the backing by extraction in chloroform. The PVC is not soluble in the chloroform. This may be useful for analysis of the backing

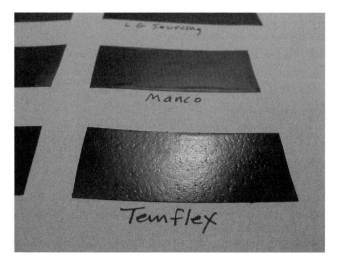

Figure 12.15 The surface features of electrical tape may vary.

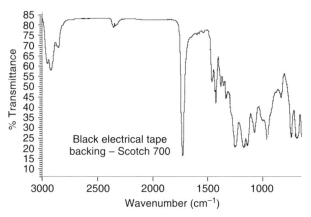

Figure 12.16 FTIR spectrum of electrical tape PVC backing with prominent contribution from the plasticizer.

with pyrolysis GC/MS, where the plasticizer content can be diminished somewhat to bring the other pyrolysis products in scale.

The adhesive of electrical tape can have a variety of natural rubber and copolymer elastomers, fillers, and cross-linkers found in other PSAs but is thinner and less tacky than tapes designed for repair and packaging. It therefore does not have a tenacious "stick" to its own backing and a release coat is generally not necessary. The adhesive may be clear or black by the addition of carbon black. Plasticizer will be present in the adhesive because it has either "migrated" from the backing or been intentionally added to counteract the expected migration.

In summary, the analysis scheme for black electrical tapes should include elemental analysis of the backing and adhesive. FTIR analysis should be

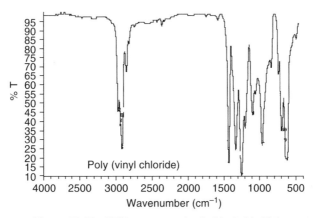

Figure 12.17 FTIR spectrum of poly (vinyl chloride).

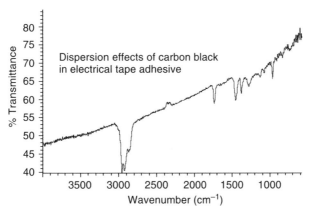

Figure 12.18 FTIR spectrum showing sloping baselines, seen in dispersion effects from high index compounds.

attempted but may show dispersion effects from the presence of carbon black. Pyrolysis GC/MS is a good way to analyze the organic components of black vinyl tape in the presence of carbon black [11, 13].

12.6 POLYPROPYLENE PACKAGING TAPE

The most common types of packaging tapes are constructed of clear isotactic polypropylene film with a clear acrylic or SIS-based adhesive and are marketed in 2 inch wide rolls. As expected, the transparency of the tape indicates that the typical inorganic fillers found in other tapes are rarely present in packaging tapes. The film, however, can be tan colored or, conversely and more commonly, the adhesive can be tan colored. In either case, iron oxides and titanium dioxide are added to impart the color. Dobney et al. [2, 14] have reported discriminating potential for colored packing tapes using laser-ablation, high resolution ICP-MS.

These tapes do not have a reinforcement fabric but find their strength in the biaxial orientation of the polypropylene backing. An understanding of how polymer films are oriented is useful to the forensic analyst. The most discriminating features of the clear film tapes arise not so much from compositional differences as from optical behavior under plane-polarized and cross-polarized light. These properties arise from the manufacturing process.

12.6.1 Oriented Films

In the normal state, polymers are amorphous; that is, there is no order, the polymer chains are random. During the manufacturing process the polymer is raised above its melting point, where the chains are unordered and relatively mobile. As it is slowly cooled, the chains will line up in an orderly fashion in packets called spherulites. When the temperature falls below the glass transition temperature (T_g) the chains become "fixed." This gives an "ordered" crystalline polymer. To produce an oriented polymer, the film is stretched as it is slowly cooled, causing the sperulites to line up in the direction of the stretch. If the polymer film is stretched in only one direction, it is "monoaxially oriented." If it is stretched in two directions simultaneously, it is "biaxially oriented." The spherulites have lattice points like regular crystals and thus a polymer with spherulites will behave as a crystal. The polymer, however, will remain amorphous in some areas and never be totally crystalline. Oriented films are found in many other applications besides tapes [15–17].

Polypropylene, polyester, and polyethylene lend themselves well to ordering and orienting, but not all polymers can produce spherulites. The acronym BOPP refers to "biaxially oriented polypropylene." MOPP refers to "monoaxially oriented polypropylene." Both MOPP and BOPP may be found in packaging tape films. MOPP is marketed as a "hand-tearable" tape; that is, it can be torn in the direction of the orientation whereas the BOPP cannot be torn by hand and requires a "nick" at the edge with a dispenser or scissors.

12.6.2 Polarized Light Microscopy Examinations of Packing Tapes

Because MOPP and BOPP tapes exhibit crystalline behavior, they lend themselves well to PLM examinations [18]. Many of the same techniques used for the examinations of synthetic fibers can be applied to polymer films. Whereas most synthetic fibers behave as uniaxial crystals (having two refractive indexes), polymer films behave as biaxial crystals (having three refractive indexes). Note that the term "biaxial" as used by a polymer chemist refers to a stretch in two directions. To a microscopist it means a crystal with three refractive indexes.

The following observations can be made on any clear polymer tape film with very little sample preparation. The tape is adhered to a glass slide adhesive side down. This should include both machine edges. The adhesive is isotropic and, if clear, does not interfere with the analysis. If adhesive must be removed, the film itself can be mounted in a liquid such as Permount or Aerochlor.

12.6.3   Is It MOPP or BOPP?

Under crossed polarizers and using the 10× objective and 10× eyepiece, rotate the film about 5 degrees off fullest extinction. The orientation marks of BOPP film will appear as multiple Xs showing the bidirectional stretching. A MOPP film will show orientation marks in only one direction. The angles of the X orientation marks in BOPP films may vary and can be measured with a 2V reticule, if available (Figure 12.19).

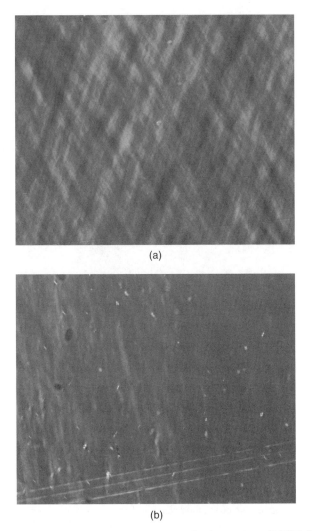

(a)

(b)

Figure 12.19 *(See color insert.) (a) Biaxially oriented polypropylene (BOPP) film. (b) Mono-axially oriented polypropylene (MOPP) film.*

12.6.4 Thickness

If a questioned and a known tape are being compared, it is useful to mount the machine edges side by side so that they may be observed in the same field. Under cross polarizers, the slide is rotated to the tape's brightest position. Note the colors of both tapes. Very small differences in thickness can be detected in the retardation colors of the two tapes observed. These differences can be amplified with the use of full- and quarter-wave plates (Figure 12.20).

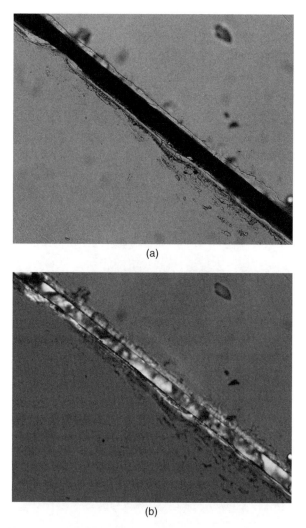

(a)

(b)

Figure 12.20 *(See color insert.) (a) A threatening note was taped to a car window using a polypropylene office tape. The questioned piece and known piece were mounted edge to edge. Under cross-polarized light, a slight difference in retardation is noted. (b) The same field as (a) but with a quarter-wave plate (147.5nm) inserted in the path accentuates the difference in retardation and shows that these two tapes have a significant difference in thickness.*

12.6.5 Degree of Offset from the Machine Edge

The slow and fast refractive indexes in the plane of the polymer film do not necessary align with the machine and cross directions of the tape. To determine this offset, align the machine edge with the north–south eyepiece graticule. Note the position on the rotatable stage, and then rotate the stage until the tape is at its fullest extinction. Subtract the difference. This may be anywhere from 0 to 35 degrees relative to the machine edge.

Other observations in clear tape might show artifacts or additives. Any irregularities in the tape surface will also be apparent under polarized light, as differences in thickness appear as different retardation colors or a mottled appearance.

An analysis scheme of clear packaging tapes should include FTIR analysis of the backing and adhesive and elemental analysis of colored tapes. While PLM is the most discriminating tool in the comparisons of clear polymer films, not every trace evidence chemist is trained in microscopy. If there is enough sample, a simple tear test may offer a clue as to whether the film is MOPP or BOPP. Can it be torn across or along the machine direction? Other methods such as X-ray powder diffraction may be employed to distinguish a MOPP from a BOPP film. The retardation colors and thus a comparison of the thicknesses may be assessed by simply viewing the tape between two pieces of cross-polarized film [19].

12.7 STRAPPING/FILAMENT TAPES

Tapes that are marketed as "filament" tape generally have glass filaments as a reinforcement fiber but may occasionally have nylon fibers. In each case the fibers are in bundles that run along the machine direction. Both of these types of reinforcements are virtually impossible to tear by hand. There is a great deal more variability in the visual/macroscopic appearance of the tapes in this class. In the previous discussion of clear polypropylene packaging tapes, all use a polypropylene film, mostly with an acrylic adhesive, the same width and colorless. There are few macroscopic features to visually discern one manufacturer from another. The strapping and filament tapes, on the other hand, do not have a standard width, have variable number and lay-down of reinforcement fibers, and are more apt to have bubbles and other anomalies in the adhesives (Figure 12.21). The backing may be polypropylene or polyester. There may be additives in both the backing and the adhesive. Adhesives may be clear or opaque. Fillers may be from the now familiar list given in the discussion of adhesive fillers (12.3.2).

The analysis and comparison of strapping or filament tapes as evidence should include elemental analysis of the backing and adhesive and FTIR analysis of the backing, adhesive, and possibly the reinforcement fiber. If synthetic fibers are present, they should be checked for

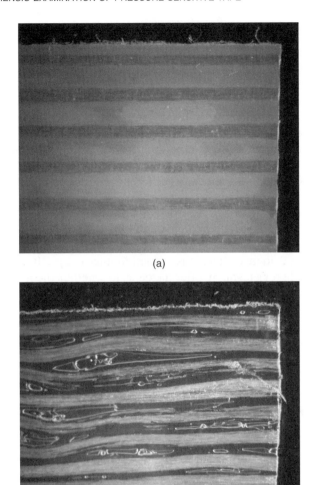

(a)

(b)

Figure 12.21 *Four different strapping tapes under 10× magnification: (a) ACE, (b) Scotch, (c) Shurtape, and (d) Universal.*

fluorescence. PLM can offer information on the nature of the inorganic fillers if present and offers the best information regarding the reinforcement fibers. In addition, the polymer film of strapping/filament tapes is likely to be an oriented film. Therefore, the same PLM examination used for the polypropylene packaging tapes is also applicable to these films (Figure 12.22).

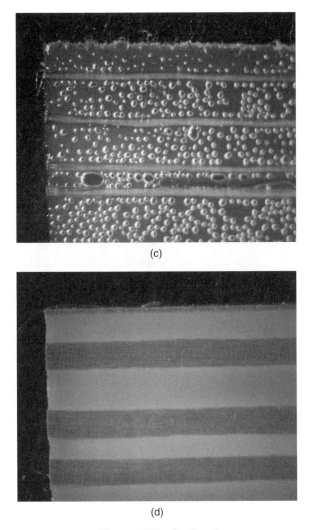

(c)

(d)

Figure 12.21 *Continued*

12.8 MASKING TAPE

Masking tape is designed for use in masking off areas during painting applications. As masking tape is so common to many households, it often appears as evidence. It has a saturated creped paper backing. The saturant serves to hold the paper fibers together, to improve the physical and barrier characteristics, and to repel moisture and solvents from paint products. Carboxylated styrene butadiene is often used as a saturant. Masking tape adhesives tend to have

Figure 12.22   *A 100× view under cross polars slightly off extinction. This strapping tape has a biaxially oriented polyester film. The fiberglass reinforcement is noted on each side.*

low tack to avoid damage to surfaces. The paper is commonly a light tan color. The adhesive can be clear or opaque and can be natural rubber based, often blended with other elastomers, or SIS based.

The surface of the saturated paper backing is originally slightly porous and so is filled and smoothed over with a backsizing. Even so, the backing surface may hold some clues when comparing masking tapes. There is a "male" and "female" side, one side having protruding crepe lines, the other side having indentations. Either side may be used as the coated side depending on the manufacturer [20].

The most discriminating examination of masking tape is the visual/macroscopic methods as there is much variation in the tan backing color and width (Figure 12.23). Studies on masking tape using fluorescence spectroscopy have shown some discrimination between tapes [21]. The discriminating potential of pyrolysis CG/MS is unknown to this author. As might be expected, the FTIR analysis shows a predominance of cellulose from the paper component. Elemental analysis should be done to address any inorganic components. Except where there are inorganic fillers in the adhesive, PLM adds little to the comparison.

12.9   INITIAL HANDLING

The initial examination of any tape's physical characteristics is quite predictive of whether a questioned tape could have come from a given source. This is particularly true of more complex products such as duct tape. If they are

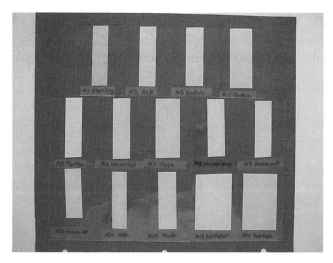

Figure 12.23 *(See color insert.) Masking tape color and width are variable.*

similar at this stage, however, instrumental testing should be conducted to assure that the chemical and microscopic compositions are consistent. The analyst must weigh these issues, keeping in mind that with each added test he/she is adding confidence to the final conclusion. The wording of a report of conclusions and subsequent testimony should reflect the limitations and strengths of those methods selected. When a questioned and known tape are different at the initial macroscopic level, no further testing is necessary.

12.9.1   Sharing Evidence with Other Sections

Examinations of tape products often involve multiple sections of the crime lab. Print processing is often requested and can present a compromise where the processing chemicals interfere with the trace examinations. When possible, the trace section should attempt to locate and sample a small area of the tape free of foreign debris prior to print processing. Only a small sample (a square centimeter) is necessary for further chemical testing, such as FTIR spectroscopy, PLM, and elemental analysis. Areas sampled should not be taken from tape ends and should always be labeled. Precautions should be taken not to compromise the recovery of latent prints or DNA and samples should be taken in consultation with the other disciplines involved. After print processing and DNA sampling, the trace analyst can complete other measurements and examinations of the tape that are not affected by the processing chemicals.

Care must be taken in the processing of tape to protect the integrity of the torn ends as these are the most likely area for latent prints and saliva. A physical match of the torn ends to a known roll might also be possible. The DNA section might test the torn ends of tape for saliva as tape users often use their teeth to cut the tape. Any cuts made to the tape by the analyst should be labeled.

12.9.2   Untangling Tape and Recovering Trace Evidence

The adhesive serves as a "trace magnet" and may provide hair, fibers, or other particles that could associate a suspect to a crime. Examination of the entire adhesive surface for trace evidence often requires untangling tape evidence. The best way to accomplish this is with gentle heat while slowly pulling the pieces apart. Alternately, the tape can be frozen, dropping its adhesion and tack to allow separation from itself. The separated pieces may then be examined under a stereoscope for trace evidence adhered to the adhesive.

12.10   METHODS

Analysis of tape products will tap into the full range of laboratory skills of the forensic scientist. On a case-by-case basis the analyst will determine the selection of tests that are most suitable. The methods recommended here are not all-inclusive. Obviously, there are methods and instruments that are not available in many forensic labs. With additional testing there is a point of diminishing returns that is reached. More testing may only rarely yield new and meaningful results.

12.10.1   Physical End Matching

A physical match of a questioned end to a known roll provides individualizing evidence and should be the first aim of tape comparisons. Most of the tapes encountered in forensic casework have a polymeric backing that stretches when torn. Several points should be noted when conducting end matches on stretchy materials:

1. A piece of polymeric tape torn by hand will never go back together the same way.
2. Some tapes are more difficult to tear by hand, stretching more, and therefore the end match will be more distorted.
3. The straighter the end cut, such as with scissors, the fewer points there are for comparison.
4. On reinforced tapes, make a note of the fiber offset from the edge and note that fiber tears must match up as well.
5. You might see a "notch" at one end of the tear where the tear started.

Where a physical end match is found between a questioned and known tape, further examinations are usually not necessary. The decision to do further testing should be decided on a case-by-case basis and depends on the strength of the match. Photographic documentation should always be included with case notes (Figure 12.24).

Figure 12.24 *An abduction/bank robbery case. Questioned and known tapes were physically matched. The questioned tape had been treated with crystal violet for latent prints prior to trace examinations.*

12.10.2 Physical Characteristics

The easiest examinations of tape products are often the most discriminating observations when comparing a known and questioned tape. Initial examinations include:

1. Macroscopic comparison of the backing color and adhesive color.
2. Use of oblique lighting to compare the surface textures.
3. Measurement of the tape width.
4. A yarn count if the tape has a fiber reinforcement.
5. Total tape thickness in mils should be assessed but may show some within-roll variability due to its compressibility and its slight coating variation. Special calipers are available, which have the slightest contact with the surfaces. Thickness may also be assessed by cross sectioning.

12.10.3 Separation of the Backing, Reinforcement, and Adhesive

To measure the thickness of the backing, the adhesive can be separated by sonicating a piece of the tape in hexane. This is also helpful in tapes with reinforcement fabric such as duct tape, allowing the fibers to be clean of adhesive for further testing. Always leave some of the tape samples intact.

12.10.4 FTIR Analysis

Infrared analysis is a valuable tool in the characterization and comparison of known and questioned tape components. The polymer class of the backing, the plasticizers, the organic and inorganic composition of the adhesive, and identification of the fabric reinforcement are all possible with FTIR analysis. Limitations occur where the heavy absorbance of an inorganic filler such as calcium carbonate masks other components in a tape adhesive. The presence of carbon black may cause dispersion effects and loss of absorbance in electrical tape backings and adhesives.

The sample preparation for FTIR analysis depends on the accessories available. The easiest and least destructive method of analysis of tape products is accomplished with an ATR (attenuated total reflectance) accessory. These are available as single reflection bench accessories, horizontal multireflection plates, microscope ATR objectives, and ATR slide-ons. Additionally, the internal reflectance element (IRE) crystal may be KRS-5, ZnSe, germanium, diamond, or silicon. Of these options, the diamond KRS-5 focusing element fits into the FTIR bench (DTGS detector) and gives the widest spectral range (4000 to 260 cm^{-1}) with a minimum of sample. The depth of beam penetration into the sample surface is 2 μm or less using any of these ATR accessories, making it unnecessary to separate the backing from the adhesive. The surface (adhesive side or backing side) of interest is simply pressed against the IRE for analysis [7, 22, 23].

Alternately, where an ATR accessory is unavailable, separation of the adhesive and backing will be necessary. Adhesives may be "pinched" from the backing and feathered out on a KBr pellet for micro-FTIR analysis (Table 12.1, Figure 12.25).

12.10.5 Elemental Analysis

Elemental analysis of tape backings and adhesive provides a complement to the FTIR and PLM data and aids in the identification of the inorganic com-

TABLE 12.1 Absorbances in cm^{-1} to Look for in FTIR Spectra of Tape Components

Natural rubber	1660, 1450, 1370, 835
Polyterpene tackifiers	1450, 1370, 835
SIS	1450, 1370, 890, 835, 760, 700
SBS	1498, 1450, 966, 910, 760, 700
Polybutadiene rubber	1450, 1309, 995, 910
Poly(butyl acrylate)	1735, 1455, 1380, 1245, 1165
Poly(vinyl chloride)	1430, 1330, 1255, 1095, 960, 690, 635
Dioctyl phthlate (DOP)	1730, 1600, 1580, 1460, 1380, 1275, 1123, 1075, 745
Calcium carbonate	1470 to 1430, 875
Kaolin clay	3620, 1030, 1007, 912
Titanium dixoide	800 to 450
Talc	1015, 670

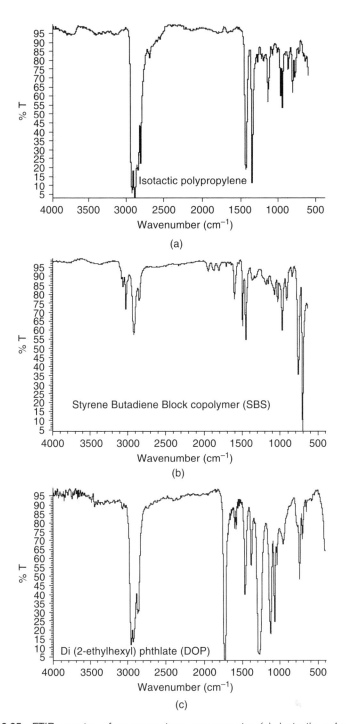

Figure 12.25 FTIR spectra of common tape components: (a) isotactic polypropylene, (b) styrene butadiene block copolymer (SBS), (c) di (2-ethylhexyl) phthalate (DOP), (d) styrene isoprene block copolymer (SIS), (e) butyl acrylate, (f) Shurtape duct tape adhesive, (g) Polyken duct tape adhesive, (h) Intertape duct tape adhesive, (i) K-Mart Performer brand duct tape adhesive, (j) Kendall–Polyken brand duct tape adhesive, and (k) unknown brand duct tape adhesive.

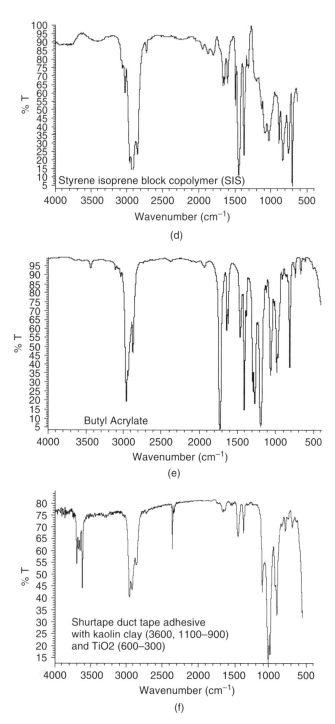

Figure 12.25 *Continued*

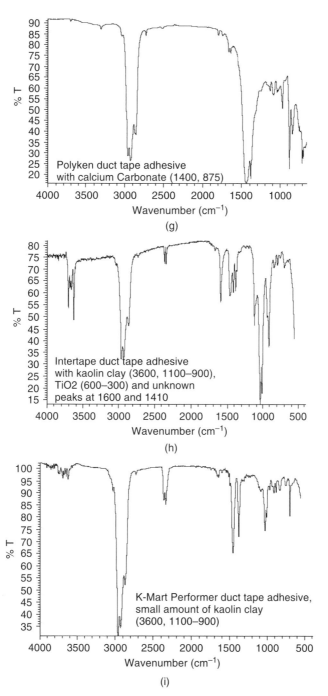

Figure 12.25 Continued

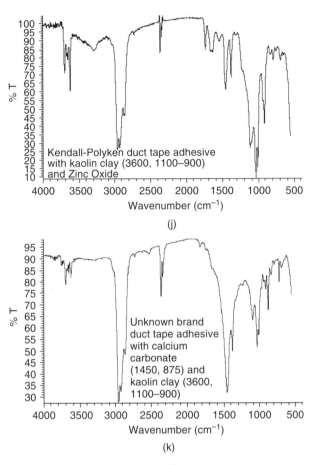

Figure 12.25 Continued

ponents of tape, that is, the fillers, extenders, colorants, and stabilizers [2, 5, 9, 24]. Several different types of instrumentation will provide this information including the scanning electron microscope (SEM) with energy dispersive X-ray spectroscopy (SEM-EDS), SEM with wavelength dispersive X-ray spectroscopy (SEM-WDS), X-ray fluorescence (XRF), and inductively coupled plasma (ICP). X-ray powder diffraction (XRD) offers identification of crystalline materials.

Adhesive formulations are often heterogeneous blends of many components, such as in duct tapes and other tapes with opaque adhesives. An approach to analysis should include multiple sample areas or a sufficiently large area to address a representative sampling.

Analysis of both sides of the tape backing may help detect multiple layers in the backings that may otherwise escape detection.

While EDX, XRF, and ICP cannot identify the minerals present they can offer some clues. If Ca is present, look for peaks in the FTIR spectrum at 1470 to 1430 and 875 cm⁻¹ (calcite or dolomite). If Al and Si are present, look for peaks at 3620,1030,1107, and 912 cm⁻¹ (kaolinite). Note that the ratios of Al and Si may vary depending on where the kaolinite is mined. Mg may be from dolomite (calcium magnesium silicate) or talc (magnesium silicate). If it is talc, there will peaks on the FTIR at 1015 and 670 cm⁻¹.

12.10.6 Polarized Light Microscopy

Microscopic examinations provide complementary information to the FTIR analysis and elemental data. The observations and results from the PLM and plane polarized light can sometimes offer stand-alone identification of tape components, such as the cotton in reinforcement fibers. Identification of the polyester fibers may also stand-alone in the hands of a skilled microscopist, as few fibers have the high birefringence of polyester. Other features of the reinforcement fibers such as delusterant, diameter, and shape cannot be assessed any other way.

As discussed elsewhere in this chapter, PLM offers the best discrimination of clear packaging tapes, since the compositions of backing and adhesive are not nearly as variable as the optical properties of the film (Figure 12.26). Other backings, even those with colored or matte surfaces or containing fillers, can still have some transparency. PLM or plane polarized light may still be able to offer some information about the film or the dispersed constituents in these semiopaque backings.

The inorganic fillers of tape adhesive or fillers extracted from tape backings may be identified by their optical properties [25]. A small pinch of adhesive

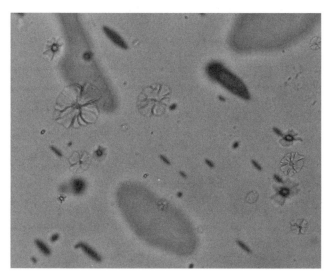

Figure 12.26 *Polypropylene film of packaging tape, 100× plane polarized light showing unidentified flower-shaped objects.*

TABLE 12.2 **Optical Properties of Commonly Encountered Inorganic Fillers**

Calcite ($CaCO_3$) (–) 1.658 (ω), 1.486 (ε)
Dolomite ($CaMgCO_3$) (–) 1.682 (ω), 1.503 (ε)
Kaolin clay (Al Si) B = (–) 0.005, 1.553–1.565 (α), 1.559–1.569 (β), 1.560–1.570 (γ)
Zn oxide (+) 2.013 (ω), 2.029 (ε)
Titanium dioxide (TiO_2) (+) 2.620 (ω), 2.900 (ε)
Talc (Mg Si) B = (–) 0.03–0.05, 1.538–1.554 (α), 1.575–1.599 (β), 1.575–1.602 (γ)

can be dispersed in xylene, dried on a microscope slide, and mounted in an appropriate refractive index (RI) medium. An RI of 1.66 is a good general purpose value for the identification of calcite, dolomite, kaolinite, zinc oxide, and titanium dioxide. A 1.56 RI medium works best for the identification of talc. Although dolomite and calcite have very similar optical properties, they will show different dispersion staining colors in a 1.66 high dispersion index liquid. FTIR analysis cannot distinguish dolomite and calcite. The presence of magnesium in the elemental data offers a clue that dolomite might be present. However, magnesium could also be from something other than dolomite. The optical properties of these common minerals can be found in most microscopy handbooks to aid in their identification (Table 12.2). The PLM techniques offer an alternative method for laboratories that do not have elemental analysis capabilities.

12.10.7 Pyrolysis GC/MS

Pyrolysis GC/MS provides a sensitive and discriminating analysis of the organic components of tapes. It is useful in the identification of components in complex mixtures. It is also useful in polymers that are not suitable for FTIR analysis. Black vinyl tape is an example of this, where the carbon black in the backing and often in the adhesive causes dispersion effects in the FTIR analysis. Also, calcium carbonate when present can mask other constituents. Pyrolysis GC/MS can identify all of the organic components without interference from the inorganic fillers. Pyrolysis GC/MS is also better than FTIR spectroscopy in distinguishing small differences in very similar compounds such as the adipates and sebecate plasticizers (Table 12.3).

Other suitable instrument conditions are provided in References 26–28.

Note that black vinyl tapes have a high concentration of plasticizer. A brief soaking of a small piece of the tape in chloroform prior to analysis will remove much of this, but leave enough for identification of the plasticizer.

The down side to pyrolysis GC/MS is the time involved and destruction of the sample.

12.10.8 Sourcing Tape Products to a Manufacturer

It may sometimes be useful to determine the manufacturer of a tape product where a questioned piece is received as evidence with no known source for

TABLE 12.3   Degradation Products of Pyrolyzed Tape Polymers

Pyrolysis Product	Likely Source
Decane, decene	Polyethylene
Dimethyl heptene	Polypropylene
Isoprene, limonene, dipentene	Polyisoprene
HCl, benzene, naphthalene	Poly(vinyl chloride)
Furans, levoglucsan	Cellulose, paper
Butadiene, vinylcyclohexene	Polybutadiene
Benzene, vinyl benzoate	Polyester

Typical Instrument Conditions	
Interface oven	300 °C
Pyrolysis temperature	700 °C
Pyrolysis interval	15 seconds
Clean temperature	1000 °C
Clean interval	20 seconds
Column	30 m × 0.25 HP-5
Inlet	Constant pressure @ 5.9 psi
Inlet temperature	300 °C
Split ratio	75 : 1
Carrier	Helium
GC oven program	Hold @ 40 °C for 2 min, ramp 6°/min to 295 °C and hold 5 min

comparison. There is currently in development a searchable elemental spectral database called SLICE. It is a collaborative effort between the FBI Laboratory and XK Inc. and can be searched with EDS, WDS, or XRF data. However, the success of this project depends on frequent input of current data, as product components change with market trends.

12.11   CASE EXAMPLE

In October 1997 an off-white refrigerator was found behind a shopping center in San Diego. The refrigerator was taped closed with white duct tape. The body of an unknown female was found inside the refrigerator (Figures 12.27–12.29). The investigation led to an apartment in Long Beach, California. During a search of the premises, a piece of white duct tape was found in the garage (no roll of tape was ever recovered). This tape was examined and found to be similar in color, size, and construction, and the layers were chemically similar to the tape around the refrigerator. This tape and other physical evidence helped convict the victim's roommate.

Figure 12.27   *Defendant (left) sits in court. He is accused of killing a woman, stuffing her body in the refrigerator at right (covered with white paper), and dumping it. Inset photo shows white duct tape at the top of the refrigerator. (Larger photo by Scott Linnett, San Diego Union-Tribune. Used with permission. Inset photo by Tanya DuLaney, Criminalist, San Diego Police Department Crime Laboratory.)*

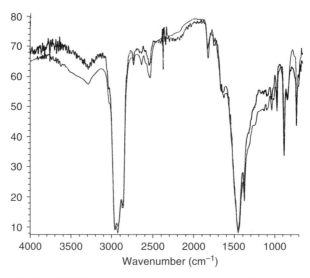

Figure 12.28   *Overlay of FTIR spectra of adhesives (reflectance) from tape from refrigerator and tape piece from garage, rubber based with calcium carbonate.*

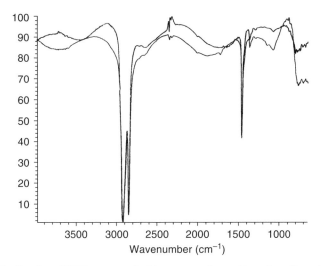

Figure 12.29 *Overlay of FTIR spectra of white outer surface of tape from the refrigerator and tape piece from the garage, showing polyethylene.*

The tape was constructed of a white plastic outer layer on top of a fiber matrix with an off-white adhesive bottom layer. Fibers ran longitudinally (worp) and horizontally (fill). The longitudinal fibers were a combination of cotton fibers and polyester fibers. The horizontal fibers were all polyester fibers. The outer layer was polyethylene and the adhesive was a rubber-based adhesive.

ACKNOWLEDGMENTS

The author wishes to acknowledge John Johnston, retired from the tape manufacturing industry, for his technical peer review of this chapter for accuracy. Thanks also to Tanya DuLaney, Criminalist, San Diego Police Department Crime Laboratory, for providing the case example.

REFERENCES

1. Blackledge R (1987). Tapes with adhesive backing; their characterization in the forensic science laboratory. In: *Applied Polymer Analysis and Characterization*, Mitchell J Jr (ed.). Hanser Publishers, Munich, Chapter III-F, pp. 413–421.

2. Dobney AM, et al. (2001). Elemental composition of packaging tapes using HR ICPMS. Presented at the 2001 meeting of the American Academy of Forensic Sciences, Seattle, WA.

3. Jenkins TL Jr. (1984). Elemental examination of silver duct tape using energy dispersive X-ray spectrometry. In: *Proceedings of the International Symposium on the*

Analysis and Identification of Polymers, FBI Academy, Quantico, VA, July 31–August 2, pp. 147–149.

4. Kee TG (1984). The characterization of PVC adhesive tape. In: *Proceedings of the International Symposium on the Analysis and Identification of Polymers*, FBI Academy, Quantico, VA, July 31–August 2, pp. 77–85.

5. Keto RO (1984). Forensic characterization of black polyvinyl chloride electrical tape. In: *Proceedings of the International Symposium on the Analysis and Identification of Polymers*, FBI Academy, Quantico, VA, July 31–August 2, pp. 137–143.

6. Maynard P, et al. (2001). Adhesive tape analysis: establishing the evidential value of specific techniques. *Journal of Forensic Sciences* **46**(2):280–287.

7. Merrill RA, Bartick EG (2000). Analysis of pressure sensitive adhesive tape: I. Evaluation of infrared ATR accessory advances. *Journal of Forensic Sciences* **45**(1):93–98.

8. Noble W, Wheals BB, Whitehouse MJ (1974). The characterization of adhesives by pyrolysis gas chromatography and infrared spectroscopy. *Journal of Forensic Sciences* **19**(3):163–174.

9. Smith J (1998). The forensic value of duct tape comparisons. *Midwestern Association of Forensic Scientists Newsletter* **27**(1):28–33. Reprinted in *CACNews* 3rd Quarter, 1998, pp. 14–17.

10. Snodgrass H (1991). Duct tape analysis as trace evidence. In: *Proceedings of the International Symposium on Trace Evidence*, FBI Academy, Quantico, VA, June pp. 69–73.

11. Williams ER, Munson TO, Munson BS (1988). The comparison of black polyvinyl-chloride (PVC) tapes by pyrolysis gas chromatography. *Journal of Forensic Sciences* **33**(5):1163–1170.

12. Teetsov A, Stellmack ML (2004). Hand-sectioning and identification of pressure-sensitive tapes (McCrone Associates) In: *Proceedings of the PSTC Tech XXVII Global Conference*, May 2004.

13. Merrill RA, Bartick EG (1989). Procedure for the forensic analysis of black plastic tape. Internal FBI Document, FBI Forensic Science Research Unit, Quantico, VA, January.

14. Dobney AM, et al. (2001). Sector field ICP-MS applied to the forensic analysis of commercially available adhesive packaging tapes. *Journal of Analytical Atomic Spectroscopy* **2002**(17):478–484.

15. Rappe R (1992). Microscopical examination of polymer films. *Microscope* **40**:93–101.

16. Carraher CE Jr (ed.) (1996). *Polymer Chemistry—An Introduction*, 4th ed. Marcel Dekker, New York.

17. Mathias LJ (1984). Fundamentals of the polymer chemistry of acrylics and vinyls. In: *Proceedings of the International Symposium on the Analysis and Identification of Polymers*, FBI Academy, Quantico, VA, See also http://www.psrc.usm.edu/macrog.

18. Smith J, Weaver R (2004). PLM examinations of clear polymer films. *Microscope* **52**(3&4):112–118.

19. Rogers B (2005). Clear adhesive tape analysis using polarizing light techniques, the Megascope. *Southwestern Association of Forensic Scientists Journal* **27**(1):53–58.

20. Johnston J (2003). *Pressure Sensitive Adhesive Tapes—A Guide to their Function, Design, Manufacture and Use.* Pressure Sensitive Tape Council, Northbrook, IL, p. 21.

21. Blackledge RD (1984). Comparison of masking tapes by fluorescence spectroscopy. In: *Proceedings of the International Symposium on the Analysis and Identification of Polymers*, FBI Academy, Quantico, VA, July 31–August 2, p. 135.

22. Merrill RA, Bartick EG (1999). Advances of infrared ATR analysis of duct Tape. Presented at the 51st Annual Meeting of the American Academy of Forensic Sciences, Orlando FL, February 15–20.

23. Coates J, Reffner J (1999). Visualization of micro-ATR infrared spectroscopy. *Spectroscopy* **14**(4):34–45.

24. Benson JD (1984). Forensic examination of duct tape. In: *Proceedings of the International Symposium on the Analysis and Identification of Polymers*, FBI Academy, Quantico, VA, July 31–August 2, pp. 145–146.

25. Randle WA (2004). Microscopical examination of duct tape adhesive fillers. Presented at INTER/MICRO, June, Chicago, IL.

26. Wampler RP, Zawodny CP (1999). Analysis of polymer packaging products using pyrolysis-gas chromatography-mass spectrometry. *CDS Analytical* **Sept.**: 31–32.

27. Wampler RP, Phair M (1997). Analysis of rubber materials by pyrolysis GC. *Rubber World* 30–34.

28. Bakowski NL, Bender EC, Munson TO (1985). Comparison and identification of adhesives used in improvised explosive devices by pyrolysis–capillary column gas chromatography–mass spectrometry. *Journal of Analytical & Applied Pyrolysis* **8**:483–492.

ADDITIONAL SOURCES

1. Agron N, Schecter X (1986). Physical comparisons and some characteristics of electrical tape. *Association of Firearms and Toolmarks Examiners Journal* **18**(31): 53–59.

2. Courtney M (1994). Evidential examinations of duct tape. *Southwestern Association of Forensic Scientists Journal* **15**(1):10–16.

3. Fanconi BM (1984). Trends in polymer development and analytical techniques. In: *Proceedings of the International Symposium on Analysis and ID of Polymers*, FBI Academy, Quantico, VA, p. 87.

4. Pizzi A, Mittal KL (eds.) (1994). *Handbook of Adhesive Technology*. Marcel Dekker, New York.

5. Rappe R (1987). Measurement of the principle refractive indices of oriented polymer films. *Microscope* **35**:67–82.

6. Sakayanagi M, et al. (2003). Identification of pressure sensitive adhesive polypropylene tape. *Journal of Forensic Sciences* **48**:68–76.

7. Satas D (ed.) (1996). *Handbook of Pressure-Sensitive Adhesive Technology*. Van Norstrand Reinhold, New York.

8. Schmitz J (2004). Taking the universal out of universal fix-it—a chemical/forensic analysis of duct tape. Thesis paper presented to Butler University, Indianapolis, IN Spring.

9. Sclademan JA (1997). Tackifiers and their effect on adhesive curing. *Adhesive Age* **Sept.**:24–26.

13

Discrimination of Forensic Analytical Chemical Data Using Multivariate Statistics

Stephen L. Morgan

Department of Chemistry and Biochemistry, University of South Carolina, Columbia, South Carolina

Edward G. Bartick

Retired, Research Chemist, FBI Academy, Counterterrorism and Forensic Science Research Unit, Quantico, Virginia. Present: Director Forensic Science Program, Department of Chemistry and Biochemistry, Suffolk University, Boston, Massachusetts

13.1 PATTERNS IN DATA

Identification of patterns in data and interpretation of observed differences is a frequent task for the forensic chemist. A fiber examiner performs UV–visible microspectrophotometry to evaluate the likelihood of associations between known and questioned fibers. A trace evidence chemist might assess correspondence between pyrolysis gas chromatograms of a paint chip left on a bicycle at a hit-and-run incident and chromatograms of paint from vehicles

Forensic Analysis on the Cutting Edge: New Methods for Trace Evidence Analysis, Edited by Robert D. Blackledge.
Copyright © 2007 John Wiley & Sons, Inc.

suspected to have been involved. An arson investigator compares headspace gas chromatograms of questioned debris taken from a suspicious fire with chromatograms of known accelerants.

Most scientists are familiar with statistics associated with a single type of measurement variable. For example, the refractive index of questioned glass fragments might be compared with that of window glass fragments found at a crime scene. The resulting data values have only one variable (refractive index) measured for a number of different objects (different glass fragments).

Statistics used for summarizing univariate measurements include sample means and standard deviations. Univariate procedures (calculations of means, standard deviations, confidence intervals, two-sample t-test, etc.) for small data sets can be conducted by hand, on small calculators, or with a spreadsheet. However, scientists are less familiar with statistics based on measurements of two or more variables for each sample. Modern spectroscopy, chromatography, or mass spectrometry (MS) routinely produces data of high dimensionality.

Microspectrophotometry, for example, produces spectral intensities (absorbance, transmittance, photon counts) at several thousand wavelengths. "Hyphenated techniques" (e.g., liquid chromatography/mass spectrometry) produce data from a single analytical sample in the form of a two-dimensional table (ion counts as a function of retention time and mass-to-charge ratio). Means and standard deviations (or variances) are also employed in multivariate statistics but must be calculated for each variable; additionally, covariances (or correlations), measuring the strength of the linear relationships between variables, are calculated. These statistics are often not single numbers, but are arrays (matrices) of numbers. Due to the size of data sets and the complexity of the procedures, statistics associated with multivariate data are often inaccessible without computer-assisted analysis. Increases in computing power have made computationally intensive data analysis feasible, and increased availability of software for multivariate statistics has made these techniques accessible.

To assess similarities and dissimilarities among several sets of chromatograms or spectra, a forensic chemist often looks for familiar features to reduce the complexity of the data. For example, a drug identification chemist might recognize a characteristic ion at a specific mass-to-charge ratio in a mass spectrum, or a trace investigator might spot a peak, usually associated with a common polymer, at a characteristic vibrational frequency in an infrared spectrum. Often, qualitative use of a "fingerprint" chromatogram or spectrum is all that is needed to make a decision. Examiners develop a practiced eye for details and are able quickly recognize features that are different or common among just a few spectra or chromatograms. In some instances, the presence or absence of major peaks or specific target analytes may resolve issues of similarity between two or more patterns. However, finding a small difference in a large data set may be virtually impossible. The desired information may be present in the data, but it is not easy to spot because of data overload.

Multivariate data can also be confusing and misleading when examined a single variable at a time. Real differences between samples might not show up as the presence or absence of a single peak, or as inflation or reduction of single peaks. A combination of small changes could be the key to recognizing a significant difference. Furthermore, the decision of whether a spectrum or chromatogram is significantly different from a previous one, or whether a sample is consistent with a specific class of materials, may depend on recognition of a subtle multivariate trend partially obscured by background features (e.g., fluorescence) or by noise. Trace evidence found at a crime scene may also involve mixtures of chemical components. While kerosene, gasoline, or diesel fuel may easily be identified from chromatograms, if a mixture of all three accelerants is used to start a fire, the analyst must recognize the more complex combined pattern. Finally, when faced with comparisons among several hundred spectra or chromatograms, simple visual inspection does not suffice.

The validity and relevance of any scientific technique applied to a particular forensic question may be challenged in court [1]. In *Frye v. United States* [2], involving admissibility of evidence from a lie detector test based on systolic blood pressure, the court ruled that scientific evidence "must be sufficiently established to have gained general acceptance in the particular field in which it belongs." *People v. Young* [3], which involved typing of blood stains by electrophoresis, required that scientific evidence be validated by published independent studies and accepted by impartial experts. *Daubert v. Merrill Dow Pharmaceuticals, Inc.* [4], a case involving whether a causal link existed between an antinausea drug and birth defects, recognized Rule 702 in the Federal Rules of Evidence [5] as the standard for admitting scientific evidence. *Daubert* also established a checklist for assessing the reliability of scientific expert testimony: (1) whether the scientific technique or theory can be or has been tested in an objective manner; (2) whether the technique or theory has been subject to peer review and publication; (3) whether the rate of error of the technique or theory has been established; (4) whether standards and controls have been maintained in the process, and (5) the degree to which the technique or theory has been generally accepted in the scientific community. The judge serves as a "gatekeeper" who analyzes what is said by experts and assures the validity of their scientific principles. *Kumho Tire Co. v. Carmichael* [6], which involved the cause of a tire blowout, made it clear that *Daubert* issues apply to all expert testimony, including testimony involving statistical analyses.

When a forensic examiner testifies in court concerning multivariate analytical data, questions may arise regarding the validity of comparisons. How should the similarity of complex patterns be evaluated? Are the patterns similar enough to be considered to have originated from similar source materials? Can differences be explained by sampling variability from the original source material or by experimental variability in the laboratory? The statistical significance of differences in patterns can be addressed with the methodology of multivariate statistics.

Multivariate statistics has a documented history of development and application in the scientific literature. The ideas underlying principal component analysis (PCA) as a dimension reduction and data display technique were originated by Pearson [7] in 1901, a time when machine computing was largely unavailable. Pearson noted that although the methods "can be easily applied," the calculations are "cumbersome" for more than four variables. Hotelling [8] described algorithms for computing principal components in 1933. Addressing the problem of comparing patterns in multivariate space, Mahalanobis [9, 10] introduced the multivariate distance measure that bears his name in 1936. Linear discriminant analysis (LDA), a method for describing and predicting the separation of groups of multivariate data representing different sample types, was first derived in 1936 by Fisher [11]. The technique was applied to discriminating between varieties of iris flowers using combinations of physical measurements.

While numerous texts discuss statistics in forensic science and in the courtroom [12–18], the primary focus is on sampling, probability, and univariate statistics. This chapter offers a descriptive introduction to principal component analysis and linear discriminant analysis, with the goal of providing forensic scientists with a basic understanding of these powerful data analysis tools. Practical examples of use and interpretation are provided, and selected forensic applications are reviewed.

13.2   EXPERIMENTAL DESIGN AND PREPROCESSING

Too often, whatever data is conveniently available is the data that is analyzed. While "happenstance" data may easily be obtainable from historical records, data sets that are not systematically designed may not represent the full range of behavior of a system under study. The well planned data set includes all possible constituents of interest (e.g., variations in chemistry), contains variation over the range of interest of the objects (e.g., types of forensic materials), and appropriately spans feature values of interest (e.g., wavelength, ion mass). Analysis of a limited data set may generalize poorly in prediction or classification of new samples. Proper experimental design can ensure representative sampling over a range of objects and variables [19–22]. Most importantly, to permit the level of experimental uncertainty and the significance of differences among objects to be assessed, replicate measurements on all variables should be made for as many objects as possible. At the planning stage, data analysts should communicate with experts to understand the nature of the evidence samples under study and the character of the data to be analyzed.

Multivariate analysis starts by considering each sample (spectrum or chromatogram) as a vector of measurements on p different variables (e.g., wavelengths or retention times). Note that the word "sample" has dual uses in chemistry and in statistics. Although a forensic chemist might refer to an analytical "sample" as an item of evidence (a fiber, an aliquot of a solution) to be analyzed, a statistician refers to a statistical "sample" as drawn from the universe of possible data. In multivariate analysis, "sample" refers to a single multivariate analysis result (e.g., a spectrum). These p-dimensional pattern vectors (or samples) constitute the rows of the data matrix, where n and p refer to the number of samples and variables, respectively.

$$p \text{ variables}$$

$$\mathbf{X} = \begin{bmatrix} x_{11} & \cdots & \cdots & x_{1j} & \cdots & \cdots & x_{1p} \\ \vdots & & & \vdots & & & \vdots \\ x_{i1} & \cdots & \cdots & x_{ij} & \cdots & \cdots & x_{im} \\ \vdots & & & \vdots & & & \vdots \\ x_{n1} & \cdots & \cdots & x_{nj} & \cdots & \cdots & x_{np} \end{bmatrix} \Bigg\} \, n \text{ samples}$$

In this array, the value x_{ij} represents the measurement for the jth variable of the ith sample, for example, the absorbance value of the jth wavelength in the ith absorbance spectrum. Figure 13.1 shows a three-dimensional view of ten replicate samples of UV–visible spectra for each of seven different purple nylon 6,6 fibers. The corresponding data matrix has 70 rows of samples (spectra) and 1527 columns of variables (wavelengths from 330 to 850 nm).

Preprocessing is used to average, normalize, smooth, or filter and remove random and systematic variation that might confound later interpretation. Transformation (e.g., by taking the logarithm of measured values) is used to change the scaling or magnitude of values to facilitate comparisons or to reexpress information (as in converting transmittance to absorbance). Multi-

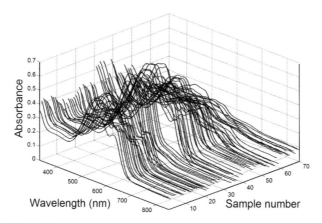

Figure 13.1 *UV–visible absorbance spectra (ten replicates) from seven purple dyed nylon 6,6 fibers.*

variate statistical analyses can be sensitive to data preprocessing, and inappropriate preprocessing may compromise the analysis. Preprocessing can be performed on one sample at a time over all feature variables, on one feature variable at a time over all samples, or on the entire data matrix.

Smoothing, background correction, and normalization treat one pattern vector at a time. If excessive noise is present, smoothing can be applied to increase the signal-to-noise ratio. One method is to use the data values in a moving window to estimate a "noise-free" value for the data point at its center by replacing it with the mean or median of the window values. Another approach, the running polynomial smooth, involves fitting a polynomial to points in the window and replacing the center value with the predicted value from the model. The Savitzky–Golay algorithm, perhaps the most common such digital filtering approach, is well documented [23, 24] and often implemented in instrument software. The amount of smoothing depends on the number of data points in the window and on the degree of polynomial fitted. Figure 13.2 shows smoothing, with a cubic polynomial and a window of 25 data points, on seven replicate (low intensity) fluorescence spectra (365 nm excitation) taken from a gray cotton fiber. High frequency noise is diminished while peak shapes are preserved. Care should be exercised because smoothing can cause distortions in peak height or width and can impair resolution of peaks [25]. Other strategies, such as signal averaging and low pass filters, are also used [26].

Constant or systematically varying background may confound clear interpretation. For example, gradient programming can cause baseline shifts in

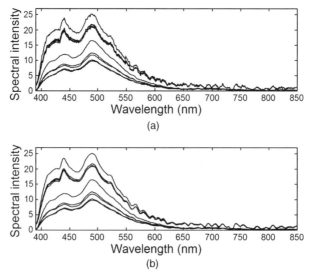

Figure 13.2   *Effect of smoothing on replicate fluorescence spectra of a single gray fiber: (a) original spectra and (b) spectra after Savitzky–Golay smoothing.*

chromatography, and background fluorescence may dominate a Raman spectrum. As with smoothing, background correction is performed separately on each data vector (chromatogram, spectrum). Some data systems permit manual subtraction of baselines between points chosen by the operator. Baseline correction can also be performed by subtracting an explicit fitted model (e.g., a straight line or polynomial) for a baseline trend from each spectrum. Ten Raman spectra of a Kevlar fiber [27] are shown in Figure 13.3a. The effect of least squares fitting of a fourth order polynomial [28] to each of these spectra is shown in Figure 13.3b. When predicted values at each wavelength are subtracted from each spectrum, the sloping background is removed. Another technique is to replace sample vectors by their first derivative. A simple way to calculate derivatives for equally spaced data is to take differences between adjacent data points. Figure 13.3c demonstrates baseline removal by taking first derivatives of the Raman spectra of Kevlar. Although derivatives remove

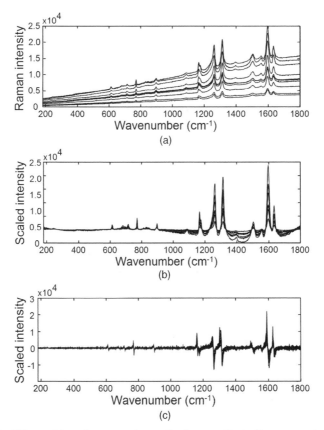

Figure 13.3 Effects of baseline correction methods on replicate Raman spectra of a Kevlar fiber: (a) original spectra, (b) spectra after subtracting a fitted fourth order polynomial from each spectrum, and (c) spectra after taking the first derivative of each spectrum.

varying baseline features, calculating differences tends to propagate noise (the most rapidly changing component) rather than averaging it. Improved noise filtering with derivatives can be achieved by Savitzky–Golay least squares smoothing [23].

Background correction or smoothing is typically performed prior to normalization. Normalization removes systematic variations associated with size, concentration, or amount effects in each sample pattern. For example, if a larger aliquot of test solution is subjected to chromatography, peak areas in the chromatogram will be higher; if spectral pathlength varies because of sample thickness, absorbance will vary. However, the pattern of relative intensities is the same. Normalization can be accomplished by dividing the value of each variable across a row of the data matrix by the sum of absolute values of all intensities across the row. Figure 13.4 shows the normalization of ten replicate spectra of the seventh fiber from Figure 13.1. Spectra were taken at different locations along the fiber and, perhaps because of varying amounts of dye or differences in sample thickness, are offset from one another in intensity. After normalization, feature intensities are expressed as a fraction of the total pattern intensity, the sum of the spectral intensities in each spectrum is unity, and variability is reduced. Other normalization procedures are also employed. For example, the multidimensional length of the pattern vector can be set to unity by dividing every intensity value by the square root of the sum of squares of the values. Mass spectra are normalized by scaling all intensities as a percentage of the intensity of the highest (base) ion peak. Normalization may not

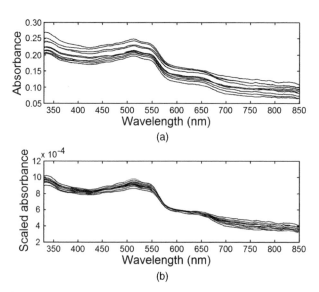

Figure 13.4 Effect of normalization on replicate UV–visible spectra of a single purple fiber: (a) original spectra and (b) spectra normalized to unit sum.

be appropriate if concentration effects are of interest, and other effects are also documented [29, 30].

Mean centering and autoscaling are typically performed on one feature variable at a time over all samples. Mean centering by columns involves calculating the mean of each variable over all sample vectors (the "centroid"), and subtracting that value from the corresponding elements of each sample vector. Figure 13.5b shows the effect of mean centering on a data set comprised of eight carbohydrate amounts, determined as alditol acetates by gas chromatography (GC), in 43 samples of legionella-like organisms [31]. Mean centering transforms rows of the data matrix into deviations from the average and removes constant background from the data without changing relative variation in the variables. Autoscaling (also known as variance scaling or standardization) transforms each column's data into a z-score, a set of values having a mean of zero and a standard deviation (or, equivalently, variance) of one

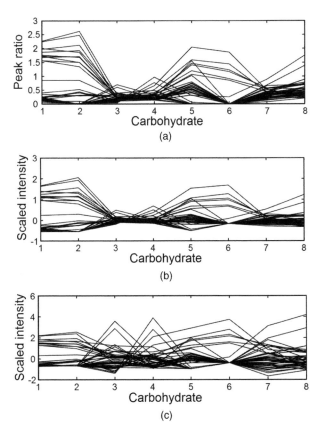

Figure 13.5 *Mean centering and autoscaling of data for eight carbohydrates measured by gas chromatography in 43 samples of legionella-like organisms: (a) original data, (b) mean centered data, and (c) autoscaled data.*

[32]. Variables with higher variance (variables 1, 2, 5, and 6 in Figure 13.5a,b) are more widely distributed in the space of the features than variables with lower variance (variables 3, 4, 7, 8). Multivariate techniques are sensitive to variance and will weight variables of higher variance more than those of lower variance. To deal with this issue, autoscaling starts with mean centering and goes one step further: the mean centered data in each column is divided by the standard deviation of the data in that column. After autoscaling, the vertical dispersions for all variables are similar (Figure 13.5c). Autoscaling is recommended when variables have different measurement units or exhibit large differences in variance. If outliers are present, the median may be used instead of the mean for robust estimation of the center of the data [33].

13.3 DIMENSIONALITY REDUCTION BY PRINCIPAL COMPONENT ANALYSIS FOR VISUALIZING MULTIVARIATE DATA

Principal component analysis is a dimensionality reduction technique that takes advantage of the fact that the variables may not be independent of one another, but are often correlated because of the underlying information. For example, spectral intensities at some wavelengths may rise and fall together as a consequence of correlation with chemical structure. In an infrared spectrum, chemical bonds often show their presence at multiple frequencies due to various bending and stretching vibrations. Correlation between two variables, x and y, is measured by the correlation coefficient, r, which is calculated as the ratio of the covariance between x and y and the product of the standard deviations of x and y. The sign of the correlation coefficient indicates whether variables are positively or negatively correlated, that is, whether y-values become larger or smaller as the corresponding x-values become larger. Values of r range in magnitude from zero (no correlation) to one (perfect correlation). The closer r is to unity, the closer the data is to a straight line [20, 32].

Hotelling [8] explained PCA by saying that there may be a smaller "fundamental set of independent variables which determine the values" of the original variables. When variables are correlated, the inherent dimensionality of multivariate data is less than the number of variables and these dominating directions of variability contain most of the information. Spectra, chromatograms, and mass spectra all contain redundant information. It is often the combination of peaks at correlated frequencies in an infrared spectrum that identify an organic compound. Identifying the molecular structure of a drug from a mass spectrum involves assessing peaks at different mass-to-charge ratios in the context of known fragmentation processes.

PCA projects multivariate data into lower dimensional subspaces by creating linear combinations of the correlated variables to form a new set of orthogonal and uncorrelated variables that have maximum variance. To illustrate PCA, the UV–visible spectra of the seven purple dyed fibers were normalized to unit area (Figure 13.6). Absorbance values taken from two of the

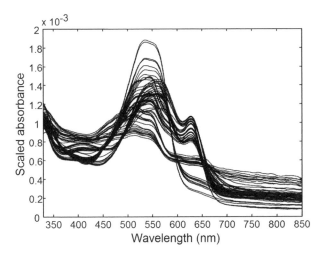

Figure 13.6 *Replicate UV–visible absorbance spectra (normalized to unit sum) from seven different purple dyed nylon 6,6 fibers.*

wavelengths are plotted in Figure 13.7a. These two measurements are negatively correlated ($r = -0.6392$) because the first wavelength (391 nm) is located near a valley, the second wavelength (541 nm) is near a peak, and spectra with a lower valley tend to have a higher peak, and vice versa.

For the two-variable data set of fiber spectra (Figure 13.7a), mean centering translates the origin of the coordinate system to the average in both dimensions (Figure 13.7b). The first principal component (PC) is the line through the centroid of the data producing the largest variance for the perpendicular projections of the data points onto that line. The first PC fits the data best, in the sense of minimizing the sum of squares of perpendicular deviations of the data points from this line. When a horizontal line through the centroid is rotated counterclockwise, the first PC is found at the angle at which the variance of the data projected onto the line is a maximum. The perpendicular projections from the data points to the line with the highest variance are shown in Figure 13.7c. The second PC is captures the next greatest amount of variance in a direction orthogonal (at 90°) to the first PC and is found in the same way, with the restriction of being perpendicular to the first PC (Figure 13.7d). The possible number of PCs is the smaller of the number of samples or variables; in this case, two variables restrict the number of PCs to two (Figure 13.8).

More efficient methods of extracting PCs are based on solving for the eigenvectors and eigenvalues from the covariance matrix (containing variances and covariances of the variables), or from the matrix of correlations between the variables [34, 35]. A robust and efficient algorithm for eigenanalysis is the singular value decomposition [26]. Analysis of mean-centered data is equivalent to eigenanalysis of the covariance matrix. The eigenvalues, sorted

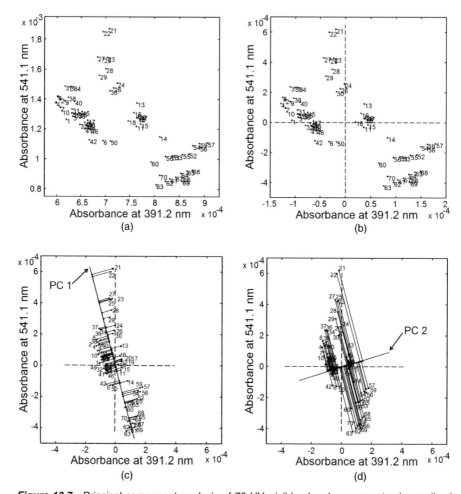

Figure 13.7 *Principal component analysis of 70 UV–visible absorbance spectra (normalized to unit sum) of seven different purple dyed nylon 6,6 fibers, measured at two wavelengths; numbered points refer to the 70 spectra: (a) absorbance; (b) mean centered absorbance; (c) the first PC, maximizing the variance of perpendicular projections onto a line through the centroid, and (d) the second PC, maximizing variance of projections onto a line perpendicular to the first PC.*

by magnitude, are the variances explained by the corresponding PCs, and the eigenvalues sum to the total variance.

Analysis of autoscaled data is equivalent to eigenanalysis of the correlation matrix, and the eigenvalues sum to the number of original variables. Eigenvectors contain the weights (linear contributions) of each variable defining the PCs. Eigenvectors can also be expressed in terms of correlations with the original variables, in which case the weights are called "loadings" [34]. When a PC and a variable are at right angles to one another, the two vectors are not correlated (r is zero). When a PC and a variable are oriented in the same (or

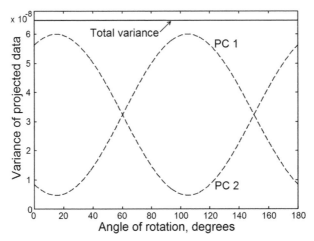

Figure 13.8 *Projected variance versus angle of rotation for two PCs (from Figure 13.7).*

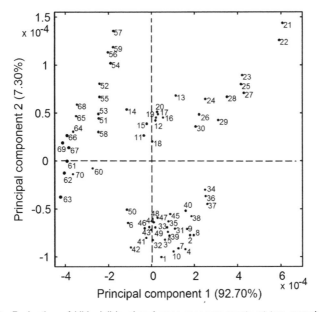

Figure 13.9 *Projection of UV–visible absorbance measurements at two wavelengths (391.2 and 541.1 nm) from seven different purple dyed nylon 6,6 fibers into the space of the first two principal components. Numbered points refer to the 70 spectra.*

opposite) direction, their correlation coefficient is equal to one (or minus one), and the PC reproduces the variation of that variable perfectly. Different computer programs may give eigenvector components with opposite signs due to arbitrary orientation of PCs in one direction, or in 180° from that direction.

Figure 13.9 shows the projections of the data (termed scores) onto the axes of the two PCs derived above. This plot shows the data, with the origin translated, and axes rotated to the orientation of the PCs. The variation of the data in Figure 13.7a is exactly reproduced because a number of PCs equal to the number of original variables has been used. If it had been decided that one PC was sufficient to explain most of the variation (92.70%), the data set could have been represented only by the projections onto the first PC. In that case, the separation of data explained by the second PC (7.30% of the variance) would have been ignored.

PCA can be used to reduce data dimensionality by retaining only PCs that capture a sufficiently large fraction of the variance and by deleting PCs associated with the remaining small proportion of the variance. "Significant" PCs have eigenvalues (variances) representing systematic variation (signal), and not just noise. If just a few PCs are adequate in this sense, the cluster of data points in p-dimensional space lies mostly in a subspace of lower dimensionality. A projection of the data into the lower dimensional space may then provide a useful graphical display of the relationships among samples and of the clustering of samples in groups. The seven groups of replicate spectra are in sets of ten numbered points, and some clustering of the replicates from each group can be recognized in Figure 13.9. Such clustering in PCA scores suggests further investigation of this behavior is warranted.

PCA was conducted on the covariance matrix for the normalized data matrix (70 spectra × 1527 wavelengths) for the seven different purple textile fibers. Various methods have been suggested for deciding how many PCs adequately represent systematic variation [35, 36], but simple inspection works well in this instance. The variance accounted for by the first 20 PCs is shown in Figure 13.10 as plots of percent variance (a "scree" plot) and of cumulative percent variance. The first two and three PCs explain 87.3% and 96.7% of the variance, respectively. Figure 13.11 shows loading plots for correlations between the first three PCs and the original wavelengths. Dashed lines are drawn on these plots at cor-

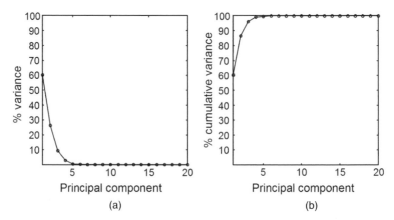

Figure 13.10 *(a) Percent variance and (b) cumulative percent variance as a function of the number of principal components for UV–visible absorbance spectra from seven different purple dyed nylon 6,6 fibers.*

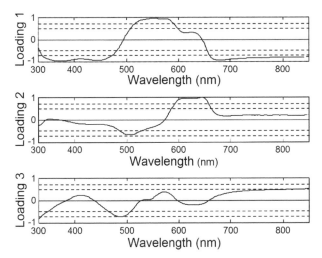

Figure 13.11 *Loading plots of correlations of principal components 1–3 with wavelengths for UV–visible absorbance spectra from seven different purple dyed nylon 6,6 fibers.*

relation coefficients of ±0.5 and ±0.707. After four PCs, correlations with variables do not exceed ±0.5. Projections of data into the space of the first two and three PCs are plotted in Figure 13.12. Grouping of replicate spectra from each of the seven fibers are indicated by 95% confidence ellipses [34].

Figure 13.13 shows the ten replicate spectra for each of the seven groups of replicate spectra for the purple dyed nylon 6,6 fibers. Fibers 1, 2, and 3 differ in spectral features just above 400 nm and around 650 nm and are separated by differences in the second PC (Figure 13.12). The loading plots (Figure 13.11) confirm that the second PC has higher correlations with wavelengths in the 500 nm and the 600–650 nm regions. Confidence ellipses for another trio of fibers (2, 4, and 5) overlap in the two-dimensional scores plot; however, adding the third PC discriminates fiber 2 spectra from those of fibers 4 and 5. The third PC is correlated with changes in the 450–525 nm region (Figure 13.11), where spectra from fiber 2 differ from those of fibers 4 and 5 (Figure 13.13). Confidence ellipses for spectra from fiber groups 4 and 5 overlap and have near identical spectra. Ellipses for fiber groups 6 and 7 are separated from one another and have different spectra (Figure 13.13).

The PCA scores plots (Figure 13.12) are consistent with the known dye formulations applied to these nylon fibers (Table 13.1). Fibers 4 and 5 have the same combination of three dyes; the overlap of spectra from these groups suggests that the dye formulations are quite similar. Fiber 2 is dyed with two of the three dyes with which fibers 4 and 5 are dyed. Fiber 2 is dyed with a red dye that is not present on fibers 4 and 5, and fibers 4 and 5 also have a different red dye not present on fiber 2. Although fibers 6 and 7 are dyed with the same three dyes, the PCA scores plots indicate that spectra of these two fibers can be discriminated. The minor differences in the spectra of these fibers around 600–650 nm suggest that the dye formulations are not identical with

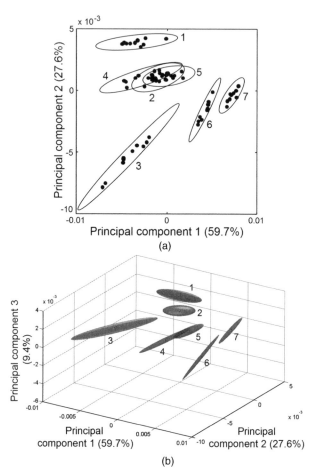

Figure 13.12   *Projection of UV–visible absorbance spectra from seven different purple dyed nylon 6,6 fibers into the space of (a) the first two PCs and (b) the first three PCs. Labels 1–7 and their 95% confidence ellipses identify the seven fibers for which ten replicate spectra were obtained.*

respect to amounts of the dyes. In summary, while some differences in spectra are obvious by visual inspection of Figure 13.13, the PCA scores plots of Figure 13.12 provide visual summaries of the relationships among the seven groups of fiber spectra.

13.4   VISUALIZING GROUP DIFFERENCES BY LINEAR DISCRIMINANT ANALYSIS

Trends that differentiate different groups of samples may not coincide with directions of greatest variability. Linear discriminant analysis (also called canonical variates analysis) constructs a new set of axes that best separates

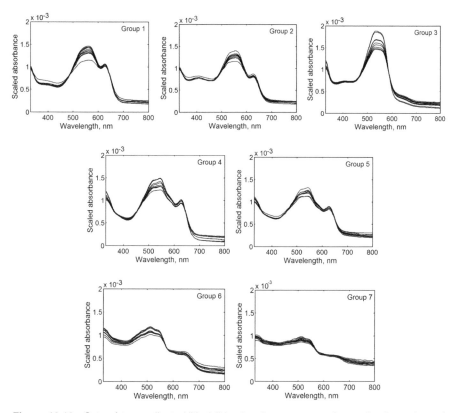

Figure 13.13 *Sets of ten replicate UV–visible absorbance spectra (normalized to unit sum) from seven different purple dyed nylon 6,6 fibers.*

TABLE 13.1 Acid dye formulations for seven purple dyed nylon 6,6 fibers (1–7). Dye names are not specified due to their proprietary nature.

	Dyes					
Fibers	Red 1	Red 2	Blue 1	Blue 2	Yellow 1	Gray 1
1	✓		✓			
2	✓		✓		✓	
3	✓	✓		✓		
4		✓	✓		✓	
5		✓	✓		✓	
6		✓			✓	✓
7		✓			✓	✓

data into groups [37–41]. These discriminant axes (canonical variates, or CVs) are linear combinations of the original features. The criterion maximized by LDA is the ratio of the variance between groups (i.e., separation between groups) divided by the variance within groups (i.e., experimental variability among spectra belonging to the same group). This criterion, first described by Fisher [11], is known as the Fisher ratio. Because knowledge of group membership (or class) for each sample is required, LDA is a supervised technique. PCA, which does not use grouping information, is an unsupervised technique.

LDA finds a reduced space representation that places members of the same group as close as possible and moves all groups as far apart as possible. Except for maximizing a different criterion, LDA is similar to PCA in that data is projected into the space of the first several canonical variates as discriminant scores. The canonical variates are determined by finding the eigenvectors of the matrix $\mathbf{W}^{-1}\mathbf{B}$, a matrix form of the Fisher ratio. The matrix $\mathbf{B}$ is the between-groups sum of squares and cross-products matrix, and $\mathbf{W}^{-1}$ is the inverse of the pooled within-groups sum of squares and cross-products matrix [37–40]. Clustering of similar samples (the within-groups variance) provides an assessment of the significance of distances between samples (the between-groups variance) judged different from one another. The between- to within-groups variances accounted for by each CV are given by the sorted eigenvalues.

The contrast between PCA and LDA is instructive for the previous GC analyses of eight carbohydrates (rhamnose, fucose, ribose, mannose, two aminodideoxyhexoses, muramic acid, and glucosamine) for 43 samples of three species of legionella-like organisms. After autoscaling (Figure 13.5), PCA explained 68.6% and 84.1% of the variation with two and three PCs, respectively. The PCA scores plot on the first two PCs (Figure 13.14a) shows overlap of groups 1 and 3, and the combination of all three PCs (Figure 13.14b) separated the groups from one another. The LDA discriminant scores plot is shown in Figure 13.15. For any data set, the number of canonical variates can never exceed either the number of groups minus one or the number of variables. Thus, three groups of data can, at most, be plotted as a function of only two CVs. Three group centroids define a plane, there are only two nonzero CV eigenvalues, and 100% of the variation between groups lies in this two-dimensional space. Although the PCA scores plot of Figure 13.14 shows the most variance possible to be shown in three dimensions, variance is not discrimination. The LDA scores plot shows better group separation than the PC scores plot. The differences in carbohydrate profiles for three legionella groups reflected in these scores plots led us to suggest that *Fluoribacter* is distinct from the other two legionella species [31]. Recognition of such differences and the ability to identify pathogenic organisms has forensic implications for detection of biological threats.

One difficulty with data of very high dimensionality is that LDA also requires the number of samples (spectra, chromatograms) to be greater than the number of variables (wavelengths, chromatographic peaks); otherwise, the within-groups sum of squares and cross-products matrix cannot be inverted. A common solution to this problem is to reduce the dimensionality of the data prior to LDA. Some applications have satisfied this problem by hand-picking

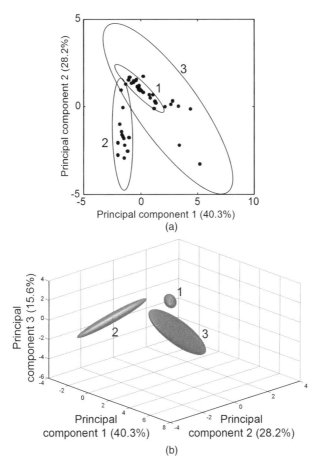

Figure 13.14 *Projection of data for eight carbohydrates measured by gas chromatography in 43 samples of legionella-like organisms into the space of (a) the first two PCs and (b) the first three PCs. Group labels represent replicate data and ellipses are 95% confidence intervals for (1) 20 samples of Legionella pneumophila, (2) 13 samples of Tatlockia micdadei, and (3) 10 samples of Fluoribacter bozemanae.*

a reduced set of variables (e.g., see References 42 and 43). While an automatic procedure for dimensionality reduction is desirable, caution should be exercised to avoid overfitting small data sets with large numbers of variables [44, 45]. PCA is a common choice for reducing the number of variables prior to LDA [27, 46–51].

The data for ten replicate UV–visible absorbance spectra from seven different purple dyed nylon 6,6 fibers (Figure 13.14), with 70 samples and 1527 wavelengths, was normalized to unit sum and subjected to PCA. The reduced matrix produced by projecting the data into the space of the first four PCs

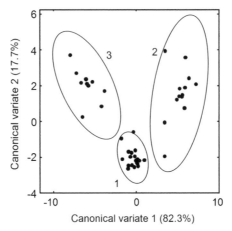

Figure 13.15 *Projection of data into the space of the first two CVs for eight carbohydrates measured by gas chromatography in 43 samples of legionella-like organisms. Group labels designate 95% confidence ellipses for the three groups: (1) 20 samples of Legionella pneumophila, (2) 13 samples of Tatlockia micdadei, and (3) 10 samples of Fluoribacter bozemanae.*

(explaining 99.5% of the variance) was analyzed by LDA (Figure 13.16). The first two CVs account for 90.3%, and the first three CVs account for 98.9% of the between-groups to within-groups variability present in the data using four PCs. Group separation relative to the variability within groups is improved over the PCA outcome in Figure 13.12. Except for groups 4 and 5, which have very similar dye formulations, LDA separates the patterns of each group of fibers. The spectra from the two fibers not dyed with a blue dye (groups 6 and 7) are at the lower right of the plot of projections onto two CVs and at the bottom of the plot onto three CVs (Figure 13.16). The three groups of spectra from fibers dyed with the red 1 dye are on the left side of the plot for two CVs and at the left and top of the plot for three CVs. Clustering of spectra for the fibers dyed with the yellow dye is not apparent in the LDA projections.

As mentioned previously, variables contributing to group separation can be identified by examining components of the LDA eigenvectors. Another way is to plot the Fisher ratio calculated for single variables at a time. Univariate Fisher ratios measure the ability of each variable by itself to discriminate among groups [49–55]. A variable has a Fisher ratio less than one when the between-groups variability is less than the within-groups variability. Larger values indicate variables that better discriminate among groups. Univariate Fisher ratios can also be used to select variables for further analysis [49, 50, 52, 53]. Figure 13.17 shows a univariate Fisher ratio plot for the seven groups of absorbance spectra for purple nylon 6,6 fibers. The regions just above 400 nm and near 600 nm, previously mentioned to discriminate among the fibers, have high univariate Fisher ratios. Identifying differences between group patterns can also lead to an understanding of the chemical basis of dis-

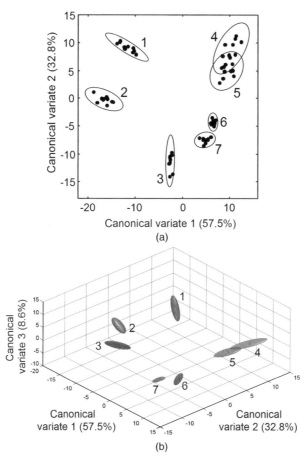

Figure 13.16 Projection of 70 UV–visible absorbance spectra from seven different purple dyed nylon 6,6 fibers into the space of (a) the first two CVs and (b) the first three CVs. Group labels 1–7 identify the seven fibers for which replicate spectra were measured.

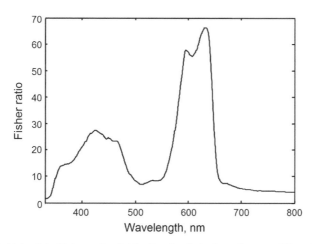

Figure 13.17 Univariate Fisher ratio plot for the discrimination of seven different purple dyed nylon 6,6 fibers based on 70 UV–visible absorbance spectra.

crimination [49–51, 54, 55]. Evaluating classification performance for prediction of groups to which samples belong is discussed in the next section.

13.5 GROUP SEPARATION, CLASSIFICATION ACCURACY, AND OUTLIER DETECTION

Although based on different criteria, PCA and LDA find lower dimensional projections of multidimensional data that facilitate visualization of relationships among groups of samples and identification of the contributions of the original variables (e.g., wavelengths) to separation of groups. Whether these perceived differences are sufficient for reliable classification of unknown samples is another question entirely. Before using PCA or LDA to discriminate among groups or to classify samples, one might consider whether it is worth doing. If the means of the groups of interest were equal, it would not make sense to try to discriminate among samples of the groups.

For univariate data from each of two groups, the significance of the difference between the two group means $\bar{x}_1$ and $\bar{x}_2$, based on n_1 and n_2 measurements, is evaluated by the two-group Student's t-test using the test statistic

$$t = \frac{|\bar{x}_1 - \bar{x}_2|}{s_p \sqrt{(1/n_1 + 1/n_2)}}$$

where s_p is the pooled standard deviation based on the combined groups (assuming equal variances) [32]. This statistic can be viewed as a signal-to-noise ratio: distance between means (numerator), expressed in units of the standard deviation of the difference (denominator). If the calculated value of t exceeds the value that could be expected to occur by chance, then the null hypothesis that the population means of the two groups are equal is rejected at the stated level of confidence, and the alternative hypothesis that the means are different (a two-sided alternative) is accepted. When more than two groups are compared, one-way analysis of variance (ANOVA) tests the significance of differences in the means by examining the ratio of among-groups to within-groups variability [56].

The multivariate generalization of the univariate Student's t-test is Hotelling's T^2 test for the equivalence of two multivariate means [57], for which the test statistic is defined as

$$T^2 = \frac{n_1 n_2}{n_1 + n_2} \mathbf{d}' \mathbf{C}_p^{-1} \mathbf{d}$$

where n_1 and n_2 are the number of samples in the two groups, $\mathbf{d}$ is the difference vector of centroids, $\mathbf{C}_p^{-1}$ is the inverse of the pooled within-groups covariance matrix, and p is the number of dimensions in which the samples are measured [37–39]. Hotelling's T^2 is the *squared* "between-to-within" variability ratio: the squared distance between the means (numerator) divided by the covariance matrix (denominator) containing variances and covariances of the variables. Although the above calculation involves matrices, the result is a single number. For univariate data, Hotelling's T^2 reduces to the square of the Student's t statistic. An assumption of Hotelling's T^2 test is that within-group sample covariance matrices are equal and can be pooled, which is similar to the assumption of equal variance in the two-group Student's t-test. Modest deviations from equality are usually not serious, but if this assumption is not valid, other tests are available [39]. In analogy with the F test for univariate data [32], test statistics are available for testing the equality of two or more covariance matrices [37–39]. Within the framework of multivariate analysis of variance (MANOVA), several tests (Pillai, Lawley–Hotelling, Wilk's lambda, and Roy's largest root) for the significance of differences among several multivariate means exist, differing slightly in interpretation but all related to Hotelling's T^2 statistic [38, 39]. One or more of these tests are often provided by discriminant analysis software.

Classification of an unknown sample as a member of a particular group requires a quantitative assessment of pattern similarity. Treating each pattern as a point in p-dimensional space, the simplest measure of similarity (or proximity) between two patterns is the Euclidean distance based on the Pythagorean theorem. Figure 13.18a illustrates Euclidean distances from a centrally located data point to four group means in a two-dimensional space. Euclidean distances between points will be small when patterns are similar and larger when patterns are dissimilar. The shortest Euclidean distance from the center point is to the group mean directly above.

Analogous to a standardized z-score for univariate data [32], the Mahalanobis distance for multivariate data [58] is calculated from the centroid of a specified group of samples. The Mahalanobis distance, D^2, is a covariance adjusted distance measure (a single number) that takes into account both the variances of the individual variables and their correlations with one another:

$$D^2 = \mathbf{d}'\mathbf{C}^{-1}\mathbf{d}$$

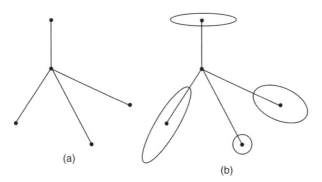

(a)

(b)

Figure 13.18 Configuration of five points illustrating (a) Euclidean distances from the center point and (b) confidence ellipses for Mahalanobis distances at equal probabilities.

The vector **d** is the vector of differences at each variable (e.g., wavelength) between a point and a group mean, and C^{-1} is the inverse of the covariance matrix for the group to which the distance is calculated [37–39, 59]. The sampling distribution of the squared Mahalanobis distance D^2 and of the Hotelling's T^2 statistic are both related to the F-distribution, and critical values of T^2 and D^2 at selected probability and degrees of freedom can be specified. To calculate covariance matrices, the number of feature dimensions employed (e.g., PCs retained for LDA) must not be larger than the smallest number of samples in any group minus two. Mahalanobis distances can be employed to plot ellipses (in two dimensions) or ellipsoids (in three dimensions) around each group cluster for a chosen level of confidence. Ellipses in previous scores plots were drawn at 95% probability. Figure 13.18b shows ellipses drawn around the group means at equivalent Mahalanobis distances. If the group covariance matrices are not the same, the ellipses tilt in different directions. The smallest Mahalanobis distance from the center sample to any of the four group centroids is to the group at the lower left. Because the confidence ellipse for the group at the top is not tilted, it would be unusual (of low probability) for a sample from that group to be located at the position of the center sample. The orientation of the ellipse at the lower left makes the sample in the center more likely to have originated from that cluster. When the probability of group membership for a sample is below a predefined acceptable threshold, the sample could be classified as "in-doubt" [39]. Soft independent modeling of class analogy (SIMCA), a technique introduced by Wold [60], goes further in developing classification models based on principal component analysis of each class separately.

Predicting group membership of samples (classification) can be done in several ways, depending on the selection of a similarity measure. Most common is the Mahalanobis distance, based either on a pooled covariance matrix from all groups or on covariance matrices estimated for each group separately. A sample is classified as a member of the group for which the Mahalanobis distance from the sample to the group centroid is smallest. Higher classification accuracy and lower error implies greater discrimination between the groups of samples. A variety of methods for estimating the accuracy of classification (and its complement, misclassification rate) have been proposed [61].

The simplest way to estimate classification accuracy is the resubstitution method. The complete data set is employed as a training set to develop a classification procedure based on the known class membership of each sample, and the class membership of every sample in the data set is predicted. For example, after preprocessing, projections of the samples into a PCA or LDA reduced-dimensional space are used to classify each sample into the nearest group using Mahalanobis distance. However, because resubstitution uses the same data to build the classification model and to evaluate its accuracy, classification accuracy is typically overestimated. When the classification model is applied to new data, the error rate would be expected to be worse than predicted by this overly optimistic method.

A second approach partitions the available data into two portions: a training set for development of the classification model, and a test set for prediction of classification. By separating the data used to build the classification model from the data used to evaluate its performance, the estimate of error is unbiased. The data could be split in various ways, such as 80% training and 20% testing. Although using independent training and test data is best, that approach is not optimal in data-limited situations because the classification model is not based on all available data. Obtaining enough data to have reasonably sized training and test data sets may be demanding of time or resources.

A third approach, leave-one-out cross-validation [62], provides a nearly unbiased estimate of classification accuracy. A sample is temporarily deleted from the data set, a classifier is built from the training set of remaining observations, and the model is used to predict the group membership of the deleted sample (e.g., classify the left-out sample into the nearest group). This process iterates through all n samples, treating each available sample as an unknown to be classified using the remaining $n - 1$ samples. Cross-validation efficiently uses all the available data for estimating classification accuracy and, although it is time consuming, is usually preferred. It is also possible to leave out more than a single sample at a time. For example, the classification accuracy might be reported as the average of multiple iterations of randomly omitting 20% of the samples for classification on the basis of the remaining 80%.

Projections into the space of the first four PCs for the absorbance spectra for the purple nylon 6,6 fibers accounted for 99.46% of the variance. These scores were subjected to LDA and classification using Mahalanobis distances, resulting in a leave-one-out classification accuracy of 94.29% (66/70). Two spectra from group 4 were misclassified in group 5, and two spectra of group 5 were mistakenly classified in group 4. This 20% error rate for groups 4 and 5 is not surprising given the overlap between these two groups in Figure 13.16. How much better is this outcome than could be expected by chance? One way to answer this question is to calculate the proportional chance criterion [39]. With seven groups of ten replicate spectra, only 10/70 or 14.3% of samples would be expected to be correctly classified by random chance. Another measure, the maximum chance classification, is the expected classification accuracy if all samples are classified as members of the largest group; for equal group sizes, proportional and maximum chance classifications are the same.

Another way to assess objectively whether classification results are statistically better than could be obtained by chance is to employ computer-intensive nonparametric methods based on resampling (bootstrapping) [63, 64]. Resampling involves sampling with replacement from the original data matrix to build test data sets of the same, or larger, size. The process of resampling and classifying is repeated many times, and the distribution of classification results is used to generate confidence intervals. Another approach uses randomization of the data matrix or of the group assignments of samples [65]. Every sample from the original data matrix is present in this resampled data, except that the connection between the samples and the group assignments

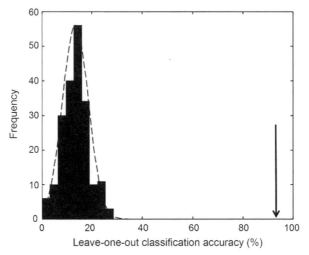

Figure 13.19 *Randomization test of LDA/Mahalanobis distance classification for UV–visible absorbance spectra of seven purple dyed fibers. The leave-one-out classification accuracy of 98.57% achieved with the original data represents a z-score of 19.2.*

has been removed. Randomization and classification is repeated many times and the statistical significance of the classification achieved on the original data matrix is judged by reference to the resulting "null" distribution for which no group structure exists. Figure 13.19 shows randomization testing results on LDA/Mahalanobis distance classification for the UV–visible absorbance spectra of the seven purple fibers. The average and standard deviation of the leave-one-out classification accuracy from 200 iterations was 13.37% and 5.24%, respectively. The 94.29% accuracy cited previously (marked by the vertical arrow in Figure 13.19) represents a z-score of 15.44. Because of this highly improbable outcome, the null hypothesis that no real grouping relationships exist in the data can be rejected.

In addition to providing recognition that a particular measurement might be unrepresentative, outlier testing [66] is often used to inspect data routinely for reliability and to identify the need to bring a measurement process into a better state of quality control [67]. Hotelling's T^2 values or Mahalanobis distances measured for single samples from the center of a group of samples can be employed diagnostically to recognize unusual samples (outliers) that deviate from the behavior of the remaining samples [68]. Figure 13.20 demonstrates the use of Hotelling's T^2 to detect outliers (samples 59 and 60) in a set of 80 UV–visible spectra. One difficulty with use of either D^2 or T^2 values for outlier detection is that multiple outliers can distort the covariance matrix and cause outliers to appear as if they belong. Outlier detection based on the distribution of sample vector lengths upon repeated sampling with replacement has been used in a forensic application involving reflectance absorbance

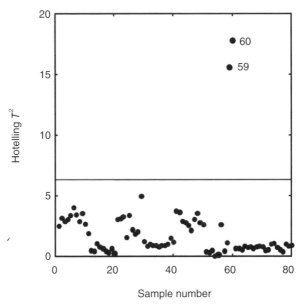

Figure 13.20 *Hotelling's T² values calculated for the projections into the space of the first two PCs for a group of 80 UV–visible absorbance spectra from eight black cotton fibers. The critical value of Hotelling's T² at the 95% level of confidence is drawn as a horizontal line at 6.31.*

IR spectra of copy toners [33]. A variety of robust methods for multivariate outliers have recently been reviewed [69].

13.6   SELECTED APPLICATIONS

Accelerants   Tan et al. [70] summed peak areas within 19 regions in GC/MS chromatograms of gasoline and analyzed the data by PCA with Mahalanobis distance classification. Perfect classification of test samples and of simulated fire debris was achieved. Johnson and Synovec [71] applied PCA to the classification of two-dimensional GC patterns of jet fuels. Univariate LDA was employed to select informative variables for analysis. Doble et al. [72] employed PCA and Mahalanobis distances to classify gasoline into regular or premium grade groups on the basis of peak areas of 44 compounds. An accuracy of 80–93% was achieved based on random splitting of the 88 available samples into ten equal-sized training and test data sets. Sandercock and Du Pasquier [73] used solid phase microextraction GC/MS to analyze 35 samples of gasoline of various grades. Selected ion monitoring produced a data set of peak areas for target analytes which was normalized and log-transformed. Projections into the first three PCs, followed by LDA, produced 96% (168/175 correct) classification into 32 different groups. The same researchers [74] also took integrated peak areas from GC/MS chromatograms of 160 samples and classified all samples correctly into 18 groups by LDA of the projections into the space of the first three PCs. Time and evaporation effects in the data were also modeled. Petroleum biomarkers determined by GC/MS have been subjected to PCA for forensic oil spill identification [75]. Principal component analysis (PCA) was also used to discriminate between evaporated samples of kerosene and diesel fuel using peak areas of 14 components. The significance of observed differences was evaluated using a likelihood ratio [76, 77].

Document Examination and Analysis of Currency   Hida et al. [78] performed elemental analysis of 277 counterfeit coins by X-ray diffraction and X-ray fluorescence. PCA on data for five elements was used descriptively to show separation in groups. PCA, LDA, and SIMCA were applied by Kher et al. [79] for classification of document papers based on their attenuated total reflectance (ATR) and diffuse reflectance IR spectra. When PCA scores were used as inputs to LDA, 68% of all possible pairs of paper could be discriminated. Kher et al. [80] also analyzed inks from black and blue ballpoint pens by high performance liquid chromatography (HPLC) with UV–visible detection. PCA of retention times and absorbances at four wavelengths discriminated different ink samples with cross-validation. Hida and Mitsui [81] performed elemental analysis by X-ray fluorescence on 200 prepaid subway cards. Samples separated into four groups by PCA using data for five elements. Calcium and barium levels were correlated with the first PC; titanium and chlorine levels were associated with changes in the second PC. Another set of 32 turnpike cards could be separated into counterfeit and real cards on the basis of projections into the space of two PCs.

Bartick et al. [82] investigated Fourier transform infrared (FTIR) microscopy of copy toners and employed canonical variates analysis for visualization of discrimination among samples from different manufacturers. Egan et al. [49] analyzed copy toner samples using reflection–absorption infrared microscopy. For 430 copy toners, 90% were correctly grouped into groups previously established by spectral matching. For two pairs of groups that were often misclassified, univariate Fisher ratio plots revealed discriminating features enabling 96% correct LDA classification. Egan et al. [50] found PCA and cluster analysis of scanning electron microscopy (SEM) data for 166 copy toner samples to discriminate 13 statistically different subgroups, with the presence or absence of a ferrite base being a major division (Figure 13.21). Using both SEM and IR data, 41% of toner samples could be assigned to specific manufacturers. With combined pyrolysis GC/MS and SEM data, 54% of samples could be differentiated. The synergy of the complementary information provided by the three analytical techniques employed narrowed matching possibilities.

Thanasoulias et al. [83] achieved 100% classification (on resubstitution) using LDA on the projections into the space of the first three PCs for 50 UV–visible spectra of blue ball point pens. Kher et al. [84] found PCA valuable for forensic classification of paper types using infrared spectroscopy. HPLC

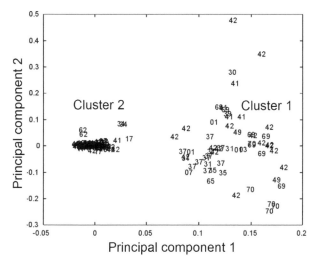

Figure 13.21 *Projections of amounts of 10 elements measured by SEM for 166 samples into the space of the first two principal components (adapted from Reference 50). Cluster 1 has low iron concentrations and no manganese. Cluster 2 contains much higher iron and manganese abundances. The first two PCs account for 56% of the variation in the data.*

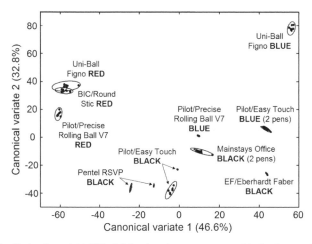

Figure 13.22 *Projection of 14 UV–visible absorbance spectra of ballpoint pen inks (ten replicate spectra each) into the space of the first two PCs [86].*

and ATR IR spectroscopy data were employed by Kher et al. [85] to characterize blue ballpoint pen inks. The data was analyzed by PCA and classified using distances from group centroids. Cross-validated classification accuracies were 62.5% and 97.9% for IR and HPLC, respectively, although the two techniques were found to complement one another. A recent unpublished study of ballpoint inks [86] used UV–visible microspectrophotometry and LDA to characterize similarities among 14 different inks (Figure 13.22). The cross-validated classification accuracy based on Mahalanobis distances to group means was 89%. All misclassifications were between pens of the same type and color (e.g., within the two Mainstays black and the two Pilot blue pens) or between pens of very similar colors (Uni-ball and BIC red pens).

Drug Analysis Jonson[87] reported using PCA for classification of amphetamine samples based on impurities detected by GC. Canonical variates analysis (LDA) was used for statistical comparisons of impurities in methyl amphetamine by Perkal et al. [88]. The country of origin in seized samples of heroin was modeled using LDA of measured concentrations of selected alkaloids and adulterants [89]. Of 505 samples from Turkey, Pakistan, India, and Southeast Asia, 83% were successfully classified. Janhunen and Cole [90] measured opiate, noscapine, and papaverine content by GC in 31 street samples of heroin. Chromatograms could be separated into eight groups by eye and LDA discriminated 91.9% of the samples, including pairs that were not discriminated visually. FTIR and GC/MS data for illicit amphetamines are compared by Praisler et al. [91] using PCA and SIMCA [60] for the purpose of drug screening. Krawczyk and Parczewski [92] selected 15 gas chromatographic peaks and performed PCA on peak area data to identify amphetamine sources. Correlations and Euclidean distances between samples were used to classify samples. Klemenc [93] applied several multivariate techniques, including PCA, to normalized GC/MS data and classified samples into three groups with 95–100% accuracy using the average of the three nearest neighbor distances. PCA, discriminant analysis, and other multivariate techniques were also discussed in the context of heroin profiling by Dams et al. [94]. Ryder [95] used PCA on Raman spectra to classify solid mixtures of narcotics. Waddell et al. [96] applied PCA to the analysis of data from 14 trace metals found in 99 ecstasy tablets. Four PCs accounted for 74% of the variance in the data and indicated potential groupings of samples from different seizures. Daéid and Waddell [97] reviewed chemometric procedures for profiling illicit drug seizures, Leger and Ryder [98] treated Raman spectra of 85 narcotics samples with polynomial modeling for baseline correction. The data was then analyzed by PCA for discrimination and PC scores were used for quantitation. Cocaine and MDMA could be determined with cross-validated errors of quantitation of several percent.

Fibers Morgan et al. [99] evaluated the discriminating power of UV–visible and fluorescence microspectrophotometry using randomly selected fibers of two colors (yellow and red) for four fiber types (cotton, acrylic, nylon, polyester). PCA and LDA were used to evaluate discrimination between similar fibers of various textile fiber types. The results suggest that, for dye formulations that fluoresce, fluorescence can be more discriminating than absorbance. Enlow et al. [51] used LDA to classify 270 ATR FTIR spectra from eight different nylon subclasses with 98.5% cross-validated accuracy. Figure 13.23 shows the average of 80 nylon 6,6 and 100 nylon 6 spectra, along with a plot

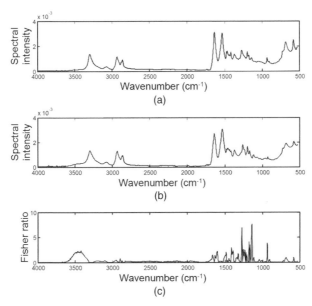

Figure 13.23 (a) Average nylon 6,6 spectrum, (b) average nylon 6 spectrum, and (c) single wavelength Fisher ratio plot. (Adapted from Reference 51.)

of single wavelength Fisher ratios. Significant differences between the nylon 6,6 and nylon 6 spectra appear at 1140 and 1273 cm^{-1}. The spectrum of nylon 6 has a valley at 1140 cm^{-1} and a peak at 1261 cm^{-1}. These differences are hard to spot by eye.

PCA of peaks derived from pyrolysis GC analysis of lignin was found to discriminate samples from different species of *Eucalyptus* trees [100]. Causin et al. [101] used PCA on integrated peak areas for eight peaks in pyrolysis GC/MS programs and examined the eigenvectors of the first three PCs to identify informative peaks differentiating acrylic fibers. ATR FTIR spectroscopy was employed for analysis of fibers including blends of polyester with cotton, lycra, wool, and rayon [102]. LDA and Mahalanobis distances were used in combination with library matching to standard spectra to confirm identity of blends with 98% classification accuracy. Raman microspectroscopy was employed by Clelland et al. [27] to characterize and discriminate among aliphatic, aromatic, and mixed aliphatic/aromatic undyed polyamide fibers. After baseline subtraction and normalization, projections of the spectra into the space of the first six PCs were analyzed by LDA. Spectral differences between most fiber subclasses were visually distinguishable and statistically significant (Figure 13.24). As expected, spectra of nylon 12 and Vestamid® (a trade name for nylon 12) were not distinguishable by LDA; similarly, spectra of Nomex® (a DuPont polymer) and Conex® (made by Teijin, in Osaka, Japan under license from DuPont) were also not discriminated.

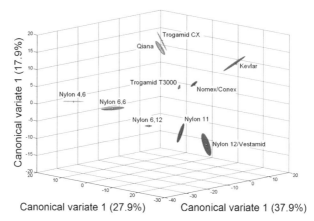

Figure 13.24 *Projections of Raman spectra from ten different polyamide and aramid subclasses into the space of the first three canonical variates, with 95% confidence ellipses for each group. (Adapted from Reference 28.)*

Glass Koons et al. [103] measured nine elements in 184 samples of sheet and container glasses by inductively coupled plasma (ICP) atomic emission spectrometry. Projections into the space of the first two PCs correctly classified all but two of the samples; the two samples classified incorrectly were bottles from the same distiller. Hicks et al. [104] compared 119 glass samples from crime scenes using refractive index (RI) measurements. Of the 16,653 possible pairs, only 223 pairs were not differentiated. Samples that could not be discriminated by RI were analyzed using nondestructive energy dispersive X-ray microfluorescence. LDA of this data produced classification error rates of 5–8%. Hotelling's T^2 test was also used to evaluate discrimination. Aeschliman et al. [105] reported that PCA provided objective and routine comparisons of forensic samples (glass, brass, steel) analyzed by laser ablation ICP MS. Bajic et al. [106] employed laser ablation ICP MS to differentiate glass samples by their trace elemental composition. Standard residential window and tempered glass samples that could not be differentiated by refractive index or density were discriminated from one another using PCA of the resulting mass spectra.

Other Trace Evidence Cohen et al. [107] employed neutron activation analysis to compare the antimony, arsenic, copper, and manganese content of ammunition from two manufacturers with fragments found at a crime scene. When subjected to PCA, the known and questioned ammunition samples formed separate clusters, indicating that the questioned fragments did not match any of the known ammunition samples. The number of elements to measure for forensic analysis of bullet lead was investigated by comparing the variance explained by different numbers of elements used in PCA of elemental composition data [108]. Mahoney et al. [109] used PCA to reduce the dimensionality of complex mass spectra of gunpowder and to visualize differences between samples of black and smokeless powders.

Kochanowski and Morgan [110] conducted a pyrolysis GC/MS investigation of 100 automobile paint samples of five different colors. PCA and canonical variates analysis were used to visualize clustering and to validate comparisons among different pyrograms. The LDA scores plot (Figure 13.25) from this work shows tight within-group variability and distinct separation between 20 groups of three replicate paint pyrograms. Gresham et al. [111] reported on characterization of nail polishes and paint surfaces by static secondary ion MS. Although dominant ions specific to coating were often observed, PCA was found to be useful in handing the multivariate nature of the data and in visualize differences in spectra. Another multivariate technique, partial least squares [112], was also found to be useful for identification of coatings.

Thanasoulias et al. [113] acquired UV–visible spectra of the acid fraction of humus from 44 soil samples from five different areas. After selecting 20 variables, PCA was used to detect and remove four outliers. Nine PCs were retained for LDA classification, which produce 85% accuracy. Chazottes et al. [114] analyzed the particle size distribution for 24 soil samples from simulated crime scenes. PCA, followed by a distance-based cluster analysis, was found to discriminate between two different geological sources.

Burger et al. [115] preprocessed capillary electropherograms of condom lubricants using PCA. The scores on five PCs were classified by LDA with Mahalanobis distances to achieve 233/263 correct classifications. Multivariate analysis has been widely applied in forensic anthropology on anthropometric and biometric measurements. One forensic analytical application by Brooks et al. [116] reported data for 21 elements in cremated remains measured by ICP optical emission spectrometry. PCA loading plots identified seven elements responsible for grouping of sample clusters. LDA was then applied

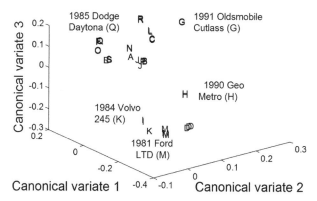

Figure 13.25 *Projections of 20 groups of white automotive paint pyrograms (denoted by alphabetic letters) into the space of the first three orthogonal canonical variates. (Adapted from Reference 110.)*

to the seven-element data set to classify questionable human remains as legitimate or contaminated. Widjaja and Seah [117] employed PCA and LDA for successful differentiation of Raman spectra of fingernails and toenails.

13.7 CONCLUSION

The forensic applications documented in this chapter illustrate the contributions that multivariate statistics and computer-assisted data interpretation can make to forensic decision-making. Multivariate techniques provide a greater ability to discriminate between groups compared to visual analysis, especially in large sample sets. Judging the significance of differences between questioned and known trace evidence items based on comparisons of single spectra or chromatograms can be somewhat subjective. Identifying discriminating features by visual inspection among more than just a few such samples can be time consuming and difficult. Multivariate analysis is performed using all available data, as opposed to spectral library matching, which might compare a single spectrum at a time to library spectra. Use of entire spectral patterns in comparisons may be daunting given their complexity. Multivariate methods utilize all information in the data set, while retaining sensitivity to minor, but discriminating, features.

Principal component and linear discriminant analysis can be employed as exploratory tools to visualize patterns in multivariate data. PCA and LDA do not create new information from existing data: they merely reexpress the information that is already present in the data, but hard to detect. Both techniques can be used to generate visually interpretable maps displaying the relationships among different groups of samples. PCA reduces the dimensionality of multivariate data by preserving the maximum amount of variability in as few dimensions as possible. On the other hand, LDA creates a reduced dimensionality map of the data that displays the best discrimination between groups (measured by the ratio of between- to within-groups variability). Because the resulting principal components or canonical variates are linear combinations of the original variables, differences between samples in these reduced dimensions can be associated with the responsible variables. With methods for assessing multivariate similarity of patterns, group membership of unknown or questioned samples can be predicted based on PCA or LDA projections. Classification accuracy and statistical measures of confidence in classification decisions can be derived by cross-validation or resampling. Hypothesis testing tools for assessing the reliability of discrimination and strategies for multivariate outlier detection are available.

Acquisition of multiple replicate analyses (spectra, chromatograms, etc.) from each individual evidence object is a prerequisite for obtaining valid conclusions from data. While failing to show a difference provides support for the null hypothesis of no difference between two groups of samples, this outcome does not conclusively establish equality of the two sets of samples. However, the dissimilarity of replicate samples establishes the extent of variability attributed to experimental uncertainty. With this information, the forensic analyst can evaluate both the statistical significance of large differences and nonsignificance of trivial differences.

The information provided by multivariate statistics, in the analysis of the reproducibility of replicate patterns and in the determination of statistical significance for observed differences of multivariate patterns, has the potential to provide support for issues raised by the *Daubert* [4] requirements of reliability and relevance for scientific evidence. Whether observed differences are significant or not in a *practical* sense cannot be answered solely by statistics. The measured characteristics of the evidence items at hand must be compared with the range of possible variations for that material and other materials that could be confused with that material. The distribution of measured characteristics across pertinent sample populations also usually depends on location, time, and environmental exposure. The validity of any assumptions and the uncertainties in population distributions might outweigh considerations given to the multivariate statistics utilized to establish groupings. Interpreting differences in patterns in terms of chemical or physical phenomena that are relevant to evidential value ultimately requires profound knowledge of the nature of the evidence, its physical and chemical nature, and, most importantly, its history.

ACKNOWLEDGMENTS

Support from the Federal Bureau of Investigation through contract J-FBI-02-132 is acknowledged. The assistance at the University of South Carolina of Alison M. Bush, Sparkle T. Ellison, Suzanna H. Hall, James E. Hendrix, Alexander A. Nieuwland, Amy R. Stefan, Heather M. Taylor, Allyson A. Wells, and Jennifer J. Yiu with this work is also appreciated.

REFERENCES

1. Foster KR, Huber W (1997). *Judging Science: Scientific Knowledge and the Federal Courts.* The MIT Press, Cambridge, MA.
2. *Frye v. United States* (1923). *54* App. D.C., 293 F. 1013.
3. *People v. Young* (1986). 425 Mich. 470; 391 N.W. 2d, 270, Supreme Ct. Mich.
4. *Daubert v. Merrill Dow Pharmaceuticals, Inc.* (1993). 509 U.S. 579, 113 S. Ct., 2786, 2796.
5. Federal Rules of Evidence (31 December 2004). http://judiciary.house.gov/media/pdfs/printers/108th/evid2004.pdf. Accessed: 20 July 2006.
6. *Kumho Tire Co. v. Carmichael* (1999). 526 U.S. 137.
7. Pearson K (1901). On lines and planes of closest fit to systems of points in space. *Philosophical Magazine* **2**(6):559–572.

8. Hotelling H (1933). Analysis of a complex of statistical variables into principal components. *Journal of Educational Psychology* **10**:69–79.

9. Mahalanobis PC (1936). On the generalized distance in statistics. *Proceedings of the National Institute of Science, Calcutta* **12**:49–55.

10. Mahalanobis PC (1948). Historical note on the D^2-statistic. *Sankhya* **9**:237.

11. Fisher RA (1936). The use of multiple measurements in taxonomic problems. *Annals of Eugenics* **7**:179–188.

12. Finkelstein MO, Levin B (1990). *Statistics for Lawyers.* Springer-Verlag, New York.

13. Aitken CGG, Stoney DA (1991). *The Use of Statistics in Forensic Science.* Ellis Horwood, New York.

14. Kaye DH, Freedman DA (1994). Reference guide on statistics. In: *Reference Manual on Scientific Evidence.* West Publishing Co., St. Paul, MN, pp. 331–414.

15. Aitken CGG (1995). *Statistics and the Evaluation of Evidence for Forensic Scientists.* John Wiley & Sons, Hoboken, NJ.

16. Gastwirth JL (2000). *Statistical Science in the Courtroom.* Springer-Verlag, New York.

17. Goode PI (2001). *Applying Statistics in the Courtroom: A New Approach for Attorneys and Expert Witnesses.* Chapman & Hall/CRC Press LLC, Boca Raton, FL.

18. Lucy D (2005). *Introduction to Statistics for Forensic Scientists.* John Wiley & Sons, Hoboken, NJ.

19. Box GEP, Hunter JS, Hunter WG (2005). *Statistics for Experimenters: Design, Innovation, and Discovery*, 2nd ed. John Wiley & Sons, Hoboken, NJ.

20. Deming SN, Morgan SL (1993). *Experimental Design: A Chemometric Approach.* Elsevier Science Publishers B.V., Amsterdam, The Netherlands.

21. Beebe KR, Pell RJ, Seasholtz MB (1998). *Chemometrics: A Practical Guide.* John Wiley & Sons, Hoboken, NJ.

22. Gemperline P (2006). *Practical Guide to Chemometrics*, 2nd ed. John Wiley & Sons, Hoboken, NJ.

23. Savitzky A, Golay MJE (1964). Smoothing and differentiation of data by simplified least squares procedures. *Analytical Chemistry* **36**(8):1627–1639.

24. Press WH, Teukolosky SA, Vetterling WT, Flannery BP (2002). *Numerical Recipes in C: The Art of Scientific Computing*, 2nd ed. Cambridge University Press, New York.

25. Marchand P, Marmet L (1983). Binomial smoothing filter: a way to avoid some pitfalls of least squares polynomial smoothing. *Review of Scientific Instruments* **54**(8):1034–1041.

26. Hamming RW (1989). *Digital Filters*, 3rd ed. Dover Publications, Mineola, NY.

27. Clelland BL, Vasser BJ, Hendrix JE, Angel SM, Bartick EG, Morgan SL (2007). Discrimination of aliphatic and aromatic polyamide fibers using Raman microspectroscopy and multivariate statistics. *Applied Spectroscopy*, in press.

28. Brennan JF, Wang Y, Dasari RR, Feld MS (1997). Near-infrared Raman spectrometer systems for human tissue studies. *Applied Spectroscopy* **51**(2):201–208.

29. Johansson E, Wold S, Sjödin K (1984). Minimizing effects of closure on analytical data. *Analytical Chemistry* **56**(9):1685–1688.

30. Sahota RS, Morgan SL (1992). Recognition of chemical markers in chromatographic data by an individual feature reliability approach to classification. *Analytical Chemistry* **64**(20):2383–2392.

31. Fox A, Lau PY, Brown A, Morgan SL, Zhu ZT, Lema M, Walla MD (1984). Carbohydrate profiling of some Legionellaceae by capillary gas chromatography–mass spectrometry. In: *Proceedings of the Second International Symposium on Legionella*, Thornsberry C, Ballows A, Feeley JC, Jakubowski W (eds.). American Society of Microbiology, pp. 71–73.

32. Moore DS, McCabe GP (2005). *Introduction to the Practice of Statistics*, 5th ed. W. H. Freeman & Co., New York.

33. Egan WJ, Morgan SL (1998). Outlier detection in multivariate analytical chemical data. *Analytical Chemistry* **70**(11):2372–2379.

34. Jackson JE (1991). *A User's Guide to Principal Components*. John Wiley & Sons, Hoboken, NJ.

35. Jolliffe IT (2002). *Principal Component Analysis*, 2nd ed. Springer-Verlag, New York.

36. Malinowski ER (2002). *Factor Analysis in Chemistry*, 3rd ed. John Wiley & Sons, Hoboken, NJ.

37. Krzanowski WJ (2000). *Principles of Multivariate Analysis: A User's Perspective*, rev. ed. Oxford University Press, New York.

38. Rencher AC (2002). *Methods of Multivariate Analysis*, 2nd ed. John Wiley & Sons, Hoboken, NJ.

39. Huberty CJ, Olejnik S (2006). *Applied MANOVA and Discriminant Analysis*, 2nd ed. John Wiley & Sons, Hoboken, NJ.

40. Campbell NA, Atchley WR (1981). The geometry of canonical variate analysis. *Systematic Zoology* **30**(3):268–280.

41. Rencher AC (1992). Interpretation of canonical discriminant functions, canonical variates, and principal components. *American Statistician* **46**(30):217–225.

42. Kawahara FK, Santner JF, Julian EC (1974). Characterization of heavy residual fuel oils and asphalts by infrared spectrophotometry using statistical discriminant function analysis. *Analytical Chemistry* **46**(2):266–273.

43. Smith CS, Morgan SL, Parks CD, Fox A (1990). Discrimination and clustering of streptococci by pyrolysis gas chromatography/mass spectrometry and multivariate data analysis. *Journal of Analytical and Applied Pyrolysis* **18**(2):97–115.

44. Kwan WO, Kowalski BR (1978). Classification of wines by applying pattern recognition to chemical composition data. *Journal of Food Science* **43**:1320–1323.

45. Rencher AC (1992). Bias in apparent classification rates in stepwise discriminant analysis. *Communications in Statistics—Series B, Simulation and Computation* **21**(2):373–389.

46. Kemsley EK (1996). Discriminant analysis of high-dimensional data: a comparison of principal components analysis and partial least squares data reduction methods. *Chemometrics and Intelligent Laboratory Systems* **33**(1):47–61.

47. Mallet Y, Coomans D, de Vel O (1996). Recent developments in discriminant analysis on high-dimensional spectral data. *Chemometric Intelligent Laboratory Systems* **35**(2):157–173.

48. Indahl UG, Sahni NS, Kirkhus B, Naes T (1996). Multivariate strategies for classification based on NIR-spectra—with application to mayonnaise. *Chemometric Intelligent Laboratory Systems* **49**(1):19–31.

49. Egan WJ, Morgan SL, Bartick EG, Merrill RA, Taylor HJ (2003). Forensic discrimination of photocopy and printer toners. II. Discriminant analysis applied to infrared reflection–absorption spectroscopy. *Analytical and Bioanalytical Chemistry* **376**(8):1279–1285.

50. Egan WJ, Galipo RC, Kochanowski BK, Morgan SL, Bartick EG, Miller ML, Ward DC, Mothershead RF II(2003). Forensic discrimination of photocopy and printer toners. III. Multivariate statistics applied to scanning electron microscopy and pyrolysis gas chromatography/mass spectrometry. *Analytical and Bioanalytical Chemistry* **376**(8):1286–1297.

51. Enlow EM, Kennedy JL, Nieuwland AA, Hendrix JE, Morgan SL (2005). Discrimination of nylon polymers using attenuated total reflection Fourier-transform infrared spectroscopy and multivariate statistical techniques. *Applied Spectroscopy* **59**(8):986–992.

52. Harper AM, Duewer DL, Kowalski BR (1977). ARTHUR and experimental data analysis: the heuristic use of a polyalgorithm. In: *Chemometrics: Theory and Application*, Kowalski BR (ed.). John Wiley & Sons, Hoboken, NJ.

53. Sharaf MA, Illman DL, Kowalski BR (1986). *Chemometrics.* John Wiley & Sons, Hoboken, NJ.

54. Sahota RS, Morgan SL (1992). Recognition of chemical markers in chromatographic data by an individual feature reliability approach to classification. *Analytical Chemistry* **64**(20):2383–2392.

55. Pierce KM, Hoggard JC, Hope JL, Rainey PM, Hoffnagle AN, Jack RM, Wright BW, Synovec RE (2006). Fisher ratio method applied to third-order separation data to identify significant chemical components of metabolite extracts. *Analytical Chemistry* **78**(14):5068–5075.

56. Harris RJ (1994). *ANOVA: An Analysis of Variance Primer.* F. E. Peacock Publishers, Itasca, IL.

57. Hotelling H (1931). The generalization of Student's *t*-ratio. *Annals of Mathematical Statistics* **2**(3):360–378.

58. Mahalanobis PC (1936). On the generalized distance in statistics. *Proceedings of the National Institute of Science of India* **12**:49–55.

59. De Maesschalck R, Jouan-Rimbaud D, Massart DL (2000). The Mahalanobis distance. *Chemometrics and Intelligent Laboratory Systems* **50**(1):1–18.

60. Wold S (1976). Pattern recognition by means of disjoint principal components models. *Pattern Recognition* **8**(3):127–139.

61. Hand DJ (1997). *Construction and Assessment of Classification Rules.* John Wiley & Sons, Hoboken, NJ.

62. Lachenbruch PA (1967). An almost unbiased method of obtaining confidence intervals for the probability of misclassification in discriminant analysis. *Biometrics* **23**(4):639–645.

63. Efron B, Tibshirani RJ (1993). *An Introduction to the Bootstrap*, Chapman & Hall, New York.

64. Smith BM, Gemperline PJ (2002). Bootstrap methods for assessing the performance of near-infrared pattern classification techniques. *Journal of Chemometrics* **16**(5):241–246.

65. Solow AR (1990). A randomization test for misclassification probability in discriminant analysis. *Ecology* **7**(16):2379–2382.

66. Barnett V, Lewis T (1994). *Outliers in Statistical Data*, 3rd ed. John Wiley & Sons, Hoboken, NJ.

67. Taylor JK (1987). *Quality Assurance of Chemical Measurements*. Lewis Publishers, Chelsea, MI.

68. Chiang LH, Russell EL, Braatz RD (2000). Fault diagnosis in chemical processes using Fisher discriminant analysis, discriminant partial least squares, and principal component analysis. *Chemometrics and Intelligent Laboratory Systems* **50**(1):243–252.

69. Møller SF, von Frese J, Bro R (2005). Robust methods for multivariate data analysis. *Journal of Chemometrics* **19**(10):549–563.

70. Tan B, Hardy JK, Snavely RE (2000). Accelerant classification by gas chromatography/mass spectrometry and multivariate pattern recognition. *Analytica Chimica Acta* **422**(1):37–46.

71. Johnson KJ, Synovec RE (2002). Pattern recognition of jet fuels: comprehensive GC×GC with ANOVA-based feature selection and principal component analysis. *Chemometrics and Intelligent Laboratory Systems* **60**(1–2):225–237.

72. Doble P, Sandercock M, Du Pasquier E, Petocz P, Roux C, Dawson M (2003). Classification of premium and regular gasoline by gas chromatography/mass spectrometry, principal component analysis and artificial neural networks. *Forensic Science International* **132**(1):26–39.

73. Sandercock PML, Du Pasquier E (2003). Chemical fingerprinting of unevaporated automotive gasoline samples. *Forensic Science International* **134**(1):1–10.

74. Sandercock PML, Du Pasquier E (2004). Chemical fingerprinting of gasoline: 2. Comparison of unevaporated and evaporated automotive gasoline samples. *Forensic Science International* **140**(1):43–59.

75. Christensen JH, Hansen AB, Tomasi G, Mortensen J, Andersen O (2004). Integrated methodology for forensic oil spill identification. *Environmental Science and Technology* **38**(10):2912–2918.

76. Borusiewicz R, Zadora G, Zieba-Palus J (2004). Application of head-space analysis with passive adsorption for forensic purposes in the automated thermal desorption–gas chromatography–mass spectrometry system. *Chromatographia* **60**:S133–S142.

77. Aitken CGG, Lucy D (2004). Evaluation of trace evidence in the form of multivariate data. *Applied Statistics* **53**(1):109–122.

78. Hida M, Sato H, Sugawara H, Mitsui T (2001). Classification of counterfeit coins using multivariate analysis with X-ray diffraction and X-ray fluorescence methods. *Forensic Science International* **115**(1–2):129–134.

79. Kher A, Mulholland M, Reedy B, Maynard P (2001). Classification of document papers by infrared spectroscopy and multivariate statistical techniques. *Applied Spectroscopy* **55**(9):1192–1198.

80. Kher AA, Green EV, Mulholland MI (2001). Evaluation of principal components Analysis with high-performance liquid chromatographic and photodiode array detection for the forensic differentiation of ballpoint pen inks. *Journal of Forensic Science* **46**(4):878–883.

81. Hida M, Mitsui T (2001). Classification of prepaid cards based on multivariate treatment of data obtained by X-ray fluorescence analysis. *Forensic Science International* **119**(3):305–309.

82. Bartick EG, Merrill RA, Egan WJ, Kochanowski BK, Morgan SL (1997). Forensic discrimination of photocopy toners by FT-infrared reflectance spectroscopy. In: *Proceedings of the International Conference on Fourier Transform Spectroscopy*, Athens, GA, 13 August 1997. American Institute of Physics Plenum Press, New York, pp. 257–259.

83. Thanasoulias NC, Parisi NA, Evmiridis NP (2003). Multivariate chemometrics for the forensic discrimination of blue ball-point pen inks based on their Vis spectra. *Forensic Science International* **138**(1–3):75–84.

84. Kher A, Stewart S, Mulholland M (2005). Forensic classification of paper with infrared spectroscopy and principal components analysis. *Journal of Near Infrared Spectroscopy* **13**(4):225–229.

85. Kher A, Mulholland M, Green E, Reedy B (2006). Forensic classification of ball-point pen inks using high performance liquid chromatography and infrared spectroscopy with principal components analysis and linear discriminant analysis. *Vibrational Spectroscopy* **40**(2):270–277.

86. Hall NO, Stefan AR, Morgan SL (2006). Characterization of ball-point pen ink by UV/visible microspectrophotometry for forensics and document conservation. Unpublished manuscript.

87. Jonson CSL (1994). Amphetamine profiling—improvements of data processing. *Forensic Science International* **69**(1):45–54.

88. Perkal M, Ng YL, Pearson JR (1994). Impurity profiling of methylamphetamine in Australia and the development of a national drugs database. *Forensic Science International* **69**(1):77–87.

89. Johnston A, King LA (1998). Heroin profiling: predicting the country of origin of seized heroin. *Forensic Science International* **95**(1):47–55.

90. Janhunen K, Cole MD (1999). Development of a predictive model for batch membership of street samples of heroin. *Forensic Science International* **102**(1):1–11.

91. Praisler M, Dirinck I, Van Bocxlaer J, De Leenheer A, Massart DL (2000). Pattern recognition techniques screening for drugs of abuse with gas chromatography—Fourier transform infrared spectroscopy. *Talanta* **53**(1):177–193.

92. Krawczyk W, Parczewski A (2001). Application of chemometric methods in searching for illicit Leukart amphetamine sources. *Analytica Chimica Acta* **446**(1–2):107–114.

93. Klemenc S (2001). In common batch searching of illicit heroin samples—evaluation of data by chemometrics methods. *Forensic Science International* **115**(1–2): 43–52.

94. Dams R, Benijts R, Lambert WE, Massart DL, De Leenheer AP (2001). Heroin impurity profiling: trends throughout a decade of experimenting. *Forensic Science International* **123**(2–3):81–88.

95. Ryder AG (2002). Classification of narcotics in solid mixtures using principal component analysis and Raman spectroscopy. *Journal of Forensic Science* **47**(2): 275–284.

96. Waddell RJH, Daéid NN, Littlejohn D (2004). Classification of ecstasy tablets using trace metal analysis with the application of chemometric procedures and artificial neural network algorithms. *Analyst* **129**(3):235–240.

97. Daéid NN, Waddell RJH (2005). The analytical and chemometric procedures used to profile illicit drug seizures. *Talanta* **67**(2):280–285.

98. Leger MN, Ryder AG (2006). Comparison of derivative preprocessing and automated polynomial baseline correction method for classification and quantification of narcotics in solid mixtures. *Applied Spectroscopy* **60**(2) 182–193.

99. Morgan SL, Nieuwland AA, Mubarak CR, Hendrix JE, Enlow EM, Vasser BJ, Bartick EG (2004). Forensic discrimination of dyed textile fibers using UV–VIS and fluorescence microspectrophotometry. In: *Proceedings of the European Fibres Group*, Annual Meeting, Prague, The Czech Republic.

100. Yokoi H, Nakase T, Ishida Y, Ohtani H, Tsuge S, Sonoda T, Ona T (2001). Discriminative analysis of *Eucalyptus camalduldensis* grown from seeds of various origins based on lignin components measured by pyrolysis-gas chromatography. *Journal of Analytical Applied Pyrolysis* **57**(1):145–152.

101. Causin V, Marega C, Schiavone S, Guardia VD, Marigo A (2006). Forensic analysis of acrylic fibers by pyrolysis-gas chromatography/mass spectrometry. *Journal of Analytical Applied Pyrolysis* **75**(1):43–48.

102. Espinoza E, Przybyla J, Cox R (2006). Analysis of fiber blends using horizontal attenuated total reflection Fourier transform infrared and discriminant analysis. *Applied Spectroscopy* **60**(4) 386–391.

103. Koons RD, Fiedler C, Rawalt RC (1988). Classification and discrimination of sheet and container glasses by inductively coupled plasma–atomic emission spectrometry and pattern recognition. *Journal of Forensic Science* **33**(1):49–67.

104. Hicks T, Monard Sermier F, Goldman T, Brunelle A, Champod C, Margot P (2003). The classification and discrimination of glass fragments using non destructive energy dispersive X-ray μfluorescence. *Forensic Science International* **137**(2–3): 107–118.

105. Aeschliman DB, Bajic SJ, Baldwin DP, Houk RS (2004). Multivariate pattern matching of trace elements in solids by laser ablation inductively coupled plasma–mass spectrometry: source attribution and preliminary diagnosis of fractionation. *Analytical Chemistry* **76**(11):3119–3125.

106. Bajic SJ, Aeschliman DB, Saetveit NJ, Baldwin DP, Houk RS (2005). Analysis of glass fragment by laser ablation–inductively coupled plasma–mass spectrometry and principal component analysis. *Journal of Forensic Science* **50**(5):1123–1127.

107. Cohen IM, Pla RR, Mila MI, Gomez CD (1988). Activation analysis for trace elements in bullet lead samples and characterization by multivariate analysis. *Journal of Trace and Microprobe Techniques* **6**(1):113–124.

108. Committee on Scientific Assessment of Bullet Lead Elemental Composition Comparison (2004). *Forensic Analysis: Weighing Bullet Lead Evidence*. National Research Council, The National Academies Press, Washington DC.

109. Mahoney CM, Gillen G, Fahey AJ (2006). Characterization of gunpowder samples using time-of-flight secondary ion mass spectrometry (TOF-SIMS). *Forensic Science International* **158**(1):39–51.

110. Kochanowski BK, Morgan SL (2000). Forensic discrimination of automotive paint samples using pyrolysis gas chromatography–mass spectrometry with multivariate statistics. *Journal of Chromatographic Science* **38**(3):100–108.

111. Gresham GL, Groenewold GS, Bauer WF, Ingram JC (2000). Secondary ion mass spectrometric characterization of nail polishes and paint surfaces. *Journal of Forensic Science* **45**(2):310–323.

112. Naes T, Isaksson T, Fearn T, Davies T (2002). *A User-Friendly Guide to Multivariate Calibration and Classification*. NIR Publications, Chichester, UK.

113. Thanasoulias NC, Piliouris ET, Kotti ME, Evmiridis NP (2002). Application of multivariate chemometrics in forensic soil discrimination based on the UV–Vis spectrum of the acid fraction of humus. *Forensic Science International* **130**(2–3):73–82.

114. Chazottes V, Brocard C, Peyrot B (2004). Particle size analysis of soils under simulated scene of crime conditions: the interest of multivariate analyses. *Forensic Science International* **140**(2–3):159–166.

115. Burger F, Dawson M, Roux C, Maynard P, Doble P, Kirkbride P (2005). Forensic analysis of condom and personal lubricants by capillary electrophoresis. *Talanta* **67**(2):368–376.

116. Brooks TR, Bodkin TE, Potts GE, Smullen SA (2006). Elemental analysis of human cremains using ICP-OES to classify legitimate and contaminated cremains. *Journal of Forensic Science* **51**(5):967–973.

117. Widjaja E, Seah RKH (2006). Use of Raman spectroscopy and multivariate classification techniques for the differentiation of fingernails and toenails. *Applied Spectroscopy* **60**(3):343–345.

14

The Color Determination
of Optically
Variable Flake Pigments

Michael R. Nofi

Senior Metrology Engineer, JDSU-Flex Products Group, Santa Rosa, California

14.1 INTRODUCTION

Optically variable pigments are commonly used for a wide range of security and decorative products. These pigments display distinctive optical characteristics that are discernible on both the microscopic and macroscopic level. As such, the forensic identification of this unique material as trace evidence would be potentially useful to the forensic worker. The changes in color and appearance of these pigments with varying illuminating and viewing angles make them difficult to positively identify. Even assigning a color name to describe these pigments is difficult in terms of the variable colors seen by observers. To assist in identifying these pigments, information on their unique properties and a comparison of how these properties differ from traditional pigment materials will be explained. Using color to identify pigments requires an understanding of how light interacts with matter, the nature of the color phenomenon (including additive color mixing), and how color is measured and calculated. An understanding of the various methods used to formulate OVP® color will also be explained. A description of microspectrophotometers, how they

Forensic Analysis on the Cutting Edge: New Methods for Trace Evidence Analysis, Edited by Robert D. Blackledge.
Copyright © 2007 John Wiley & Sons, Inc.

operate, and the problems associated with measuring OVP color are discussed, including the recommended practices and methods to help the forensic worker identify these materials as trace evidence.

14.2 OVP: FORM, CHARACTERISTICS, AND FUNCTION

OVP is a specialty pigment product that was developed and patented by Flex Products, Incorporated in 1988. When viewed at different angles, the color appearance of this unique pigment changes quite dramatically (i.e., from cyan to purple, magenta to gold, etc.). As an anticounterfeiting security ink, it is in use for banknote security in over 94 countries and in brand protection security labels for pharma and consumer electronic industries. It is also used for decorative special effects (since 1996) in a wide range of consumer products in the form of thermoplastic extrusions used to make cell phones, radios, custom effect paints for high end cars, motorcycles, and protective sports equipment.

The interactions of light with colorants determine their color and appearance. Colorants are pigments and dyes that change the absorption characteristics of another material. Pigments are natural or artificial materials that both absorb and scatter light and are insoluble in the application medium or substrate. Dyes are similar to pigments in that they absorb light and can be natural and artificial, but unlike pigments they do not scatter light and are soluble in their application medium or substrate. The similarities between traditional pigments and light interference OVP are that both materials can be artificial, inorganic, and insoluble. Unlike traditional pigments, OVP does not generate color by absorption or scatter, but through light interference.

Mica-based pigments, also known as pearl luster pigments, also generate color through the physics of light interference. These pigments have been available in North America since 1900. The small particles of mica coated with titanium dioxide (TiO_2) act as microscopic interference filters and change in luster, hue, and chroma as the illumination or angle of view changes. They are translucent, showing different colors when viewed in both reflection and transmission. Color characterizations of a mica-based pigment include both reflection and transmission to capture both the reflected interference color and transmitted body color. The result is a complex combination of additive color mixing for the interference color and subtractive color mixing for the absorption color. These pigments are popular in numerous decorative applications such as packaging, cosmetics, and auto finishing since they cause a change in luster or apparent color, especially on curved surfaces with changes in the angle of illumination and view.

OVP products are similar to mica-based pigments, but with important differences, having smoother, mirror-like surfaces with high diameter to thickness ratios. These materials are best described as optically variable micromirrors (Figure 14.1). Since the pigment flakes are opaque, only the reflected color is measured.

Figure 14.1   *SEM (1000×) of opaque color shift pigment flake.*

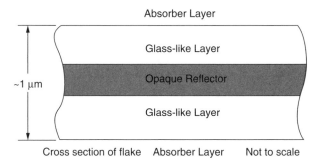

Figure 14.2   *Flake construction diagram.*

The phenomenon in which the appearance of a pigment changes with change in illumination angle or viewing angle is called gonioappearance. Gonioappearance pigments include metal flake pigments and light interference color shift pigments. Metal flake pigments change from light to dark with changes in viewing angle. Although mica-based interference pigments are gonioapparent, the change in color is less dramatic than OVP.

To better understand OVP, we should first review its construction (Figure 14.2). Each flake is made up of an opaque central reflective metal core, dielectric spacer layers in contact with both surfaces of the metal core, and semitransparent, nonselective absorbing layers on the two exterior surfaces. The reflective core, the spacer layers, and the absorber layers have no spectral peaks (that could give rise to color) across the entire visible electromagnetic spectrum. However, when placed in intimate contact with one another in the structure described, very dramatic metallic colors are made possible.

The core layer serves two purposes. First, it provides the reflective surface necessary for light interference, which produces the highly chromatic color shift characteristic of gonioapparent pigments. Second, the reflective core layer provides opacity/hiding by reflecting light prior to reaching the substrate.

The total thickness of the entire structure is on average just 1 µm—that's about 40 millionths of an inch. The typical face dimension of a pigment flake is 20 µm or three-quarters the thickness of a human hair. In order to control the color of these pigments, the whole manufacturing process must be controlled to a thickness tolerance of a few atoms [1].

Color produced by color shift pigments is not a bulk material property; instead it is based on the laws of physics and the science of light interference. In fact, highly transparent materials with no intrinsic color whatsoever can be made to exhibit specular and nonspecular colors through the physics of light interference. Examples in nature include the color seen in soap bubbles and oil spots on wet asphalt. These interference colors are also seen in butterflies and on the feathers of peacocks and hummingbirds. These colors are usually called iridescent colors and differ from colors produced by absorption and scattering in that they are usually purer and metallic in appearance.

In OVP, the phenomenon of optical thin film interference is created where there are layers of materials with differing indices of refraction and absorption. When light interacts at these layers it is temporarily split into two parts. These parts are later recombined and, depending on the optical pathlength differences between the two beams, the light waves may be in or out of phase when they recombine. For those wavelengths where the peaks and valleys match (in phase) the intensity of the resulting wavelengths is the sum of the two components. For wavelengths where the peaks and valleys are mismatched (out of phase) the sum of the two components cancels each other out. This has the effect of selectively enhancing and canceling color throughout the visible spectrum. Figure 14.3 shows the spectral characteristics of Silver-to-Green

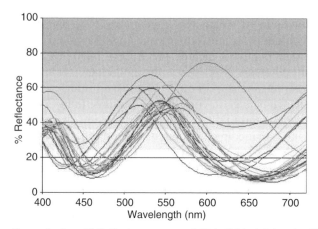

Figure 14.3   *(See color insert.) Reflectance scans of 25 individual flakes for Silver-to-Green OVP. Note: Effective angle of incidence is 35° (50× objective).*

OVP flakes as measured with a microspectrophotometer. Note the characteristic sinusoidal structure characteristic of Fabry–Perot interference coatings.

What causes the color change when viewed at different angles? The optical thickness of the film changes as a function of the illuminating angle of incidence. Films become optically thinner when viewed at higher angles of incidence (as measured from the normal to the surface) due to the optical path differences between the two interfering beams of light. This effect shifts the spectral wavelengths where optical interference occurs and results in a change in color. The color shift is always from the longer to the shorter wavelengths, that is, green to blue with increasing angles of incidence. On a chromaticity diagram the color shift trajectory moves counterclockwise with increasing angle of incidence.

If the OVP flakes are suspended in a nonabsorbing medium, the reflectance peak wavelength at one measurement angle of incidence can be used to predict the location of the reflectance peak wavelength at any other angle of incidence using the following equation:

$$m\lambda_{pk} = 4nd\cos\theta \qquad (14.1)$$

where λ_{pk} is the peak wavelength (in nm), m is the order of interference, n is the index of refraction of dielectric material, d is the physical thickness (in nm) of dielectric material, and θ is the angle of incidence (from the normal). Because of changes in the optical constants of the coating materials as a function of wavelength (optical dispersion), this formula is not exact.

Through optical thin film design, color shift pigments can be designed and made to emphasize the reflectance in certain portions of the spectrum while suppressing in the others. A wide range of pigment designs can be constructed with different color ranges created by precisely controlling the thickness of each glassy layer from a physical thickness of 0.4–1.6 μm. Of course, some colors are easier to obtain than others, but subject to the laws of optical physics, there is virtually no limit to the number of colors that can be achieved. With certain interference color designs, the change in color can be subtle or large.

14.3 COLOR MEASUREMENT

Light is electromagnetic radiant power that is visually detectable by a normal human observer. Known as the visible spectrum, it occupies a very small range of the electromagnetic spectrum bounded by the ultraviolet on the blue end and the infrared on the red end. It is that portion of the spectrum to which the human eye is sensitive, approximately 380–780 nm in wavelength. The wavelength unit generally used in spectrophotometry related to color is the SI unit nanometer (nm). A nanometer is a unit of length and is 10^{-9} meter. Unless otherwise specified, values of wavelength are generally those measured in air. This spectrum extends from violet, from 380 to 450 nm, blues from 450 to 490 nm, greens between 490 and 560 nm, a very narrow band of yellow from 560 to 590 nm, orange from 590 to 630 nm, and red between 630 and 780 nm.

We normally see color as a result of the impact of light rays on the receptor cells in the retina. The perception of color is dependent on the interrelationship of three factors: (1) *light*—to illuminate an object; (2) *object*—to reflect or transmit light to an observer; and (3) *observer*—to "sense" the reflected light.

For color measurement a combination of these three is considered on a spectral wavelength basis throughout the visible spectrum.

CIE 1931 Standard Colorimetric System   CIE (Commission Internationale de l'Eclairage), also known as the Commission on Illumination, is the main international organization concerned with problems of color and color measurement. The CIE 1931 standard colorimetric system is used to quantify the color of objects for specification and measurement of color in the lighting industry, pigment manufacturing, and psychological research in color perception. The CIE procedure for calculating color includes the math functions: illuminant, observer, and the reflection or transmission spectral profile of the object of interest.

Illuminant   The illuminant is the incident luminous energy specified by its spectral energy distribution. For illuminated objects it can affect the perceived colors. A specified illuminant is required in color measurement and is a factor in the tristimulus color calculation. The illuminant should not be confused with the light source of the microspectrophotometer. The recommended illuminant for forensic work is illuminant A, which is incandescent tungsten with a color temperature of 2856 K. CIE has published spectral output data for this illuminant and others to help standardize colorimetric computations for calculating the color of illuminated objects.

Color Matching Functions   The color matching functions are the sensitivity curves of the three receptors in the eye. This is known as the 2° standard observer. The standard observer is an ideal human observer with a visual response described by the color matching functions. Tristimulus values are calculated from the reflectance spectral profile of a material object and give the amount of red, green, and blue light required to match a color. For this measurement we use spectrophotometers, which are described in detail later.

The tristimulus values are calculated as follows:

$$X = K \int_{380}^{780} S(\lambda)\bar{x}(\lambda)R(\lambda)\,d\lambda \tag{14.2}$$

$$Y = K \int_{380}^{780} S(\lambda)\bar{y}(\lambda)R(\lambda)\,d\lambda \tag{14.3}$$

$$Z = K \int_{380}^{780} S(\lambda)\bar{z}(\lambda)R(\lambda)\,d\lambda \tag{14.4}$$

$$K = \frac{100}{\int_{380}^{780} S(\lambda)\bar{y}(\lambda)\,d\lambda} \tag{14.5}$$

where $S(\lambda)$ is the relative spectral power distribution of the illuminant; $\bar{x}(\lambda)$, $\bar{y}(\lambda)$, and $\bar{z}(\lambda)$ are color matching functions for 2° standard observer; and $R(\lambda)$ is the spectral reflectance of the specimen.

Chromaticity values are calculated from the tristimulus values and define the color quality of a color stimulus. It is defined by its chromaticity coordinates, or by its dominant (or complementary) wavelength and its spectral purity taken together and can be plotted two-dimensionally in a chromaticity diagram (Figure 14.4). Dominant wavelength is that wavelength in the spectral profile that is strongest in the measured color. It is the wavelength of light that would be subtracted to bring the color back to white. The complementary wavelength is the wavelength of light that is weakest in the color and if added to color would bring the color back to white. Spectral purity is a measure of the strength or purity of the color.

The x, y, z chromaticity coordinates are calculated from the X, Y, Z tristimulus values according to the following formulas:

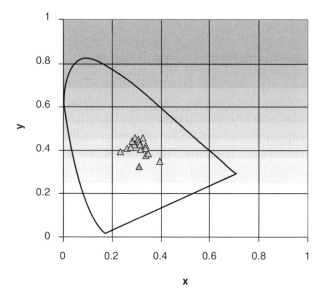

Figure 14.4 *(See color insert.) 1931 x, y, CIE chromaticity diagram of the color of 25 individual flakes of Silver-to-Green OVP (calculated from the spectra in Figure 14.3).*

$$x = \frac{X}{X+Y+Z} \tag{14.6}$$

$$y = \frac{Y}{X+Y+Z} \tag{14.7}$$

$$z = \frac{Z}{X+Y+Z} = 1-x-y \tag{14.8}$$

Differences in color can be expressed in terms of chromaticity differences and change in the reflected luminance (Y).

14.4 COLOR BLENDING

OVP flake pigments that may be encountered during forensic examination will likely consist of a blend of different materials (including nongonioapparent adulterant pigments). In order to identify these materials it is important to understand how these pigment colors are formulated.

Colors are made and matched by mixing colorants. To achieve the desired color effect, a fundamental understanding of the laws of color mixing is required. The two basic laws of color mixing are additive mixing and subtractive mixing. These laws provide a quantitative approach for calculating and predicting color formulations.

The laws are based on an understanding of how materials modify light. When light interacts with a material, different proportions of the light are reflected, transmitted, absorbed, and scattered. The color and appearance of the material depends on the proportion of light and how each proportion of light is spatially distributed.

Additive mixing does not involve mixing colorants that selectively absorb and scatter light, but rather the mixing of colored lights. The mixing of light is the superimposition of spectrally different color stimuli or, more accurately, is the *apparent* superimposition due to the inability of the eye to resolve small areas of different stimuli from the flake acting as mirrors. This occurs in the case of OVP, where the flake sizes are below the resolving power of the human eye. At a viewing distance of 10 inches, flakes smaller than 75 μm are unresolvable. To illustrate: if equal quantities of green colored flakes are mixed with magenta colored flakes, the color stimulus the observer sees is silver, that is, a neutral metallic color. Light beams from each of these flakes combine additively, creating a new color.

The applications for additive color mixing are more limited than applications for subtractive color mixing. Some examples of additive color mixing include color television and stage lighting. Recent developments include the use of colored light emitting diodes (LEDs) to create color balanced white light. For paints and inks, additive color mixing is used in the blending of OVP. This is in contrast to the blending of traditional pigments, where the subtractive mixing rule is used to selectively remove light from the material through absorption. This process can be quite complex, especially when the colorant also scatters light. Broad applications for subtractive mixing include color printing, color photography, and the coloring of absorbing inks and paints (including transparent mica-based pearlescent pigments).

14.5 ADDITIVE COLOR THEORY

The additive primary colors are red, green, and blue (Figure 14.5). Mixing these three color lights in different proportions can result in a wide range of colors. If light from three additive primaries are mixed in the correct proportion, the sum of the three colors produces white. The addition of two additive primary colors produces a secondary color. Mathematically, we can express the additive relationship between these colors using the following formulas:

Blue + Green = Cyan
Green + Red = Yellow
Blue + Red = Magenta
Red + Green + Blue = White

(*Note*: For metallic color the colors yellow and white are often referred to as gold and silver, respectively. It is interesting to note that there are no naturally occurring blue, green, or red metals, although copper is close to red.)

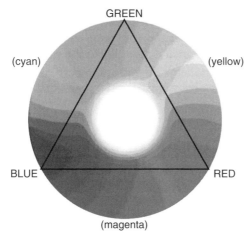

Figure 14.5 *(See color insert.) Primary colors for additive color mixing shown in capital letters. The lower case letters in parenthesis are the secondary colors.*

The methods for color mixing described previously are based on the simple addition and subtraction of colors. The formal mathematical theory for additive color mixing is described next. This is the basis for computer color formulation using additive color theory and can be used by the forensic scientist to determine the macrocolor of paint from individual colored pigment flakes.

1. The color of any material may be expressed in terms of its tristimulus values X, Y, Z. The tristimulus values are

$$X = \sum_{380}^{780} R(\lambda)S(\lambda)\bar{x}(\lambda) \tag{14.9}$$

$$Y = \sum_{380}^{780} R(\lambda)S(\lambda)\bar{y}(\lambda) \tag{14.10}$$

$$Z = \sum_{380}^{780} R(\lambda)S(\lambda)\bar{z}(\lambda) \tag{14.11}$$

where S, a function of wavelength, represents the power distribution of the illuminant; R represents the reflectance (or transmittance) of the material; $\bar{x}$, $\bar{y}$, and $\bar{z}$ are the observer's color matching functions; and X, Y, and Z are the tristimulus values for the spectrum colors.

2. In an additive color mixture, the light from each component reaches the eye in an unmodified state. Additive color mixture is in contrast to absorptive color mixture in which the light passes through two or more components in series, and each component absorbs some of the light.

One method of obtaining an additive mixture takes advantage of the limited resolving power of the eye. Thus, a roofing material composed of red and green particles may appear a dark yellow (brown) when viewed from such a distance that the individual particles cannot be resolved. We assume that the OVP flakes are so small that the blends form an additive color mixture.

3. If the colored components of equal area per unit mass have masses m_1, $m_2, \ldots, m_n$ in an additive color mixture, the tristimulus values of the mixture are

$$X = \frac{m_1 X_1 + m_2 X_2 + \cdots + m_n X_n}{m_1 + m_2 + \cdots + m_n} = \frac{\sum_{i=1}^{n} m_i X_i}{\sum_{i=1}^{n} m_i} \tag{14.12}$$

$$Y = \frac{m_1 Y_1 + m_2 Y_2 + \cdots + m_n Y_n}{m_1 + m_2 + \cdots + m_n} = \frac{\sum_{i=1}^{n} m_i Y_i}{\sum_{i=1}^{n} m_i} \tag{14.13}$$

$$Z = \frac{m_1 Z_1 + m_2 Z_2 + \cdots + m_n Z_n}{m_1 + m_2 + \cdots + m_n} = \frac{\sum\limits_{i=1}^{n} m_i Z_i}{\sum\limits_{i=1}^{n} m_i} \qquad (14.14)$$

where X_1, Y_1, Z_1 through X_n, Y_n, Z_n are the tristimulus values of the colors of each of the components.

Equations (14.12)–(14.14) follow from the definition of the tristimulus values in Equations (14.2)–(14.4) and from the fact that the integral of a sum is the sum of the integrals.

Equations (14.12)–(14.14) can be extended to any number of components with different colors.

From the tristimulus values the chromaticity coordinates can be calculated using Equations (14.6)–(14.8). For graphing purposes, only Equations (14.6) and (14.7) are required.

4. By a similar argument for additive color mixtures, reflectance values for two components at a particular wavelength can be averaged by the following, assuming equal area per unit mass. The apparent reflectance of a mixture is given by

$$R = \frac{m_1 R_1 + m_2 R_2 + \cdots + m_n R_n}{m_1 + m_2 + \cdots + m_n} = \frac{\sum\limits_{i=1}^{n} m_i R_i}{\sum\limits_{i=1}^{n} m_i} \qquad (14.15)$$

where R_1 through R_n are the reflectance values of single pigment components at a single wavelength, and m_1 through m_n are the equal area per unit masses of individual components.

14.6 METHODS OF FORMULATING OVP

OVP is commonly blended with similar materials (including other traditional pigments) to achieve a specific color effect. It is important that the forensic worker is aware of the commonly used methods and materials employed to fabricate these blends in order to more accurately discriminate/identify their occurrence during the course of an investigation. The blending process is relatively simple and straightforward and is discussed in detail later.

A Simple Graphical Method A simple method for blending using additive color mixing can be performed by plotting the colors of two different batches of OVP on the 1931 CIE chromaticity diagram. Plotting the chromaticity coordinates (x, y) of each batch on the diagram in Figure 14.6 shows the location of these colors in color space. If the two batches B_1 and B_2, with individual

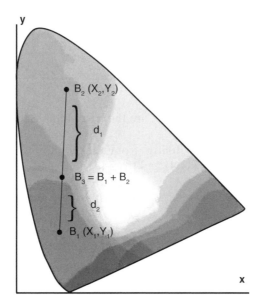

Figure 14.6 *(See color insert.) 1931 CIE chromaticity diagram showing "straight line" and "center of gravity" rules for additive color mixing.*

chromaticity coordinates (x_1, y_1) and (x_2, y_2), are mixed additively, the resulting color will have coordinates (x_3, y_3) that specify a point B_3 lying on the straight line connecting B_1 and B_2. This is called the straight line rule. Where this point lies along the line will depend on the pigment quantity (weight) from each batch mixed together. This can be solved geometrically [2]. If the distance between B_1 and B_3 is (d_1) and the distance from B_3 and B_2 is (d_2), a simple ratio and proportion formula can solve this problem: $d_1/d_2 = m_2/m_1$. If m_1 and m_2 are the masses of the individual blend components and B_1 and B_2 are located at coordinates (x_1, y_1) and (x_2, y_2), then the "center of gravity" of the pair is blend B_3 at (x_3, y_3). This is called the center of gravity rule. This method works fine for simple blending involving two pigments but is less effective when working with three or more pigments.

Creating a Color Gamut A gamut of colors can be created by using a set of three primary OVP colors (Figure 14.7). The color boundaries are defined by the straight line rule between pairs of primaries (P_1) and (P_2). If we add a third primary (P_3), we now have three points in color space. This third point can be connected to the other two primaries, thus forming a color triangle. With a mixture of these three primaries any color can be created within the boundaries of this triangle. Colors outside the triangle cannot be achieved using these three primary pigments. Perhaps the best example of additive color mixing is a color television, where combinations of electron beam guns (R, G, B) are used with different screen phosphors to create colors.

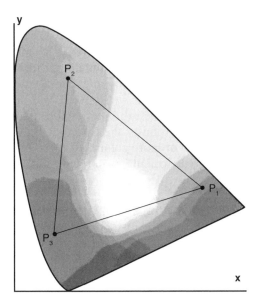

Figure 14.7   *(See color insert.) Color gamut showing available colors within the color triangle bounded by three primary colors.*

Blending by Weight   This pigment loading method uses simple ratios and proportions and does not require the use of spectrophotometers and additive color mixing software to create new colors. This technique works by mixing OVP with other colorants (including other gonioapparent pigments) on a weight basis. It is important, however, to remember that additive color mixing works by area and not weight. For example, equal weights of aluminum flake and carbon black when mixed with an OVP do not have the same areas; therefore, formulating on a weight basis only can be misleading. The surface area per unit weight must be factored into the blending calculations.

Interesting color effects can be achieved by formulators by varying the pigment loading in the polymer system. The loading level can be varied to achieve the desired appearance. Depending on the results this process may be performed iteratively. Loading to full opacity provides the most spectacular color shift effect. Lighter pigment loading will result in more subtle color shift effects [3].

A starting point for the amount of OVP to be added in a polymer system varies from 0.5% to 6.0%. The thickness of the flake determines the color produced. Thus, the number of flakes per unit weight varies with color. A larger number of flakes per unit weight improve hiding power (coverage) at the same pigment loading.

14.7   BLENDING OF PIGMENTS

OVP products have been found to be compatible in a wide range of standard formulations. They can be blended with aluminum, carbon black, and other inorganic and organic pigments, including mica-based interference pigments to create a wide range of color effects.

These pigments were designed to exhibit dramatic hue shifts at high levels of chroma. A method for extension of the color palette can be achieved by simply blending together two or more color shift pigments to produce a new color. By using additive color formulations, hundreds of new colors can be produced.

Blending Two OVP Colors Together   Examples include blending two similar OVP colors to produce hues in the midrange of the two starting colors. Samples can be produced by blending pigments of complementary colors to produce more subtle shades with distinctive color shifts.

Blending with Achromatic Colors   The use of achromatic, or colorless, pigments to modify OVP colors follows the additive mixing rule. The addition of a desaturant, such as aluminum flake, will decrease chroma with an increase in lightness, creating a more subdued pastel color. Adding carbon black also desaturates the color, but conserves considerably more chroma than aluminum. Lightness does decrease from the masstone (unadulterated) color creating rich, dark colors when adding carbon black. The hue travel (color shift) remains unaffected with the additions of carbon black and aluminum. Addition of either of these pigments will improve hiding power, as both aluminum and carbon black are opaque pigments, thus reducing the total amount of OVP required.

Blending with Chromatic Absorbing Pigments   Additions of traditional absorbing colored pigments similar to the near normal or face color of OVP deepen the saturation of the face color (moving the color travel trajectory toward the dominant hue of the absorbing pigment) while shifting the high angle color to a less chromatic position. Addition of other colored pigments similar to the high angle color to an OVP desaturates the face color while intensifying the chromaticity of the high angle color. High loading additions of colored pigments may overwhelm the OVP effect. This is because the colored pigments restrict the interaction of light with the OVP. However, noticeable color shifts will still be evident at lower loadings.

Blending with Transparent and Semitransparent Pigments or Dyes
These formulations can confuse the investigator since the semitransparent pigments permit the interaction of light with the surface of the OVP. Acting as a selective color filter to opaque color shift pigments, the addition of these pigments will result in color that is a combination of additive and subtractive color theory.

14.8 MICROSPECTROPHOTOMETRY

A spectrophotometer is an invaluable tool in forensics for identifying materials through their spectral properties or signatures. An understanding of what these instruments measure will help the forensic worker to identify OVP materials.

Basically, a spectrophotometer measures how light, as a function of wavelength, interacts with a material. For color measurement, the spectral variable is wavelength and the measured property is reflected, transmitted, or absorbed radiant power. The wavelength range for color work is usually specified as 380–780 nm.

When light interacts with matter, some loss in intensity occurs. Starting with a unit amount of light, we calculate this loss as the ratio of outgoing light to incoming light. The three processes by which light interacts with materials include:

Transmission—the process whereby radiant energy passes through a material or object. Rectilinear transmission is transmission of light through the medium without diffusion. Diffuse transmission is transmission in which diffusion occurs within the medium, independently, on a macroscopic scale, or the laws of refraction. Mixed transmission is a combination of diffuse and rectilinear transmission.

Reflection—the process by which radiant energy is returned from a material or object. Specular reflection is reflection without diffusion, in accordance with the laws of optical reflection, as in a mirror. Diffuse reflection is reflection in which flux is scattered in many directions by diffusion at or below the surface. Materials with diffuse reflectance have a frosted or matte appearance. Mixed reflection is a combination of partly specular and partly diffuse reflection.

Absorption—the transformation of radiant energy by interaction with matter, where light is converted to heat. Selective wavelength absorption of visible light is one mechanism for creating color.

A *micro*spectrophotometer is similar in function to a spectrophotometer but combines the capabilities of a spectrophotometer with those of a microscope for measuring the spectra of pigment samples too small for measurement with conventional spectrophotometers. The following example illustrates the difference between a macroscopic and microscopic color measurement. A sample of OVP appearing silver to the unaided eye is scanned using a spectrophotometer. The spectral profile shows the color to be silver. However, when individual flakes are measured with a microspectrophotometer, half of the flakes measure green and half magenta. What is the color of the flakes? Since the area of an individual flake is typically a million times smaller than the spectrophotometer measurement aperture, a spectrophotometer can only measure the average macroscopic color of flakes. Microspectrophotometers measure the microscopic color of single flakes. A microspectrophotometer can approximate a spectrophotometer measurement by averaging the colors of the individual flakes using additive color law. It is approximate, because differences in color interpretation can arise when the flakes do not have 100% surface coverage. This is a macroscopic effect measurable only with a spectrophotometer.

14.9   MEASUREMENT GEOMETRY

The microspectrophotometer has two different illumination geometries, reflected brightfield illumination and transmitted brightfield illumination. For measuring interference pigments the reflected brightfield geometry is used.

The nature of the sample determines how it will be measured for color. For example, to measure the color of a transparent sample requires a microspectrophotometer measurement in a transmission mode. Opaque or reflecting samples would be measured in a reflection mode.

The various types of samples for measurement can be classified as follows:

1. *Transparent* samples transmit radiant energy without difficulty.
2. *Translucent* samples transmit light diffusely, but not permitting a clear view of objects beyond the specimen and not in contact with it.
3. *Opaque* or *reflecting* samples transmit no radiant energy, the energy being either absorbed or reflected.

In a microspectrophotometer, the microscope objective serves two functions. It collects light from the object and forms a real image of the object for the eyepiece to magnify for viewing. It also functions as the illuminator and viewing optics for the microspectrophotometer, defining the conditions by which a sample is illuminated and viewed. Most microscopes have multiple objectives (from three to five) featuring different powers (typically 5×, 10×, 20×, and 50×). In combination with a 10× eyepiece this gives a visual magnification of 50×, 100×, 200×, and 500× respectively.

To provide a starting point for color analysis, it is necessary to define the measurement angles of incidence and view. In Kohler illumination, the microspectrophotometer angles of incidence and view are defined by the numerical aperture (N.A.) of the microscope objective. N.A. is a measure of the light gathering capacity of a lens system and determines its resolving power and depth of field. Refer to Figure 14.8.

The angle of incidence is the angle of the light as measured from the normal (perpendicular) of the flake. For a microscope, this is the cone angle α of the objective. A cone consists of many different angles from the vertex to the terminated angle, so there is not one angle of incidence but a series of different angles of incidence; therefore, the effective incident angle formula is defined as

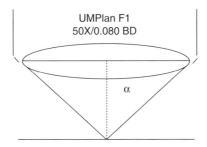

Figure 14.8 *Numerical aperture (N.A.)* = *n* sin α, *where* n *is the index of refraction of the* *medium and* α *is the half-cone angle.*

TABLE 14.1 Typical Microspectrophotometer Objectives and Aperture Sizes Used to Measure OVP

Objective	Magnification	Aperture Size	N.A.	α	E.I.
MPLan	5×	50 μm	0.1	5.7°	3.8°
MPLan	20×	12.5 μm	0.4	23.6°	15.7°
UMPLan F1	50×	5 μm	0.8	53.1°	34.9°
MPLan	100×	2.5 μm	0.9	64.2°	44.9°

$$E.I. = \frac{\displaystyle\int_{0}^{\alpha_2} \alpha \sin(\alpha)\, da}{\displaystyle\int_{0}^{\alpha_2} \sin(\alpha)\, da} \qquad (14.16)$$

where α is the half-cone angle of objective.

For best performance, most microscopes match the numerical aperture of both the illuminator and objective, so the effective angles of illumination and view are the same. The measurement of gonioapparent materials with different microscope objectives results in different spectral profiles, so it is important to specify the effective incidence angle when making measurements on these materials. Effective angle of light incidence is less important when working with solid colors or nongonioapparent colors.

14.10 SWITCHING OBJECTIVE MAGNIFICATIONS

In addition to changing the magnification and the size of the measurement aperture, switching objectives changes the measurement geometry and effective incidence angle (Table 14.1). The different objectives act as a multiangle color measurement system and can be used to confirm color shift of the flake particle. However, all things are not equal. Since the measurement aperture size also changes when switching objectives, you cannot measure the same area of the flake. This is not a problem provided the color of the flake is the same over both measurement areas.

Measurement Area   A measurement aperture defines the area of the sample that is measured. The size of the aperture varies with objective magnification, for example, a 5× objective measures an area 50 μm square, whereas a 50× objective measures an area 5 μm square. The objective is selected based on the area of the flake you wish to measure. For small samples, a good rule of thumb is to select a measurement area that is about three-quarters of the area of the sample you wish to measure.

14.11   DETERMINING SAMPLE SIZE

How many flakes must be measured to get a representative sample? The concept of accuracy relates how well the test average compares with the true value of the tested material.

$$n = \left(\frac{z_\alpha + z_\beta}{d/\sigma} \right)^2 \tag{14.17}$$

where σ is the best estimate of the standard deviation of the test; α is the risk of concluding the test is biased when it is really accurate; d is the minimum detectable bias—the minimum amount of bias that is important to detect; β is the risk of concluding the test is accurate when it is biased by d or more; and z_α and z_β are z values from the Standard Normal Table that correspond to α and β in the upper tail of the distribution, respectively [4].

14.12   MEASUREMENT UNCERTAINTY

It is important to establish the level of measurement uncertainty in the spectral and color measurements. A statement of measurement uncertainty should accompany all reported data. Repeatability and reproducibility should be determined for the entire measurement system by performing formal gauge repeatability and reproducibility studies [5]. ASTM has guidelines for confirming instrument repeatability and reproducibility including how to perform interlaboratory agreement studies between forensic labs.

Calibration Standards   Wavelength calibration is performed using either a didymium or holmium oxide glass standard. These materials have strong absorption bands in the visible region. The minimas of these bands occur at discrete wavelengths and are used to verify wavelength accuracy. These are available from most national labs such as National Institute of Science and Technology and the National Physical Laboratory.

For photometric calibration, front surface aluminum coatings with certified reflectance values are used to calibrate reflectance measurements. For transmission, calibration is done using air (%T value of 1.0 at all wavelengths). There are certified neutral density filters that are available to confirm photometric accuracy and instrument linearity. These are also available at national laboratories and various companies that supply optical standards.

Stability of the Light Source Most microspectrophotometers are a single beam type, requiring a stable light source for repeatable measurements. The instrument is calibrated against a known dark and a known light state. This establishes a calibrated span of measurement. Any changes in the intensity of the light source from the time the instrument is calibrated to the time when the measurements are made will change the dimensions of this span, resulting in measurement error. Referred to as instrumental drift, it usually increases over time, limiting the time between recalibrations. Understanding the effect of drift on measurement will determine how often you must rereference or calibrate the instrument. A double or dual beam system has a second beam that corrects for light source drift, greatly extending the time between recalibrations.

14.13 SAMPLE PREPARATION AND MEASUREMENT

When preparing samples, the objective is to eliminate as many sources of error as possible. Use a preparation method whose repeatability and reproducibility are proven. Care should be taken to ensure that the whole measurement process is defined in sufficient detail to avoid errors and can be followed by all analysts.

It is difficult to select a representative sample for spectral measurement. Material inhomogeneity can be a particular challenge when measuring OVP flake pigments. Many of these products are coated on a plastic web and subsequently stripped off and ground into pigment flake. The coating deposition process, both across and down the web, is variable resulting in a spatial profile of color changes. Compounding the matter is the fact that these products are coated on both sides at different times in the coating process. Process drift results in a different color between the top and bottom of the flakes. When the pigment is stripped off it can show color differences in individual flakes that can extend in color across the visible spectrum, in the form of a Gaussian distribution about a mean color (Figure 14.4). Without measuring a lot of individual flakes, it is difficult to definitively identify these types of materials with microspectrophotometry. A workaround for nonhomogeneous samples would be to use statistical sampling and take average measurements. Additive color blending as described earlier can be used to determine the average color or spectral profile. This information is used for color comparison to a reference material. In choosing a representative sample which flakes do you select to measure? A microscope field of view consists of hundreds of varied color flakes. A microscope reticle in the eyepiece can be used to select flakes at random. A number is selected in the center of the reticle. The x-y stage is moved without looking into the eyepiece and the flake that is closest to that number is selected for measurement.

When measuring the reflected color of interference pigments it is important that the flake is flat and normal to the optical axis of the microscope objective. Measuring unflat flakes results in invalid color measurements. Any deviation in flatness results in flake tilt. Tilt causes vignetting of the reflected light intensity and does not change the angle of incidence of the color measurement. Sensitivity to tilt is inversely proportional to numerical aperture (N.A.). The effect is more pronounced at lower N.A. values: that is, a 5× objective is more sensitive to tilt than a 100× objective. A tilt experiment with 5× objective showed complete loss of color for flakes with a tilt greater than 6°. Under the microscope, flakes with tilt appear dark in brightfield illumination. A method to determine the flatness of the flake is to focus on the flake at the highest magnification. If all four corners of the flake come into focus at the same time the flake is sufficiently flat to measure. Microscope interferometry techniques can also be used to quantitatively determine flake flatness using a Michelson two-beam interferometer microscope objective. This method is used to screen flat flakes for color measurement. However, this method may statistically bias your selection of flakes for color measurement.

To facilitate zero tilt it is helpful to disperse the flakes onto a microscope slide with as little flake interaction as possible. Physical flake interactions lead to flake tilt, the angle of which is related to the cotangent of α, which equals the flake length divided by the physical thickness of the interfering flake. When an OVP flake lays tilted on another flake, a 1 μm thick flake will lie at a theoretical angle of approximately 3°. Use of water to disperse flakes can change the color of the flake. The dielectric coating layers can take up some water, thereby changing the optical constants of the dielectric material and resulting in a color shift to a lower hue angle and therefore creating measurement errors. A recommended method to disperse the flakes onto a microscope slide is to use anhydrous alcohol. Another approach is to lightly dust the flakes onto the surface of the slide. This is done by putting a small amount of pigment on a spatula and tapping the edge of the spatula on the microscope slide. A proper dusting is just barely visible to the unaided eye. When working with these pigments cleanliness is very important to avoid cross contamination where mystery flakes appear at the microscope level that were not visible to the unaided eye.

14.14   SPECTRAL PROFILING

Color is not unique to spectra. It is possible to have two different spectral profiles that produce the same color. Materials that exhibit this property are call metamers, which are spectrally different objects or color stimuli that have

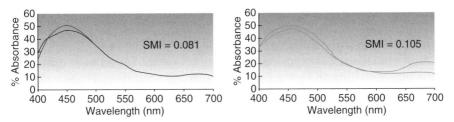

Figure 14.9 *(See color insert.) Examples of different spectral profiles and associated spectral matching indices. An SMI of zero indicates an exact spectral match.*

the same tristimulus values. In forensic work, confirmation of material matches must be done by matching spectral profiles and not based on color (or tristimulus) matching. A recommended method to measure how close a spectral match is to a known reference material is the spectral matching index (SMI) function [5]. This is defined as

$$SMI = \frac{\sum\limits_{\lambda_i = \lambda_1}^{\lambda_n} \left| R_{\lambda_i}^{std} - R_{\lambda_i}^{smpl} \right|}{\dfrac{\lambda_n - \lambda_1}{m} + 1} = \frac{\sum\limits_{\lambda_i = \lambda_1}^{\lambda_n} \left| \Delta R_{\lambda_i} \right|}{\dfrac{\lambda_n - \lambda_1}{m} + 1} \tag{14.18}$$

where m is the wavelength measurement increment and R is the reflectance factor of the reference standard and sample. λ_1 and λ_n represent the starting and ending wavelengths of the evaluation range.

Some examples of how SMI can be applied to different spectral profiles are shown in Figure 14.9. To establish the statistical significance for the SMI index it is first necessary to determine the degree of uncertainty of the microspectrophotometer measurements. SMI indicies outside this interval are statistically significant [6].

14.15 STATISTICAL METHODS OF EVALUATION

Statistical testing should be employed to determine whether or not you have a statistically significant reason to reject a null hypothesis. Is specimen A the same as specimen B? Various statistical test methods are available to determine differences and include F and t tests, chi-squared, and ANOVA [7]. Analysis can be performed using color differences, spectral matching indicies, and frequency distributions of color and spectral parameters.

14.16 CHALLENGES FOR THE FUTURE

Some challenges for the forensic scientist in measuring the color of gonioapparent materials include the correlation of microscopic spectral measurements to macroscopic spectral measurements, the development of a measurement method for removing flake tilt (through the design of a microscope tilt stage), and identifying a method for precisely locating the flakes being measured, so that a return to a particular flake is possible for re-evaluation.

New gonioapparent pigments are being introduced at an ever increasing pace, making it difficult for the forensic scientist to keep abreast of new materials. Samples of new materials should be secured as reference materials for forensic libraries. A list of other goniapparent pigments not mentioned in this chapter, but that the forensic scientist is likely to encounter, include holographic pigments, diffractive (structured) pigments, and liquid crystal pigments. Each of these materials has a unique structure and mechanism for creating color.

14.17 OTHER FORENSIC METHODS

Other forensic methods can be used with microspectrophotometry as collaborative evidence to help positively identify a material. A forensic scientist should not hesitate to employ other methods in order to make a positive identification. Some methods using material science analytical techniques include energy-dispersive X-ray (EDX) in combination with scanning electron microscopy (SEM) for compositional analysis and morphology comparisons. These techniques work quite well when reference materials are used. For a more detailed look into the structure of the pigment materials, transmission and/or scanning transmission electron microscopy (S/TEM) can be used by cross-sectioning individual flakes using ultramicrotomy and scanning/imaging them. This can shed light on the internal structures such as coating design, layer thickness, and stoichiometry of the pigment flakes specific to different methods of coating deposition [8].

ACKNOWLEDGMENTS

The author wishes to thank the many Flex Products colleagues whose knowledge and experience made this chapter possible. Special thanks to Lyuda Hall, Shirley Morikawa, Tom Markantes, Roger Phillips, Barbara Parker, Vicky Tucker, Jennifer Riedel of the Oregon Department of State Police, Forensic Services Division, and Dr. Paul Martin of Craic Technologies for their technical assistance and support.

REFERENCES

1. Phillips R (2005). Flex Products, Inc., personal communication.
2. Markantes T (2005). Flex Products, Inc., personal communication.

3. Parker B (2001). Color sells—color shift pigment formulation 101. Color and Appearance Division of the Society of Plastics Engineers, RECTEC 2001 Conference, Marco Island, Florida, September 23–25.

4. Frost P, Gutoff E (1994). *The Application of Statistical Process Control to Roll Products*, 3rd ed. PJ Associates, New Quincy, MA.

5. MSA Work Group (1995). Measurement Systems Analysis, Chrysler Corp., Ford Motor Co, General Motors Corp.

6. Hall L (2005). Flex Products, Inc., personal communication.

7. Ambrose H III, Ambrose K, Emlen D, Bright K (2002). *A Handbook of Biological Investigation*, 6th ed. Hunter Textbooks Inc., Winston-Salem, NC.

8. Mansour S (2005). JDSU, personal communication.

ADDITIONAL SOURCES

ASTM E308 (2004). *Practice for Computing the Colors of Objects by Using the CIE System.* ASTM International, West Conshohocken, PA.

ASTM E2194 (2004). *Standard Practice for Multiangle Color Measurement of Metal Flake Pigmented Materials.* ASTM International, West Conshohocken, PA.

ASTM E284 (2004). *Terminology of Appearance.* ASTM International, West Conshohocken, PA.

ASTM Standards on Precision and Bias for Various Applications, 5th ed. ASTM International, West Conshohocken, PA, 1997.

Billmeyer F Jr, Saltzman M (1981). *Principles of Color Technology*, 2nd ed. Wiley-Interscience, Hoboken, NJ.

Hunter R, Harold R (1987). *The Measurement of Appearance*, 2nd ed. Wiley-Interscience, Hoboken, NJ.

Kuehni R (1997). *Color: An Introduction to Practice and Principles.* Wiley-Interscience, Hoboken, NJ.

MacAdam D (1981). *Color Measurement.* Springer-Verlag, New York.

Malacara D (2002). *Color Vision and Colorimetry: Theory and Applications.* SPIE Press, Bellinghom, WA.

Thornton J (1997). Visual color comparisons in forensic science. *Forensic Science Review* **9**(1): 37–57.

Williamson S, Cummins H (1983). *Light and Color in Nature and Art.* John Wiley & Sons, Hoboken, NJ.

15

Forensic Science Applications of Stable Isotope Ratio Analysis

James R. Ehleringer

IsoForensics, Inc., and Biology Department, University of Utah, Salt Lake City, Utah

Thure E. Cerling

IsoForensics, Inc., and Professor of Biology and Geology and Geophysics, University of Utah, Salt Lake City, Utah

Jason B. West

Research Assistant Professor, Biology Department, University of Utah, Salt Lake City, Utah

Stable isotope ratio analyses have been commonplace in the environmental, biological, and geological fields for many decades [1, 2]. The use of stable isotopes in forensic studies is relatively new, but is now rapidly expanding because of the many ways this analytical approach can help with law enforcement investigations [3]. Stable isotope analyses can complement and link with other analytical approaches to chemical identification in an investigation (e.g., HPLC, GC/MS, LC/MS), because stable isotope analyses provide an additional "fingerprint" that further characterizes a piece of forensic evidence. Stable isotope

Forensic Analysis on the Cutting Edge: New Methods for Trace Evidence Analysis, Edited by Robert D. Blackledge.

analyses provide a means of relating or distinguishing two pieces of evidence that have exactly the same chemical composition (e.g., TNT explosive material found at a crime scene and at the suspect's residence). The study of stable isotopes as a forensic tool is based on the ability of an instrument to precisely measure very small, naturally occurring differences in the amounts of the heavy stable isotopes in evidence material and to relate that composition to other samples or other evidence. That evidence can be in the form of specific compounds (e.g., TNT), mixtures (e.g., heroin), tissues (e.g., bird feathers, hair), or other materials (e.g., packaging tape, food items). In order to conduct a stable isotope measurement, a special type of mass spectrometer called an isotope ratio mass spectrometers is used to separate the light from heavy isotopes of an element. In this chapter, we will focus on the applications of gas isotope ratio mass spectrometers, which are routinely used to analyze the stable isotope ratios of hydrogen (H), carbon (C), nitrogen (N), oxygen (O), chlorine (Cl), and/or sulfur (S) in evidentiary materials.

15.1 WHAT ARE STABLE ISOTOPES?

Elements are identified on the basis of the number of protons in their nucleus. Atoms of an element share a common number of protons; however, they may differ in the number of neutrons contained inside the nucleus. Different forms of an element are based on the numbers of neutrons within the nucleus—each form is called an isotope. Stable isotopes are those isotopes of an element that are stable: that is, they do not decay through radioactive processes over time. Most elements consist of more than one stable isotope. For instance, the element carbon (C) exists as two stable isotopes, ^{12}C and ^{13}C, while the element hydrogen (H) exists as two stable isotopes, ^{1}H and ^{2}H (also known as deuterium). Table 15.1 provides the average stable isotope abundances of those elements applicable to forensic investigations.

TABLE 15.1 Abundances of Stable Isotopes of Light Elements Typically Measured with an Isotope Ratio Mass Spectrometer

Element	Isotope	Abundance (%)
Hydrogen	^{1}H	99.985
	^{2}H	0.015
Carbon	^{12}C	98.89
	^{13}C	1.11
Nitrogen	^{14}N	99.63
	^{15}N	0.37
Oxygen	^{16}O	99.759
	^{17}O	0.037
	^{18}O	0.204
Sulfur	^{32}S	95.00
	^{33}S	0.76
	^{34}S	4.22
	^{36}S	0.014

Stable isotopes should not be confused with radioactive isotopes of an element, such as ^{14}C (also referred to as radioactive carbon) or ^{3}H (also called tritium). Radioactive isotopes have limited lifetimes and undergo a decay to form a different element, although the time required for this decay may vary widely ranging from nanoseconds to many thousands of years. For instance, carbon has five very short-lived radioactive isotopes (^{9}C, ^{10}C, ^{11}C, ^{15}C, and ^{16}C) with lifetimes of seconds to minutes and one longer-lived radioactive isotope (^{14}C) with a half-life of 5710 years. Radioactive ^{14}C is perhaps best known because of its utility in dating biological materials that are less than 50,000 years old and as a tracer in metabolic studies.

15.2 WHAT ARE THE UNITS FOR EXPRESSING THE ABUNDANCE OF STABLE ISOTOPES?

Stable isotope contents are expressed in "delta" notation as δ values in parts per thousand (‰), where $\delta‰ = (R_s/R_{Std} - 1) \times 1000‰$, and R_s and R_{std} are the ratios of the heavy to light isotope (e.g., $^{13}C/^{12}C$) in the sample and the standard. We denote the stable isotope ratios of hydrogen, carbon, nitrogen, and oxygen as $\delta^{2}H$, $\delta^{13}C$, $\delta^{15}N$, and $\delta^{18}O$, respectively.

R values have been carefully measured for internationally recognized standards. The standard used for both H and O is Standard Mean Ocean Water (SMOW), where $(^{2}H/^{1}H)_{std}$ is 0.0001558 and $^{18}O/^{16}O$ is 0.0020052. The original SMOW standard is no longer available and has been replaced by a new International Atomic Energy Agency (IAEA) standard, V-SMOW. The international carbon standard is PDB, where $(^{13}C/^{12}C)_{std}$ is 0.0112372 and is based on a belemnite from the Pee Dee Formation. As with SMOW, the original PDB standard is no longer available, but IAEA provides V-PDB with a similar R value. Atmospheric nitrogen is the internationally recognized standard with an R value of $(^{15}N/^{14}N)_{std}$ of 0.0036765. Lastly, the internationally recognized standard for sulfur is CDT, the Canyon Diablo Troilite, with a value of $(^{34}S/^{32}S)_{std}$ of 0.0450045. Typically, during most stable isotope analyses, investigators would not use IAEA standards on a routine basis. Instead, laboratories establish secondary standards to use each day that are traceable to IAEA standards and that bracket the range of isotope ratio values anticipated for the samples.

15.3 WHAT IS THE BASIS FOR VARIATIONS IN STABLE ISOTOPE ABUNDANCES?

The abundances of heavy and light stable isotopes of an element vary in nature because of both physical and biological processes. Isotopic enrichment is defined as the difference in the isotope ratio of a reactant (R_r) and a product (R_p) as $\alpha = R_r/R_p$. An approximation in delta notation that is often used is $\varepsilon = (\alpha - 1) \times 1000$, where $\delta_p = \delta_r - \varepsilon$. Three specific processes tend to affect a distribution in the abundances of stable isotope concentrations, resulting in an isotope effect (α) between substrate and product. These are equilibrium fractionation events, kinetic fractionation events, and diffusion fractionation events. Equilibrium fractionation events reflect a difference in the stable isotope ratio of a compound where there is reversible movement of the molecule between two phases, such as between water as vapor and water as a liquid. In such cases, there is a clear tendency for the heavier stable isotopes to remain in the lower energy form (liquid). Kinetic fractionation events are considered as irreversible steps, such as enzymatic reactions between a substrate and product. Again, heavier sable isotopes are more thermodynamically stable, and thus less reactive. Consequently, there is a tendency of the heavier isotopes to react less, resulting in the product being isotopically lighter than the substrate. Lastly, in the gas phase, there is a diffusion fractionation whereby molecules with heavier stable isotopes tend to diffuse more slowly than molecules with light stable isotopes.

15.4   WHAT INSTRUMENTATION IS NEEDED FOR HIGH-PRECISION STABLE ISOTOPE MEASUREMENTS?

High-precision measurements of the stable isotope abundance in a known compound or material are made by converting that substance into a gas and introducing the gas into a mass spectrometer for analysis (Figure 15.1). The

A mass spectrometer allows identification
of individual isotopes

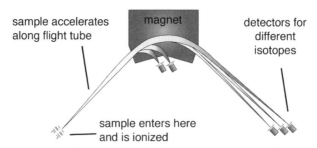

Figure 15.1 *Diagram of a stable isotope ratioing mass spectrometer. Prior to entering the mass spectrometer, the sample is typically completely combusted and the resulting gas is introduced into a gas chromatography, system, where the gases produced during combustion are separated. The pure gas sample (e.g., CO_2 or H_2) then enters the ionizing chamber, is accelerated along the flight tube where it encounters a magnetic field from a fixed magnet, and is deflected into mass-specific Faraday cups. Offline gas preparations can also be introduced to the mass spectrometer.*

instrument for the measurement of the stable isotopes of H, C, N, O, Cl, and S is referred to as an isotope ratioing mass spectrometer (IRMS). Typically, the elements of interest are introduced as the following gases: H as H_2, C as CO_2, N as N_2, O as CO or CO_2, Cl as CH_3Cl, and S as SO_2. At the inlet to the mass spectrometer, the purified gas is ionized and the ion beam is then focused and accelerated down a flight tube, where the path of the ion species is deflected by a magnet. Based on the different isotopic compositions of the ions, they are differentially deflected by the magnet. For instance, for measurements of the carbon isotope composition of a material, the carbon is oxidized to produce CO_2 and the primary species formed are $^{12}C^{16}O^{16}O$ (mass 44), $^{13}C^{16}O^{16}O$ (mass 45), and $^{12}C^{18}O^{16}O$ (mass 46). In contrast to a traditional mass spectrometer, where the strength of the magnet is varied and the ionic species are measured by a single detector, in the IRMS the magnetic field is fixed and the different isotopic ionic species (three in this example) are deflected into separate detector cups at the end of the flight tube, allowing for greater sensitivity in the measurement of the ratio of $^{13}C/^{12}C$. While a traditional mass spectrometer may be able to detect a 0.5% difference in the R_s value of $^{13}C/^{12}C$ in a sample such as might occur in ^{13}C-enriched biochemical studies, an IRMS has the capacity to resolve a 0.0002‰ difference in the R_s value at the low end of the naturally occurring $^{13}C/^{12}C$ range. In the case of CO_2, this is because R_s is measured in an IRMS as the ratio of the simultaneous currents in the two cups: $^{12}C^{16}O^{16}O$ (mass 44) and $^{13}C^{16}O^{16}O$ (mass 45) detectors.

An elemental analyzer and/or gas chromatograph are often coupled to the front end of the IRMS. With such an arrangement, it is then possible to oxidize the sample for C, N, or S isotope analyses using the elemental analyzer, to separate the different combustion gases using a gas chromatograph, and to analyze the sample gases as they pass into the inlet of the IRMS (so-called continuous-flow IRMS). In this case, helium is used as a carrier gas, transporting the combustion gas products from the elemental analyzer to the mass spectrometer. Similarly, it is possible to pyrolyze the sample in the elemental analyzer for hydrogen and oxygen isotope ratio analyses. In addition, samples can also be isolated, purified, and combusted offline and then introduced into the mass spectrometer for stable isotope analyses.

In a stable isotope analysis, a standard of known isotopic composition is analyzed before or after the sample is analyzed. This improves the accuracy of a measurement by directly comparing the isotope signals of the sample and the working standard with each other. As a consequence, a daily calibration is not used with the instrument, because essentially every sample is compared to a standard treated to the same preparation and analysis conditions. These working standards are identified by the stable isotope community and are exchanged among different laboratories in order to determine the best estimate of the actual stable isotope composition of the working standard. Under the best of conditions, there will be many working standards, reflecting a range of stable isotope ratios and a range of material types (e.g., water, plant material, animal tissue, explosive).

15.5 HOW CAN STABLE ISOTOPE ANALYSES ASSIST FORENSICS CASES?

Stable isotope ratio analyses yield potentially unique information for forensic investigations, complementing other approaches by adding a "stable isotope signature" to compounds or materials that are identified as identical by other methods (see Figure 15.2). Key applications of stable isotope ratios include sample matching, sample processing information, and source location identification. Several samples of a material of interest (e.g., illicit drugs, counterfeit money, toxins) may be obtained by investigators with the intent of identifying related groups or sources of the material [4]. Biological samples obtained from the same location and time will have experienced the same environmental conditions, they will therefore have similar stable isotope ratios [5]. Chemical processing of materials can also result in distinctive stable isotope ratios, as can impurities left behind during the production of derived products [6, 7]. Because of this, like samples may readily be grouped based on their stable isotope ratio signatures.

In addition to matching samples, forensic workers can compare individual samples to databases of stable isotope ratio information obtained from

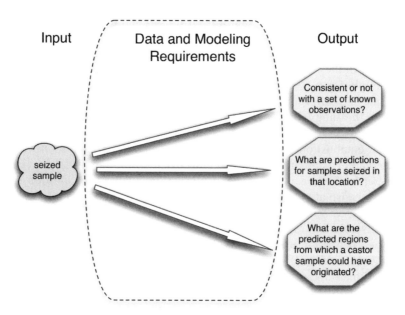

Figure 15.2 *Stable isotope ratio analysis of forensic samples. Three primary classes of information can be provided for seized samples based on stable isotope ratio data. Samples can be matched either to other samples seized or to samples previously obtained and recorded in a database. If the database is sufficiently extensive or if other modeling has been completed (see text), then two additional classes of questions can be addressed: What are the expected stable isotope ratios of samples from a given location and given a sample stable isotope ratio, from where might that sample have originated?*

authentic samples [8]. In this case, a sample of unknown origin could be assigned probable source location by comparison with an authentic stable isotope ratio database, or a previously made assignment of origin could be corroborated by database comparison. A novel approach to geographical assignment may yet yield even greater information, especially in the case that an incomplete database is available. By using first-principles models of stable isotope fractionation in organisms and the spatial modeling capacity of Geographic Information Systems, spatial maps of predicted stable isotope ratios can be constructed [9]. A sample stable isotope ratio can then be compared to the map predictions, and potential source locations can be identified. These advances represent cutting edge applications of stable isotope ratios to forensics that will continue to develop over the coming years. As databases and fundamental understandings of stable isotope ratios grow, so will their capacity to aid in forensic work.

15.6 STABLE ISOTOPE ABUNDANCES IN FORENSIC EVIDENCE

Natural isotope fractionation processes and fractionation steps associated with the synthesis of different products often lead to a wide range of stable isotope ratio values. Consider three kinds of biological samples that might be analyzed in a forensic case: human hair, drugs, and food. There is often a wide range of values in these materials, allowing for an opportunity to detect stable isotope ratio differences among samples. For example, the carbon isotope ratios of human hair can range from $-25‰$ to $-10‰$, depending on the human's diet. On the other hand, carbon isotopes in food can range from $-30‰$ to $-10‰$, depending on the plant's photosynthetic pathway. However, if plants are grown indoors with a carbon dioxide supply (such as indoor grown tomatoes), the carbon isotope ratios can be as low as $-50‰$. In drugs, the biologically derived drugs often have carbon isotope ratio values that overlap with the carbon isotope values of foods, but when synthetic drugs are included, the range of values can be much greater and is dependent on the starting materials that are used for the synthesis. It is this variation of stable isotope ratios in forensic, biological, and commercial samples that allows stable isotope analyses to add additional useful information in characterizing a sample or set of samples under consideration in a forensic case. A significant concern here is within-sample heterogeneity. This variability can make it difficult to distinguish among samples that have similar stable isotope ratios. It is normally the case that the precision of any single measurement from the mass spectrometer is more precise than that of the entire sample because of heterogeneity. This heterogeneity is often greater in biological samples than synthetic samples, because of small differences in the chemical make-up of different biological tissues. An understanding of the expected heterogeneity in a set of samples is therefore necessary prior to the interpretation of any stable isotope ratio data.

15.6.1   Food Products, Food Authenticity, and Adulteration

Plants can be divided into two photosynthetic pathway groups: C_3 and C_4 photosynthesis [10]. These two pathways exhibit distinctive differences in their carbon isotope ratio values, with C_3 plants tending to have carbon isotope ratios of about –27‰ and C_4 plants having a value of about –12‰ [11]. Among the foods we eat, most tend to be C_3 plants, including most grains, fruits, and starchy foods. In contrast, the most common C_4 plants are warm-season grasses, which include corn, sorghum, millet, and sugarcane.

Products of high commercial value are often adulterated to increase profits through substitution using a low-cost alternative. A common example of adulteration would be the dilution of honey (a product ultimately derived from C_3 plants) by the addition of low-cost fructose corn syrup (a C_4 plant product). Even though the corn-sugar substitution cannot be detected chemically through traditional GC/MS or HPLC analyses of a honey sample, IRMS analyses can distinguish between real and adulterated honeys by carbon isotope analyses [12]. Here isotope ratio analyses have also proved quite useful in detecting adulteration of imports, such reconstituted fruit juices, where water has been added back to concentrate [13] and artificial vanilla sold as true vanilla from vanilla beans [14].

Adulteration of food products can easily be observed in common U.S. foods. Consider two commonly consumed alcoholic beverages: beer and sparkling wines. Beer should be brewed with barley or wheat grain, hops, and water. Each of the biological ingredients is from C_3 plants and therefore we might expect the carbon isotope ratio of beer to be about –27‰. However, it is clear from an examination of the carbon isotope ratios of beer that many of the domestic beers have carbon isotope ratios that are consistent with a 50:50 $C_3:C_4$ mixture (Figure 15.3). In Europe, where there are authenticity laws, the carbon isotopes of beer tend to be C_3-like, consistent with barley or wheat as the primary ingredient. In contrast, the U.S. NAFTA partners appear to be producing both C_3 beers (expected) and C_3/C_4 mixture beers (unexpected). As a second example of beverage adulteration, wines should be produced from grapes, another C_3 plant, and therefore the carbon isotope ratios should be approximately –27‰. To their surprise, Martinelli et al. [15] observed that many of the sparkling wines from the United States, Brazil, and Australia were mixes of a C_3 component (grapes) and a C_4 component (most likely corn or sugarcane fructose). This was particularly true for wines from Australia, Brazil, and the Untied States. In contrast, sparkling wines from Argentina, Chile, and France tended to exhibit only C_3 signals. While the legality of these practices might be questionable, the absence of ingredient labeling requirements for these food products certainly opens the door for incorporation of corn or sugarcane sugar as a mechanism to boost profits. Where adulteration of a food product has been taken seriously has been in cases of U.S. product protection from foreign sources. For example, U.S. ATF regulations on the carbon isotope ratios of acceptable honeys has resulted in a shift in

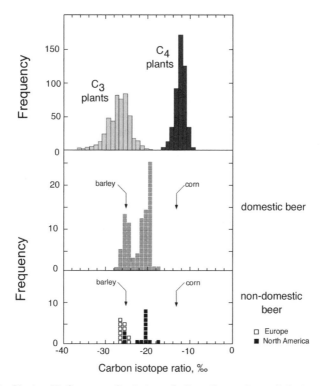

Figure 15.3 Plants with C_3 versus C_4 photosynthetic pathways have distinct, nonoverlapping carbon isotope ratio distributions. Since barley is a C_3 plant and corn is C_4, the detection of corn addition to beer is readily made. In this example, the European beer had a strong C_3 stable isotope ratio signature, whereas the North American beer showed clear addition of corn, with an average calculated proportion of 50% corn, 50% barley.

carbon isotope ratios of the received samples away from C_4 adulteration values [12].

Concerns over illnesses such as mad cow disease, testing compliance with food production regulations, and verification of claims of origin, especially for high-value products such as cheeses or wines, are all contributing to an increased use of stable isotope analysis in food authenticity [16, 17] (see also http://trace.eu.org). One of the early applications of isotopes to food authenticity was the identification of origin and adulteration in honey using carbon isotope ratios [18]. Carbon isotope ratios of plants that use the C_3 photosynthetic pathway differ significantly from plants that use the C_4 pathway. Adulteration of honey with C_4 sugars is therefore easily detected with carbon isotope analysis. Carbon isotope ratios have also been applied to maple syrups, other fruit juices, jams, vanillin, olive oil, and other sugars as a method of detecting adulteration [19, 20]. In addition to the detection of sugar additions, other applications exist such as for the detection of synthetic steroids in meat production with the same protocols for detection of human doping [21].

In addition to $\delta^{13}C$, both $\delta^{18}O$ and δ^2H have been utilized extensively for food authentication. Since, in addition to internal biochemical transformations, climate and geography affect plant $\delta^{18}O$ and δ^2H values, these have been used to authenticate both food sources and processing [17]. As an example, wine oxygen and hydrogen stable isotope ratios have been quite extensively explored as useful markers of geographic source, production methods, and even vintage [22–26], although the current methods require large databases of authentic samples for comparison with suspect samples. The combination of multiple isotopes can yield a rich array of information on food production methods and geographic sources. For example, although the existing research is currently fairly limited, the combination of hydrogen, oxygen, nitrogen, and sulfur isotopes has been explored for the detection of methods of farming beef, as well as its geographic source [27]. Combining light isotope ratio analysis with heavy isotope ratios (e.g., lead or strontium [28–30]), or with other chemical or elemental analyses [31], also promises to yield rich information about food sources and processing methods.

15.6.2 Doping and Drugs of Abuse

Illicit drugs have been shown to have stable isotope ratio values that characterize the region from which a drug sample originated. An "isotopic fingerprint" has been detected in marijuana, heroin, and cocaine, usually as a combination of the carbon and nitrogen isotope ratio values. Earlier studies were constrained by a limited number of authentic exhibits but clearly established the comparative value of isotope analyses in comparing different heroin samples [32–34]. Ehleringer et al. [35] showed that the carbon and nitrogen isotope ratios of heroin from the major poppy growing regions of the world clustered into four distinct groupings, allowing for the identification of the region of origin of different heroin exhibits.

However, it should be noted that heroin is a synthetic molecule, as it is the product of acetylation of morphine with acetic anhydride. As a consequence, any isotopic fingerprint is recorded in the morphine (biological product), which contributes 17 of the 21 carbon atoms and all of the nitrogen atoms in heroin. The carbon isotope ratios of acetic anhydride will vary with the source of carbon used to synthesize the molecule (e.g., fossil fuel, biogenic, and syngas sources). Table 15.2 shows the carbon isotope ratios of a number of different DEA exhibits acquired in one region of the world. Note the similarity in values among different exhibits. In this case, the stable isotope ratio information was useful in further characterizing the acetic anhydride and allowing investigators to link different exhibits to indicate a common source.

Ehleringer et al. [4] showed that the major growing regions of coca for cocaine in South America could be characterized by different combinations of carbon and nitrogen isotopes (Figure 15.4). In this case, these two isotopes alone explained more than 80% of the observed variation among samples originating from different growing regions. This information is useful in law

TABLE 15.2 δ^{13}C Values (‰) of Morphine Base Obtained Under Controlled Conditions from Illicit Production by Clandestine Chemists in Colombia [8]

Chemist	Morphine Base Starting Material
Colombian—1	−32.1
Colombian—2	−32.0
Colombian—3	−32.2
Colombian—4	−32.3
Colombian—5	−32.1
Colombian—6	−32.2
Colombian—7	−32.2
Colombian—8	−32.3
Colombian—9	−31.9
Colombian—10	−31.9
Colombian—11	−32.4
Colombian—12	−32.2
Average	−32.2

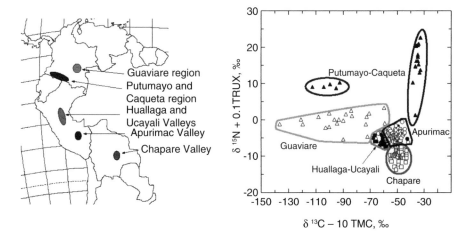

Figure 15.4 *Carbon and nitrogen stable isotope ratios allowed the correct identification of country of origin of 90% of 200 cocaine samples analyzed (from Bolivia, Colombia, or Peru). Including the abundance of minor alkaloid compounds allowed further separation of growing regions within countries and increased the accuracy of country identification to 96%. After [4]*

enforcement for strategic and intelligence purposes. Today stable isotope analyses are a key measurement in the U.S. DEA's Cocaine Signature Program, providing critical information about the origins of seized samples.

Recently, stable isotope ratio analyses were used to comment on the origins of heroin samples seized on the freighter *Pong Su*, a ship sailing under North Korean registry [8, 36]. Australian police seized 50 kg of heroin hydrochloride from the freighter and another 75 kg of heroin hydrochloride from the offload site on Australian soil. Authorities believed that it was "highly likely" that

North Korea was dealing in illegal drugs [37]. Stable isotope ratio analyses of the samples were conducted to examine their region of origin. Using authentic heroin exhibits, Ehleringer et al. [35] had earlier established that the four major growing regions for heroin (SE Asia, SW Asia, Mexico, and South America) could be distinguished on the basis of their carbon and nitrogen isotope ratios. Casale et al. [8] completed stable isotope ratio analyses of both the seized samples and deacetylated seized samples; they concluded that the seized samples were unlike any previously known heroin exhibits and that the samples were unlikely to have originated from any of the four major known growing regions.

Denton et al. [38] and Shibuya et al. [39] have shown that variations in the carbon and nitrogen isotope ratios have been useful in detecting regions of origin for marijuana seizures in Australia and Brazil, respectively. The observations by Shibuya et al. [39] were particularly useful as they provided police with information about the geographic origins of street sales in Sao Paulo, Brazil, allowing law enforcement to better focus their efforts and reconstruct trafficking routes.

Synthetic drugs such as ecstasy can also exhibit stable isotope ratio differences that are useful to law enforcement for determining relationships among batches of seized samples. Carter et al. [40] showed that, by plotting different combinations of the hydrogen, carbon, and nitrogen isotope ratios of seized ecstasy tablets, it was possible to identify clusters or groupings reflecting different production batches. This "isotopic fingerprint" allowed investigators to be able to link specific batches of the illicit drug with particular manufacturers. Palhol et al. [41] showed that is was possible to distinguish among and relate seizure samples of ecstasy originating from different geographic locations based on their nitrogen isotope ratio values. In both of these cases, the combinations of stable isotope ratios does not provide information about the origins of the seized samples, but instead allows the investigator to know how many different "cooks" or batches contributed to the samples that were seized.

Doping is a significant concern in both professional and amateur sports [42] and has resulted in the formation of organizations such as the World Anti-Doping Agency (WADA; www.wada-ama.org) with the intent to fight doping in athletics [43]. The ratio of testosterone to epitestosterone in urine can indicate synthetic steroid use and has been used to test for their use. However, because of the identification of naturally high T/E ratios in some individuals and the need to monitor athletes over time to confirm a return to a baseline condition, carbon isotope ratio analysis has been explored as an additional tool for detection [44]. Carbon isotope ratios of synthetic steroids provide a marker of steroid use because synthetic steroids are derived directly from plant sources and therefore tend to have lower $\delta^{13}C$ values [45, 46]. The use of synthetic steroids by an athlete would therefore be detected by lower $\delta^{13}C$ values in urinary steroids than those for one who has not. However, this value alone can be affected by diet and potentially other factors [47]. By comparing the $\delta^{13}C$ of endogenous steroids unaffected by synthetic inputs with those potentially derived from synthetic sources in urine samples, it is possible to unequivocally identify exogenous steroids since, if no synthetic steroids were used, their $\delta^{13}C$ should not be substantially different [47, 48].

15.6.3   Sourcing of Humans, Animals, and Animal Products

The isotopic composition of an organism will reflect the results of its dietary inputs through food and drinking water [2, 49, 50]. Large differences in the carbon isotope ratios of C_3 plants ranging from −30‰ to −24‰ (e.g., wheat, barley, potatoes) and C_4 plants ranging from −14‰ to −10‰ (e.g., corn, millet, sorghum, sugarcane) result in contrasting dietary inputs [10]. Assimilated dietary source inputs get laid down in proteins, lipids, carbonates, and carbohydrates of the muscle, bones, teeth, and hair of organisms, providing a record of the diet of that organism. Hair provides a chronological recording of the diet of an organism, be it a traveling human [51] or a migrating animal such as an elephant [52]. Naturally occurring variations in both light (H, C, N, and S) and heavy (Pb, Sr) isotopes have been used to associate humans and animals with specific geographic regions of the world. Variations in C, N, and S isotopes largely reflect dietary factors [2, 52, 53], while H isotope variations have been related to geographic patterns of water isotopes across the landscape [2, 9, 54, 55]. Within the heavy isotopes, both Pb and Sr have been shown to provide provenance and source information because these isotopes are picked up from local sources via dietary inputs [56–58].

   Geographic variations in the hydrogen and oxygen isotope ratios of water form a basis of a geographic signal recorded in the hydrogen and oxygen isotope ratios of organic matter in animals. Bowen and colleagues [9, 59, 60] have been able to extrapolate from the available location-specific data of stable isotopes in water to provide spatial maps of the predicted isotopic composition of water throughout the world. Analyses of the spatial distributions of hydrogen isotopes of waters across the North America and Europe continents reveal substantial variations in stable isotope ratios (Figure 15.5), making it possible to distinguish many geographical locations. There are not unique stable isotope ratio values for waters in a specific geographic location, but rather gradients or bands of different isotope ratio values allowing one to distinguish between locations if they were sufficiently far apart from each other.

   Kreuzer-Martin et al. [61, 62] have applied this known variation in the isotope ratios of water to show that it is possible to source microbial spores, such as the anthrax spores mailed in letters in late 2001 (http://en.wikipedia. org/wiki/2001_anthrax_attacks). The hydrogen and oxygen isotope ratios of spore cell wall materials record the isotopic composition of the water in which these microbes were cultured, providing a geolocation piece of information. This source-location information is preserved as long as the original spores remain intact. In *Bacillus* spores, approximately 74% of the oxygen atoms in a spore were derived from its source water (Figure 15.6). The utility of the stable isotope approach here is to allow investigators to both eliminate

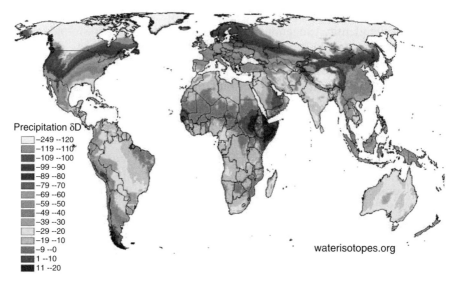

Figure 15.5 *(See color insert.) Predicted long-term annual average precipitation hydrogen isotope ratios for the land surface. This continuous layer is produced with a combination of empirical relationships between measured precipitation δ^2H and latitude and elevation, and a geostatistical smoothing algorithm for variation not explained by that relationship. Measured precipitation values are those maintained in the International Atomic Energy Agency (IAEA) water isotope database. Methods and grids are available at http://waterisotopes.org.*

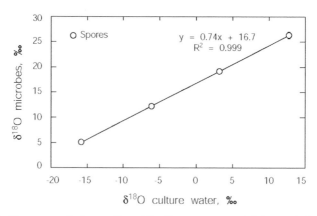

Figure 15.6 *The oxygen isotope ratios of Bacillus spores are tightly coupled to the oxygen isotope ratios of the water in which they are cultured. As can be seen in the fitted regression equation, approximately 74% of the oxygen in these bacterial spores was derived from the culture water. This oxygen isotope ratio signal allows clear matching between spores and their origins. After [61]*

possible sources by showing that stable isotope values are not consistent with a region and to further guide investigations by indicating that stable isotope ratio values are consistent with a different region. Similarly, the carbon and nitrogen isotope ratios of the spore provide information about the nature of the nutrient media in which a microbe is grown [62–64].

15.6.4 Humans: Bones, Hair, and Teeth

Trapped within the enamel of teeth are a number of trace components, including lead, strontium, and carbonates. Each of these components has isotope ratio variation resulting in four isotopes (C, O, Pb, Sr) that then provide geographic and dietary information ·to the forensic investigator. The carbon isotopes reveal dietary information from the time period when the tooth was originally formed (preadult), allowing one to gather general information about the role of C_3 versus C_4 plants in the diet [65, 66]. The oxygen in the carbonate reflects the oxygen of the water in blood, which is related to the geographical variations in drinking water [67]. Once deposited, the carbon and oxygen isotope ratios of teeth do not shift during a human's lifetime.

Carbon isotope and nitrogen isotope analyses of collagen in bone have been used to reconstruct dietary histories of both recent and prehistoric humans [68–71]. Collagen is a protein preserved in bones that reflects dietary inputs over a several year period, because of its slow turnover rate. Once preserved, collagen degrades very slowly over time, allowing for dietary reconstructions of both modern and ancient bone samples. In the anthropological arena, Ambrose and colleagues (as well as other investigators) have used carbon isotope analyses to reconstruct the importance of corn in ancestral Native American populations over the past millennium [72–74]. Richards et al. [75] recently examined dietary changes in Iron Age, Roman, and post-Roman bodies recovered from grave sites in Dorchester, England using stable isotope analyses.

The heavier element Sr shows geographic variation related to soil differences in different regions and countries [58]. Sr is taken up by plants and makes its way into the food chain, becoming incorporated into humans as part of our diet. The Sr signal then provides a tool for reconstructing the geographic origins or movements of individuals across terrains. Evans et al. [76] used this information to trace origins of individuals. Price et al. [77] found that Sr in both teeth and bones could be used to distinguish migrants versus locals in a population. Gulson et al. [57] found that the lead isotope composition of human teeth from individuals living in eastern and southern Europe was significantly different from those individuals raised in Australia. Moreover, they were able to distinguish individuals raised in different portions of the Middle East, eastern Europe, and western Europe based on the lead isotope ratios of an individual's teeth. Pye and Croft [78] review the utility of additional geological heavy isotope signals in a number of different forensic investigations.

The stable isotope ratios of human hair provide a recorder of human diet. Keratin is the dominant protein in hair and records information about both the carbohydrate and protein sources consumed by individuals. Perhaps most well documented are the changes in the carbon and nitrogen isotope ratios of hair related to geographically distinct dietary preferences [79]. Marine versus terrestrial diets are apparent [80] as are C_3-dominated versus C_4-dominated diets. European hair is distinct from U.S. hair, as is Chilean, Canadian, and German hair [81, 82].

Human hair and fingernails record dietary and water source information, and in so doing provide geographic information for forensic studies. Several recent studies have suggested that isotope ratio differences in human hair can be used to distinguish individuals of different geographic origin [83–85]. A recent study of public interest involving stable isotopes was the case of the Ice Man discovered on the border between Austria and Italy [86]. Fingernails are composed of keratin, the same protein found in hair. Recently, Nardoto et al. [79] have shown that citizens from Europe, the United States, and Brazil can be distinguished based on the isotope ratios within their fingernails. Although there is a tendency among modern human societies for a global supermarket that would homogenize isotopes in fingernails, the diets in these countries are sufficiently distinct so as to create differences in the isotope ratios of human fingerprints.

15.6.5 Stable Isotope Abundances of Manufactured Items

Manufactured or synthesized products of forensic interest can exhibit significant variation in their isotope ratios because of two manufacturing factors: differences in the substrates used to synthesize the product and differences in the manufacturing process. These materials span a broad array of items, including security papers, counterfeit currencies, plastic tapes, packaging materials, explosives, clothing, and synthetic drugs.

Consider the example of explosive compounds. PETN, RDX, TNT, and HMX are among the most common high-energy military explosives associated with terrorist events. On the other hand, ammonium nitrate, fuel oil, and pyrotechnic materials are often more common explosive materials used at the local levels. Both of these classes of explosive materials have been considered in forensic investigations in both the United States and the European Union in terms of all elements measured with an IRMS. Of significance is that explosive materials should be investigated through analysis of individual explosive compounds and not as analysis of the bulk explosive mixture (e.g., separate explosive compound binders, fillers, etc.). The left plate of Figure 15.7 shows two examples of common military explosives developed and manufactured in different countries, consisting of either PETN and RDX or TNT and RTX; the names of the manufacturers are eliminated to maintain anonymity. In both

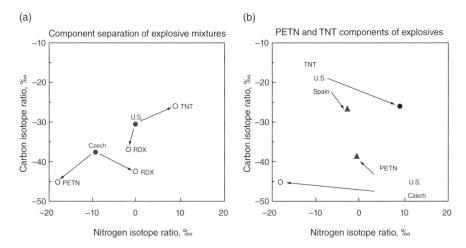

Figure 15.7 *Explosives have distinct carbon and nitrogen isotopic signatures, depending on the manufacturing process and materials used. This can be seen in the left panel showing Czech and U.S. explosives clearly separated by their $\delta^{13}C$ and $\delta^{15}N$ values. Separating these explosives into their component parts and analyzing the component isotope ratios yields even greater distinction based on differences in these component isotope ratios.*

cases, the explosive mixture is composed of two explosive compounds with very different carbon–nitrogen isotope ratio combinations. When considered as individual compounds, the isotope ratio combinations comprise a "finger-print" that characterizes the explosive material, allowing that explosive to be distinguished from other mixtures of the same chemical composition. Applications include comparisons of seized materials and efforts to link explosive materials through a series of connected sources. This difference in isotope ratio "fingerprints" is made even more evident when investigators look at the carbon–nitrogen isotope ratio combinations of pure compounds originating from different factories (here noted as different countries in Figure 15.7). PETN manufactured in the United States or Czech Republic are isotopically identifiable, whereas using traditional GC and LC preparation techniques the explosives would not be distinguishable from each other. The same analytical approach applies to distinguishing TNT and RDX originating from different factories (expressed as countries in Figure 15.7). This new analytical approach opens new opportunities in forensic science that allow investigators to distinguish among compounds that might otherwise appear as identical using traditional analytic methods.

Consider next the possibilities associated with multiple origins of or the counterfeiting of commercial products. Pharmaceutical drugs, security paper, perfumes, and other profitable items fall into this category. This situation is analogous to adulteration of biological products discussed in an earlier section.

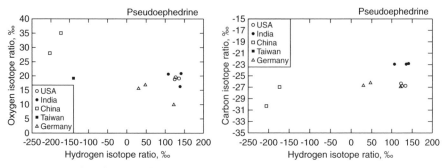

Figure 15.8 $\delta^{13}C$, δ^2H, and $\delta^{18}O$ values of pseudoephedrines of different geographic origins. Samples provided courtesy of Mexican Customs and reported by Lott et al. [7].

In this case, the forensic investigation may require an understanding of the origins of a particular material, such as with explosives as just discussed. Here stable isotopes might provide sufficient insights that allow an investigator to trace the manufacturing origins of the material at hand. Figure 15.8 shows the isotope ratio combinations for pseudoephedrines that might be imported without duties into Mexico from the United States as part of NAFTA [7]. The importing of such materials that did not originate from the United States could be detected through isotope ratio analysis. Figure 15.8 shows the carbon, hydrogen, and oxygen isotope ratios of materials manufactured in different countries using different synthetic processes. Note that the geographic origins of the materials can be detected in this case on the basis of certain isotope ratio combinations. Although the pseudoephedrines of different origins do cluster on some axes, they are separated and identifiable using other axes. In this case, stable isotope analyses can be used in screening to distinguish legitimate from illegitimate samples; or possibly to track down the origins of counterfeited sample materials.

There is no exhaustive list yet of the classes of studies for which isotope ratio analyses will or will not be useful for forensic investigations. However, it is clear that in some cases, direct sample comparisons will be used to determine whether or not two samples have a common origin (i.e., possibly coming from the same production batch). In most cases, law enforcement interests will need to develop databases of observations from different sources and repeated sampling over time in order to determine the general usefulness of stable isotopes in forensic studies.

REFERENCES

1. Sharp Z (2006). *Principles of Stable Isotope Geochemistry*. Pearson Prentice Hall, Upper Saddle River, NJ.

2. West JB, Bowen GJ, Cerling TE, Ehleringer JR (2006). Stable isotopes as one of nature's ecological recorders. *Trends in Ecology & Evolution* **21**:408–414.

3. Benson S, Lennard C, Maynard P, Roux C (2006). Forensic applications of isotope ratio mass spectrometry—a review. *Forensic Science International* **157**:1–22.

4. Ehleringer JR, Casale JF, Lott MJ, and Ford VL (2000). Tracing the geographical origin of cocaine. *Nature* **408**:311–312.

5. Ehleringer JR, Rundel WP (1989). Stable isotopes: history and units. In: Stable Isotopes in Ecological Research, Rundel PW, Ehleringer JR, Nagy KA (eds.). Springer, New York.

6. Toske SG, Cooper SD, Morello DR, Hays PA, Casale JF, Casale E (2006). Neutral heroin impurities from tetrahydrobenzylisoquinoline alkaloids. *Journal of Forensic Sciences* **51**:308–320.

7. Lott MJ, Howa J, Ehleringer JR, Jauregui JF, Douthitt C (2002). Detecting the manufacturing origins of pseudoephedrines through stable isotope ratio analysis. Poster, Forensics Isotope Ratio Mass Spectrometry Conference, 17 September. http://www.forensic-isotopes.rdg.ac.uk/conf/conf.htm.

8. Casale J, Casale E, Collins M, Morello D, Cathapermal S, Panicker S (2006). Stable isotope analyses of heroin seized from the merchant vessel *Pong Su*. *Journal of Forensic Sciences* **51**:603–606.

9. Bowen GJ, Wassenaar LI, Hobson KA (2005). Global application of stable hydrogen and oxygen isotopes to wildlife forensics. *Oecologia* **143**:337–348.

10. Farquhar GD, Ehleringer JR, Hubick KT (1989). Carbon isotope discrimination and photosynthesis. *Annual Review of Plant Physiology and Plant Molecular Biology* **40**:503–537.

11. Cerling TE, Harris JM, MacFadden BJ, Leakey MG, Quade J, Eisenmann V, Ehleringer JR (1997). Global vegetation change through the Miocene/Pliocene boundary. *Nature* **389**:153–158.

12. White JW, Winters K, Martin P, Rossman A (1998). Stable carbon isotope ratio analysis of honey: validation of internal standard procedure for worldwide application. *Journal of AOAC International* **81**:610–619.

13. Brause AR, Raterman JM, Petrus DR, Doner LW (1984). Fruits and fruit products: verification of authenticity of orange juice. *Journal of the Association of Official Analytical Chemistry* **67**:535–539.

14. Hoffman PG, Salb M (1979). Isolation and stable isotope ratio analysis of vanillin. *Journal of Agricultural and Food Chemistry* **27**:352–355.

15. Martinelli LA, Moreira MZ, Ometto JP, Alcarde AR, Rizzon LA, Stange E, Ehleringer JR (2003). Stable carbon isotopic composition of the wine and CO_2 bubbles of sparkling wines: detecting C_4 sugar additions. *Journal of Agricultural and Food Chemistry* **51**:2625–2631.

16. Rossmann A (2001). Determination of stable isotope ratios in food analysis. *Food Reviews International* **17**:347–381.

17. Kelly S, Heaton K, Hoogewerff J (2005). Tracing the geographical origin of food: the application of multi-element and multi-isotope analysis. *Trends in Food Science & Technology* (regular ed.). **16**:555–567.

18. Ziegler H, Stichler W, Maurizio A, Vorwohl G (1977). The use of stable isotopes for the characterization of honeys, their origin and adulteration. *Apidologie* **8**(4):337–347.

19. Krueger HW, Reesman RH (1982). Carbon isotope analyses in food technology. *Mass Spectrometry Reviews* **1**:205–236.

20. Hammond DA (1996). Authenticity of fruit juices, jams and preserves In: *Food Authenticity.* Ashurst PR, Dennis MJ (eds.). Blackie Academic & Professional, London.

21. Hebestreit M, Flenker U, Buisson C, Andre F, Le Bizec B, Fry H, Lang M, Weigert AP, Heinrich K, Hird S, Schanzer W (2006). Application of stable carbon isotope analysis to the detection of testosterone administration to cattle. *Journal of Agricultural and Food Chemistry* **54**:2850–2858.

22. Martin GJ, Guillou C, Martin ML, Cabanis MT, Tep Y, Aerny J (1988). Natural factors of isotope fractionation and the characterization of wines. *Journal of Agricultural and Food Chemistry* **36**:316–322.

23. Arvanitoyannis IS, Katsota MN, Psarra EP, Soufleros EH, Kallithraka S (1999). Application of quality control methods for assessing wine authenticity: use of multivariate analysis (chemometrics). *Trends in Food Science & Technology* **10**: 321–336.

24. Kosir IJ, Kocjancic M, Ogrinc N, Kidric J (2001). Use of SNIF-NMR and IRMS in combination with chemometric methods for the determination of chaptalisation and geographical origin of wines (the example of Slovenian wines). *Analytica Chimica Acta* **429**:195–206.

25. Gremaud G, Pfammatter E, Piantini U, Quaile S (2002). Classification of Swiss wines on a regional scale by means of a multi-isotopic analysis combined with chemometric methods. *Mitteilungen aus Lebensmitteluntersuchung und Hygiene* **93**:44–56.

26. Christoph N, Rossmann A, Voerkelius S (2003). Possibilities and limitations of wine authentication using stable isotope and meteorological data, data banks and statistical tests. Part 1: Wines from Franconia and Lake Constance 1992 to 2001. *Mitteilungen Klosterneuburg* **53**:23–40.

27. Boner M, Forstel H (2004). Stable isotope variation as a tool to trace the authenticity of beef. *Analytical and Bioanalytical Chemistry* **378**:301.

28. Almeida CM, Vasconcelos MTSD (2001). ICP-MS determination of strontium isotope ratio in wine in order to be used as a fingerprint of its regional origin. *Journal of Analytical Atomic Spectrometry* **16**:607–611.

29. Bentley RA (2006). Strontium isotopes from the Earth to the archaeological skeleton: a review. *Journal of Archaeological Method and Theory* **V13**:135.

30. Mihaljevic M, Ettler V, Sebek O, Strnad L, Chrastny V (2006). Lead isotopic signatures of wine and vineyard soils—tracers of lead origin. *Journal of Geochemical Exploration* **88**:130–133.

31. Almeida CMR, Vasconcelos MTSD (2003). Multielement composition of wines and their precursors including provenance soil and their potentialities as fingerprints of wine origin. *Journal of Agricultural and Food Chemistry* **51**:4788–4798.

32. Desage M, Guilluy R, Brazier JL (1991). Gas chromatography with mass spectrometry or isotope-ratio mass spectrometry in studying the geographical origin of heroin. *Analytica Chimica Acta* **247**:249–254.

33. Besacier F, Chaudron-Thozet H, Rousseau-Tsangaris M, Girard J, Lamotte A (1997). Comparative chemical analyses of drug samples: general approach and application to heroin. *Forensic Science International* **85**:113–125.

34. Besacier F, Chaudron-Thozet H, Lascaux F, Rousseau-Tsangaris M (1999). Application of gas chromatography–nitrogen isotopic mass spectrometry to the analysis of drug samples. *Analusis* **27**:213–218.

35. Ehleringer JR, Cooper DA, Lott MJ, Cook CS (1999). Geo-location of heroin and cocaine by stable isotope ratios. *Forensic Science International* **106**:27–35.

36. Collins M, Casale E, Hibbert DB, Panicker S, Robertson J, Vujic S (2006). Chemical profiling of heroin recovered from the North Korean merchant vessel *Pong Su*. *Journal of Forensic Sciences* **51**:597–602.

37. CNN.COM (2006). North Korean drug ship to be sunk.

38. Denton TM, Schmidt S, Critchley C, Stewart GR (2001). Natural abundance of stable carbon and nitrogen isotopes in *Cannabis sativa* reflects growth conditions. *Australian Journal of Plant Physiology* **28**:1005–1012.

39. Shibuya EK, Sarkis JES, Neto ON, Moreira MZ, Victoria RL (2006). Sourcing Brazilian marijuana by applying IRMS analysis to seized samples. *Forensic Science International* **160**:35–43.

40. Carter JF, Titterton EL, Murray M, Sleeman R (2002). Isotopic characterisation of 3,4-methylenedioxyamphetamine and 3,4-methylenedioxymethylamphetamine (ecstasy). *Analyst* **127**:830–833.

41. Palhol F, Lamoureux C, Naulet N (2003). 15 N isotopic analyses: a powerful tool to establish links between seized 3, 4-methylenedioxymethamphetamine (MDMA) tablets. *Analytical and Bioanalytical Chemistry* **376**:486–490.

42. Vogel G (2004). A race to the starting line. *Science* **305**:632–635.

43. Trout GJ, Kazlauskas R (2004). Sports drug testing—an analyst's perspective. *Chemical Society Reviews* **33**:1–13.

44. Catlin DH, Hatton CK, Starcevic SH (1997). Issues in detecting abuse of xenobiotic anabolic steroids and testosterone by analysis of athletes' urine. In: Cerling TE, Cook CS (1998). You are what you eat: a traveler's diet in Mongolia. *Analytical News* **2**:4–5.

45. Coppen JJW (1979). Steroids: from plants to pills—the changing picture. *Tropical Science* **21**:125–141.

46. Becchi M, Aguilera R, Farizon Y, Flament MM, Casabianca H, James P (1994). Gas chromatography/combustion/isotope-ratio. *Rapid Communications in Mass Spectrometry* **8**:304–308.

47. Aguilera R, Catlin DH, Becchi M, Phillips A, Wang C, Swerdloff RS, Pope HG, Hatton CK (1999). Screening urine for exogenous testosterone by isotope ratio mass spectrometric analysis of one pregnanediol and two androstanediols. *Journal of Chromatography B Analytical Technologies in the Biomedical and Life Sciences* **727**:95–105.

48. Shackleton CHL, Phillips A, Chang T, Li Y (1997). Confirming testosterone administration by isotope ratio mass spectrometry analysis of urinary androstanediols. *Steroids* **62**:379–387.

49. Schoeninger MJ, DeNiro MJ, Tauber T (1983). Stable nitrogen isotope ratios of bone collagen reflect marine and terrestrial components of prehistoric human diet. *Science* **220**:1381–1383.

50. Peterson BJ, Fry B (1987). Stable isotopes in ecosystem studies. *Annual Review of Ecology and Systematics* **18**:293–320.

51. Cerling TE, Cook CS (1998). You are what you eat: a traveler's diet in Mongolia. *Analytical News* **2**:4–5.

52. Cerling TE, Wittemyer G, Rasmussen HB, Vollrath F, Cerling CE, Robinson TJ, Douglas-Hamilton I (2006). Stable isotopes in elephant hair document migration patterns and diet changes. *Proceedings of the National Academy of Sciences of the United States of America* **103**:371–373.

53. Lott CA, Meehan TD, Heath JA (2003). Estimating the latitudinal origins of migratory birds using hydrogen and sulfur stable isotopes in feathers: influence of marine prey base. *Oecologia* **134**:505–510.

54. Hobson KA, Wassenaar LI (1999). Stable isotope ecology: an introduction. *Oecologia* **120**:312–313.

55. Hobson KA, Bowen GJ, Wassenaar LI, Ferrand Y, Lormee H (2004). Using stable hydrogen and oxygen isotope measurements of feathers to infer geographical origins of migrating European birds. *Oecologia* **141**:477–488.

56. Vogel JC, Talma AS, Hall Martin AJ, Viljoen PJ (1990). Carbon and nitrogen isotopes in elephants. *South African Journal of Science* **86**:147–150.

57. Gulson BL, Jameson CW, Gillings BR (1997). Stable lead isotopes in teeth as indicators of past domicile—a potential new tool in forensic science? *Journal of Forensic Sciences* **42**:787–791.

58. Beard BL, Johnson CM (2000). Strontium isotope composition of skeletal material can determine the birth place and geographic mobility of humans and animals. *Journal of Forensic Sciences* **45**:1049–1061.

59. Bowen GJ, Wilkinson B (2002). Spatial distribution of $\delta^{18}O$ in meteoric precipitation. *Geology* **30**:315–318.

60. Bowen GJ, Revenaugh J (2003). Interpolating the isotopic composition of modern meteoric precipitation. *Water Resources Research* **39**(10), 1299. DOI:10.1029/2003WR002086.

61. Kreuzer-Martin HW, Lott MJ, Dorigan J, Ehleringer JR (2003). Microbe forensics: oxygen and hydrogen stable isotope ratios in *Bacillus subtilis* cells and spores. *Proceedings of the National Academy of Sciences of the United States of America* **100**:815–819.

62. Kreuzer-Martin HW, Chesson LA, Lott MJ, Dorigan JV, Ehleringer JR (2004). Stable isotope ratios as a tool in microbial forensics—Part 1. Microbial isotopic composition as a function of growth medium. *Journal of Forensic Sciences* **49**:954–960.

63. Kreuzer-Martin HW, Chesson LA, Lott MJ, Dorigan JV, Ehleringer JR (2004). Stable isotope ratios as a tool in microbial forensics—Part 2. Isotopic variation among different growth media as a tool for sourcing origins of bacterial cells or spores. *Journal of Forensic Sciences* **49**:961–967.

64. Kreuzer-Martin HW, Chesson LA, Lott MJ, Ehleringer JR (2005). Stable isotope ratios as a tool in microbial forensics—Part 3. Effect of culturing on agar-containing growth media. *Journal of Forensic Sciences* **50**:1372–1379.

65. Kohn MJ, Cerling TE (2002). Stable isotope compositions of biological apatite. *Reviews in Mineralogy and Geochemistry* **48**:455–488.

66. Ambrose SH, Krigbaum J (2003). Bone chemistry and bioarchaeology. *Journal of Anthropological Archaeology* **22**:193–199.

67. Passey BH, Robinson TF, Ayliffe LK, Cerling TE, Sponheimer M, Dearing MD, Roeder BL, Ehleringer JR (2005). Carbon isotope fractionation between diet, breath CO_2, and bioapatite in different mammals. *Journal of Archaeological Science* **32**:1459–1470.

68. O'Connell TC, Hedges REM, Healey MA, Simpson AHR (2001). Isotopic comparison of hair, nail and bone: modern analyses. *Journal of Archaeological Science* **28**:1247–1255.

69. Richards MP, Fuller BT, Hedges REM (2001). Sulphur isotopic variation in ancient bone collagen from Europe: implications for human palaeodiet, residence mobility, and modern pollutant studies. *Earth and Planetary Science Letters* **191**: 185–190.

70. Honch NV, Higham TGF, Chapman J, Gaydarska B, Hedges REM (2006). A palaeodietary investigation of carbon (C-13/C-12) and nitrogen (N-15/N-14) in human and faunal bones from the Copper Age cemeteries of Varna I and Durankulak, Bulgaria. *Journal of Archaeological Science* **33**:1493–1504.

71. Losch S, Grupe G, Peters J (2006). Stable isotopes and dietary adaptations in humans and animals at Pre-Pottery Neolithic Nevali Cori, southeast Anatolia. *American Journal of Physical Anthropology* **131**:181–193.

72. Ambrose SH, Buikstra J, Krueger HW (2003). Status and gender differences in diet at Mound 72, Cahokia, revealed by isotopic analysis of bone. *Journal of Anthropological Archaeology* **22**:217–226.

73. Joyce M (2003). Recent advances in Maya archaeology. *Journal of Archaeological Research* **V11**:71.

74. Bousman CB, Quigg M (2006). Stable carbon isotopes from archaic human remains in the Chihuahuan Desert and central Texas. *Plains Anthropologist* **51**:123–139.

75. Richards MP, Hedges REM, Molleson TI, Vogel JC (1998). Stable isotope analysis reveals variations in human diet at the Poundbury Camp Cemetery site. *Journal of Archaeological Science* **25**:1247–1252.

76. Evans J, Stoodley N, Chenery C (2006). A strontium and oxygen isotope assessment of a possible fourth century immigrant population in a Hampshire cemetery, southern England. *Journal of Archaeological Science* **33**:265–272.

77. Price T, Burton JH, Bentley RA (2002). The characterization of biologically available strontium isotope ratios for the study of prehistoric migration. *Archaeometry* **44**:117–135.

78. Pye K, Croft DJ (eds.) (2004). *Forensic Geoscience: Principles, Techniques, and Applications*. The Geological Society, London.

79. Nardoto GB, Silva S, Kendall C, Ehleringer JR, Chesson LA, Ferraz ESB, Moreira MZ, Ometto J, Martinelli LA (2006). Geographical patterns of human diet derived from stable-isotope analysis of fingernails. *American Journal of Physical Anthropology* **131**:137–146.

80. Keenleyside A, Schwarcz H, Panayotova K (2006). Stable isotopic evidence of diet in a Greek colonial population from the Black Sea. *Journal of Archaeological Science* **33**:1205–1215.

81. Bol R, Pflieger C (2002). Stable isotope (C-13, N-15 and S-34) analysis of the hair of modern humans and their domestic animals. *Rapid Communications in Mass Spectrometry* **16**:2195–2200.

82. McCullagh JSO, Tripp JA, Hedges REM (2005). Carbon isotope analysis of bulk keratin and single amino acids from British and North American hair. *Rapid Communications in Mass Spectrometry* **19**:3227–3231.

83. O'Connell TC, Hedges REM (1999). Investigations into the effect of diet on modern human hair isotopic values. *American Journal of Physical Anthropology* **108**:409–425.

84. Sharp ZD, Atudorei V, Panarello HO, Fernandez J, Douthitt C (2003). Hydrogen isotope systematics of hair: archeological and forensic applications. *Journal of Archaeological Science* **30**:1709–1716.

85. Fraser I, Meier-Augenstein W, Kalin RM (2006). The role of stable isotopes in human identification: a longitudinal study into the variability of isotopic signals in human hair and nails. *Rapid Communications in Mass Spectrometry* **20**:1109–1116.

86. Macko SA, Lubec G, Teschler-Nicola M, Andrusevich V, Engel MH (1999). The Ice Man's diet as reflected by the stable nitrogen and carbon isotopic composition of his hair. *FASEB Journal* **13**:559–562.

Index

AATCC Test Method 118–2002, surface-modified fibers, 224

Absorption properties, optically variable pigments, 389

ABS plastic strip, 2,4,6-Trinitrotoluene (TNT), fingerprint analysis, 180–181

Accelerant analysis, multivariate statistical techniques, 359–360

"Accelerated aging" approach, ink analysis, 67–69

Accurate Arms 3100 smokeless powder:
FTIR spectrum, 261–262
GS/MS spectrum, 263–264

AccuTOF spectrometer, Direct Analysis in Real Time (DART) mass spectrometry, 176

Achromatic colors, optically variable pigments, 388

Active® Glass, surface-modified fibers, scanning electron microscopy/energy dispersive spectroscopy, 231–232

Additive color theory, optically variable pigments, 383–385

Additive mixing, optically variable pigments, color blending, 382–383

Adhesives, pressure-sensitive tapes, 296–298
duct tape, 300–307
homicide case study, 327–329
masking tape, 315–316
polarized light microscopy analysis, 325–326

Aggregate materials, cathodoluminescence, 168

Alexander v. State, glitter as evidence in, 24–31

Alliant Powder Power Pistol characterization, 253–256

Aluminum compounds, cathodoluminescence, feldspar group, 159–160

5-Aminotetrazole, automotive airbags, 35

Forensic Analysis on the Cutting Edge: New Methods for Trace Evidence Analysis, Edited by Robert D. Blackledge.
Copyright © 2007 John Wiley & Sons, Inc.

Amphetamines, multivariate statistical analysis, 362
Analysis of variance (ANOVA):
multivariate statistical analysis, 354–359
optically variable pigments, 395
Analyte target molecules:
fabric color analysis, 211–214
ink analysis, 63–64
Anatase, cathodoluminescence, 168–171
Animal products, stable isotope ratio analysis, 411–414
Anionic dyes:
fabric color analysis, ESI-MS techniques, 204–207
ink analysis, 69–71
Anthropogenic materials, cathodoluminescence, 168–171
cement and concrete, 168
duct tape, 170–171
glass, 169
paint, 170
slag, fly ash, and bottom ash, 168–169
Antimalaria mosquito nets, gas chromatrography/mass spectrometry analysis, 234
Antioxidants, condom trace evidence, 86
case studies, 108–111
Apatite, cathodoluminescence, 163
Aragonite, cathodoluminescence, 156–158
Area measurement, optically variable pigments, 392
Aromas, condom trace evidence, 87
Arson accelerants, Direct Analysis in Real Time (DART) mass spectrometry, 188–189
Art forgeries, ink analysis, 74–75
Associative evidence, fibers as, 221–222
Atmospheric pressure chemical ionization (APCI) techniques, fabric color analysis, 203–204
liquid chromatography-mass spectometry comparisons, 211–214
mass spectral analysis protocol, 216–217
Attenuated total reflectance (ATR):

document examination and currency analysis, 360–362
glitter characterization:
criminal case studies of, 25–31
infrared spectroscopy (FTIR), 15–19
Raman microspectroscopy, 19–22
surface-modified fibers, 223
Attenuated total reflectance/Fourier transform infrared spectroscopy (ATR/FTIR):
fiber analysis, 362–364
pressure-sensitive tapes, 320–324
smokless powder identification, 242
brand identification, 261–262
Authentication procedures, cathodoluminescence, 166
Automotive accident investigations:
airbags forensic analysis and, 39–54
glitter as evidence in, 30–31
hit and run accidents, glass cuts in, 274–276
Automotive airbags:
equipment and technology, 34–35
forensic analysis:
applications, 33–34
case reports and examples, 39–54
classification, 35–39
design and manufacturing changes, 54–55
future research issues, 55
history of, 34
Autoscaling:
multivariate statistical analysis, 341–342
principal component analysis, 344–348
Azo dyes:
fabric color analysis, 211–214
pepper spray detection, 133–138

Background correction, multivariate statistical analysis, 338–342
Background fluorescence, latent bloodstains, 118–125
Backing materials:
masking tape, 315–316
pressure-sensitive tape, 296–297
electrical tape, 307–309

pressure-sensitive tapes, polarized light microscopy analysis, 325–326

Backscatter electron (BSE) imaging, cathodoluminescence, calcium carbonate, 158

Backsize materials, pressure-sensitive tape, 294–295

Ballistite, smokeless powder history, 245

Ballpoint pens, ink analysis, 63–64
laser desorption mass spectrometry, 64–74

Ball powder morphology, smokeless powder characterization, 248–249

Band gap energy, cathodoluminescence, 144–147

Baseline corrections, multivariate statistical analysis, 339–342

Battle-dress uniforms (BDUs), gas chromatrography/mass spectrometry analysis, 234

Beam intensity, cathodoluminescence spectral collection, 154

Beer's law, cathodoluminescence, 147

Biaxially oriented polypropylene (BOPP), polypropylene packaging tape, 310–313

Bis(carbocyanines), latent trace analysis, fingerprints, 127–130

Bis(Heptamethine Cyanine) (BHmC), latent trace analysis, fingerprints, 129–130

Blade characteristics, glass cut analysis, 280–281

Bleached fabrics, pepper spray detection, 134–138

Blending technology, optically variable pigments, 388

Block copolymers, pressure-sensitive tapes, adhesive formulas, 297–299

Bloodstains:
automotive airbag forensics, 39–54
latent detection, 117–125
tearing patterns in fabric, 278–279

Body fluids, Direct Analysis in Real Time (DART) mass spectrometry, 181–182

Bone materials, stable isotope ratios, 413–414

Bottom ash, cathodoluminescence, 168–169

Butylated hydroxytoluene (BHT), condom lubricant residue analysis, case studies, 108–111

BUZZ OFF™ fabrics, gas chromatograph/mass spectrometry, 232–234

Calcite, cathodoluminescence, 156–158

Calcium carbonates, cathodoluminescence, 156–158

Calibration standards, optically variable pigment measurements, 392–393

Camera equipment, cathodoluminescence, 149–150

Canister powders:
brand identification, 243, 247–257
improvised explosive devices, 241–242

Canonical variates (CVs):
drug analysis, 362
linear discriminant analysis, 349–354

Capillary column gas chromatography, surface-modified fibers, 238

Capillary electrophoresis (CE), fiber color analysis, 199–203

Capsaicinoid molecules, pepper spray detection, 130–138

CAP-STUN pepper spray, latent trace evidence, 130–138

Carbonate materials, cathodoluminescence, 156–158

Carbon isotope ratios:
drugs of abuse, 408–411
food product authenticity and adulteration, 407–408
human bone, hair, and teeth, 413–414

Carpet fibers:
arson accelerants, Direct Analysis in Real Time (DART) mass spectrometry, 188–189

color analysis, capillary
electrophoresis, 199–203
Car theft, automotive airbag forensics,
case studies, 39, 46–54
Cathode configurations, 144
hot and cold configurations, 147–149
sample preparation and preservation,
152–153
spectral collection, 153–154
Cathodoluminescence (CL):
anthropogenic materials, 168–171
cement and concrete, 168
duct tape, 170–171
glass, 169
paint, 170
slag, fly ash, and bottom ash,
168–169
camera equipment, 149–150
defined, 141–143
electron source, 143–144, 147–149
forensic applications, 164–166
authentication, 166
identification, 165
provenance, 166
screening and application, 165
future applications and research, 171
geological soil and sand samples, 167
instrumentation, 147–150
limitations, 147
mechanisms, 144–147
microscope selection, 149
mineral sources, 156–164
accessory minerals, 163–164
calcium carbonate group, 156–158
feldspar group, 159–160
quartz, 160–163
SEM-CL spectrometers, 150
spectrometer, 150
techniques and forensic analysis,
151–156
image collection, 153
instrumental conditions, 151–152
luminescence fading, 155
sample alteration, 155–156
sample preparation and
preservation, 152–153
spectral collection, 154–155
terminology, 143
theory, 143

Cationic dyes:
fabric color analysis, ESI-MS
techniques, 204–207
ink analysis, 65–66, 69–71
Cement, cathodoluminescence, 168
Charge-coupled device (CCD) camera:
cathodoluminescence, 150
image collection, 153
latent trace analysis, fingerprints,
126–130
Chemical characteristics:
condom trace evidence, 95–106
of glitter, 5
multivariate statistical analysis:
accelerants, 359–360
data patterns, 333–336
document examination and currency
analysis, 360–362
drug analysis, 362
experimental design and
preprocessing, 336–342
fibers, 362–364
glass, 364
group separation, classification
accuracy, and outlier detection,
354–359
linear discriminant analysis, 348–354
principal component analysis
visualization, 342–348
trace minerals, 364–366
pepper spray detection, 133–138
trace evidence visualization and
imaging, 116–117
Chemical warfare agents, Direct Analysis
in Real Time (DART) mass
spectrometry, 189–190
Chi-squared analysis, optically variable
pigments, 395
Chlorine atoms, ink analysis, 72
Chromatic absorbing pigments, optically
variable pigment blending, 388
Chromaticity values, optically variable
pigments, color matching
functions, 381–382
diagram for, 385–386
Chromatographic separation protocol:
fiber color analysis, 216
multivariate statistical analysis,
334–336

CIE 1931 Standard Colorimetric System, optically variable pigments, color measurement, 380
chromaticity diagram, 385–386
Class evidence, glitter as, 3–8
Classification accuracy, multivariate statistical analysis, 354–359
CL-20 compound, Direct Analysis in Real Time (DART) mass spectrometry, 188
C18 liquid chromatography, fabric color analysis, 211–214
Coating materials:
  lubricant coatings, condom trace evidence:
    Direct Analysis in Real Time (DART) mass spectrometry, 182–184
    forensic analysis, 104–105
    materials characterization, 87
    residue analysis, 92–96
  pressure-sensitive tapes:
    primer coat, 296
    release coat, 293–294
Cocaine, stable isotope ratio analysis, 408–411
Codeine, Direct Analysis in Real Time (DART) mass spectrometry, 178–179
Collection techniques, glitter traces, 8–9
Collision energy (CE), fiber color analysis, 209–210
Collision induced dissociation (CID), fabric color analysis, 204–205
  structural elucidation, 208–210
  tandem mass spectrometry, 215–216
Colorants:
  ink analysis, 63–64
  laser desorption mass spectrometry, 75–76
  optically variable pigments, 376–379
    blending technologies, 382–383
  pressure-sensitive tapes:
    backing materials, 296–297
    duct tape, 299–307
Color matching functions, optically variable pigments, 380–382
Color properties:
  fiber dyes, 200–203

glitter, 5–6, 11–12
optically variable pigments:
  additive color theory, 383–385
  basic principles, 375–376
  blending protocols, 388
  color blending, 382–383
  color gamut creation, 386–387
  color measurement, 379–382
  form, characteristics, and function, 376–379
  future research issues, 395–396
  geometric measurement, 390–391
  graphical methods, 385–386
  magnification switching, 391–392
  measurement uncertainty, 392–393
  microspectrophotometry, 389–390
  sample preparation and measurement, 393–394
  sample size, 392
  spectral profiling, 394–395
  statistical analysis, 395
  weighted color blending, 387
  smokeless powder, 257–258
Color shift pigments, peroperties, 378–379
Concentration analysis, glitter traces, 8–9
Concrete, cathodoluminescence, 168
Condom trace evidence:
  Direct Analysis in Real Time (DART) mass spectrometry, 182–184
  forensic evaluation, 96–105
  forensic significance, 81–82
  lubricant coating:
    production, sale, and use, 87
    residues, 92–96
  packaging characteristics, 106–111
  physical examination protocols, 96–105
  powdering process:
    production, sale, and use, 86–87
    residues, 88–92
  production, 85–86
  residue traces, 88–96
  rough condom vulcanization:
    production, 85–86
    residues, 88, 99–101
  sales and market share and, 82–84
  sexual crimes and usage patterns, 83–84

silicone treatment, production, sale, and use, 86
usage patterns and, 82–84
Confirmatory analysis, pepper spray detection, 136–138
Confiscated sample characteristics, drug/pharmaceutical analysis, Direct Analysis in Real Time (DART) mass spectrometry, 178
Confocal depth mapping, glitter characterization, 20–22
ContactIR device, glitter characterization, 17–19
Contact trace evidence:
  automotive airbags:
    applications, 33–34
    case reports and examples, 39–54
    classification, 35–39
    future research issues, 55
    manufacturing changes, 54–55
  case studies, 105–111
  glitter, 210
    collection, separation and concentration, 8–9
    color, 5–6
    computerized database capability, 9–10
    cutting machine characteristics, 6–8
    film and particle manufacturers, 6
    individual characteristics, 3–8
    invisibility, 2–3
    layer characteristics, 4–5
    morphology, 4
    particle analysis, 9
    size properties, 3
    specific gravity, 4
    structural characteristics, 3
    transfer and retention, 3
    transport vehicles for, 8
Copper isotopes, ink analysis, 72
Cordite, smokeless powder history, 245
Cornstarch:
  condom trace evidence, 101–102
  rough condom powdering, 86
  residue analysis, 88–91
Covariance matrix, principal component analysis, 343–348

CRAIC 1000 microspectrophotometer, automotive airbag forensics, 46–54
Criminal case studies, glitter as evidence in, 24–25
Cross polarization techniques, condom powdering residue analysis, 88–91
Cross sectioning techniques, glitter characterization, 15–16
Cross-validation, multivariate statistical analysis, 357–359
Crystal field strength, cathodoluminescence, 145–146
Crystallina 321/421 systems, glitter characterization, 20–22
Crystal violet, ink analysis, 66–69
Currency analysis:
  multivariate statistical analysis, 360–362
  pack dyes, Direct Analysis in Real Time (DART) mass spectrometry, 185
Cutting machine characteristics, glitter, 6–8
Cyanoacrylate fuming, latent trace analysis, fingerprints, 126–130

Data partitioning, multivariate statistical analysis, 357–359
Data storage and analysis:
  glitter traces, 9–10
  principal component analysis, 343–348
Daubert v. Merrill Dow Pharmaceuticals, Inc., 335
Delta notation, stable isotope ratio analysis, 401
  food product authenticity and adulteration, 408
Derivatization reactions, pepper spray detection, 136–138
"Designer pens," ink analysis, 71–72
Desorption/ionization methods:
  condom lubricant residue analysis, 92–96
    case studies, 106–111
  ink analysis, 59–62, 65–69
Diazo dyes, fabric color analysis, 211–214

Diazonium reagents, pepper spray detection, 136–138

Diethylene glycol dinitrate (DGDN), smokeless powder, 245

Diffuse relectance infrared Fourier transform spectroscopy (DRIFTS), condom lubricant residue analysis, 92–96

Dihydrocapsaicin, pepper spray detection, 136–138

Dimensionality reduction, multivariate data, 342–348

Dinitrotoluene (DNT), smokeless powder, 245

Dioctyl phthalate (DOP), electrical tape, 307–309

Diopside, cathodoluminescence, 163–164

Direct Analysis in Real Time (DART) mass spectrometry:
arson accelerants, 188–189
basic principles, 175–176
body fluids, 181–182
chemical warfare agents, 189–190
condom lubricants, 182–184
drug/pharmaceutical analysis, 177–180
  confiscated samples, 178
  endogenous drugs, 178–179
  surface residues, 179–180
dyes, 184–185
experimental applications, 176
explosives, 185–188
fibers, 192–193
fingerprints, 180–181
future research, 194
glues, 191
ink analysis, 193
materials identification, elevated-temperature techniques, 190–191
plastics, 191–192

"Dirty crystals," ink analysis, 61–62

Discontinous cut patterns, glass cut analysis, 282–286

Disc-shaped smokeless powders, 253–256

Dithiocarbamates, condom residue analysis, 100

DNA analysis:
automotive airbag forensics, 35–36, 45–54

condom trace evidençe, case studies, 106–111

glitter characterization, criminal case studies, 30–31

latent trace evidence, bloodstains, 124–125

Document examination, multivariate statistical analysis, 360–362

Dolomite, cathodoluminescence, 156–158

Dopant materials:
Direct Analysis in Real Time (DART) mass spectrometry, explosives, 187–188
stable isotope ratio analysis, 408–411

Drug abuse evidence:
automotive airbag forensics, 43–54
Direct Analysis in Real Time (DART) mass spectrometry, 177–180
  confiscated samples, 178
  endogenous drugs, 178–179
  surface residues, 179–180
multivariate statistical analysis, 362
stable isotope ratio analysis, 408–411

Drunk driving investigations, automotive airbag forensics, 39–54

Duct tape:
cathodoluminescence, 170–171
forensic analysis of, 299–307
variability in, 291–292

DuPont IMR 4350 smokeless powder:
FTIR spectrum, 261–262
GS/MS spectrum, 264–265

Dyes. *See also* Pigments
Direct Analysis in Real Time (DART) mass spectrometry, 184–185
fabric analysis:
  basic principles, 197–198
  conventional comparison techniques, 198–199
  direct ESI-MS analysis, 204–207
  direct infusion MS/MS protocol, 214–216
  generalized LC-MS and LC-MS/MS protocol, 216–217
  ionization techniques, 203–204
  limitations of UV-VIS-based analysis, 199–203

liquid chromatography-mass spectrometry analysis, 210–214
negative ESI-MS analysis, 207–208
tandem mass spectrometry, 208–210
linear discriminant analysis, 349–354
optically variable pigments, 388
pressure-sensitive tapes, backing materials, 296–297

EA2192 agent, Direct Analysis in Real Time (DART) mass spectrometry, 189–190
Ecstasy, drugs of abuse, 410–411
Eigenvectors/eigenvalues:
linear discriminant analysis, 352–354
principal component analysis, 343–348
Elastomers, pressure-sensitive tapes, adhesive formulas, 297–299
Electrical tape, forensic analysis, 307–309
Electron beam interaction, cathodoluminescence, 143–144
sources, 147–149
Electron capture detector (ECD), surface-modified fibers, 238
Electron hole pairs, cathodoluminescence, 144–147
Electron impact mass spectrometry (EIMS):
ink analysis, 58–62
smokeless powder identification, GS/MS spectrum, 264–265
surface-modified fibers, pyrolysis gas chromatography/mass spectrometry, 237–238
Electron source, cathodoluminescence, 143–144
Electrospray ionization (ESI), fabric color analysis, 202–204
liquid chromatography-mass spectometry comparisons, 211–214
mass spectral analysis protocol, 216–217
Electrospray ionization mass spectrometry (ESI–MS):
fabric color analysis, 204–207
negative ion analysis, nylon windings, 207–208

quadrupole ion trap instrumentation, 211–214
structural elucidation, 208–210
pepper spray detection, 136–138
Elemental analysis:
pressure-sensitive tapes, 320, 324–325
sampling protocols, 317
stable isotopes, 400–401
Elevated-temperature Direct Analysis in Real Time (DART) mass spectrometry, material identification, 190–191
End matching techniques, pressure-sensitive tapes, 318–319
Energy dispersive spectroscopy (EDS). *See also* Scanning electron microscopy/energy dispersive spectroscopy (SEM/EDS)
surface-modified fibers, pyrolysis gas chromatography/mass spectrometry, 235–238
Energy dispersive x-ray analysis (EDXA):
glitter analysis, 22–23
criminal case studies, 25–31
optically variable pigments, 395
Enzyme-linked immunosorbent assay (ELISA), condom residue analysis, 88
latex proteins, 99–100
Ethanol acetic acid, latent bloodstain analysis, 120–125
Ethylene glycol dinitrate (EGDEN), Direct Analysis in Real Time (DART) mass spectrometry, 187–188
Euclidean distance, multivariate statistical analysis, 355–359
Explosives:
Direct Analysis in Real Time (DART) mass spectrometry, 185–188
smokeless powders:
ATR-FTIR spectroscopy, 261–262
basic properties and classification, 241–242
brand determination, 242–243, 246–257
ball-shaped powders, 248–249
disc shaped powders, 253–256

flattened ball powders, 256–257
lamella shaped powders, 258
tubular powders, 249–253
color, 257–258
Fourier Transform infrared
spectroscopy, 260–262
gas chromatograph/mass
spectrometry, 262–265
historical background, 243–245
identification, 242, 245–246
kernel/dot configuration, 258
liquid chromatography, 265–266
luster, 258
mass, 259–260
micrometry, 258–259
micromorphology, 247–257
morphology, 246–247
transmission micro-FTIR,
260–261
stable isotope ratios, 414–416
Extrinsic luminescence,
cathodoluminescence and,
145–147

Fabric analysis. *See also* Fiber analysis
Direct Analysis in Real Time (DART)
mass spectrometry, carpet
remnants, 188–189
dyes:
basic principles, 197–198
conventional comparison
techniques, 198–199
direct infusion MS/MS protocol,
214–216
ESI-MS analysis, 204–207
generalized LC-MS and LC-MS/MS
protocol, 216–217
ionization techniques, 203–204
limitations of UV-VIS-based
analysis, 199–203
liquid chromatography-mass
spectrometry analysis, 210–214
negative ESI-MS analysis, 207–208
tandem mass spectrometry, 208–210
pepper spray detection, 134–138
pressure-sensitive tapes, reinforcement
materials, 296, 299–307
Fading phenomena,
cathodoluminescence, 154

Fast atom bombardment (FAB), ink
analysis, 59–62
Feldspar group, cathodoluminescence,
159–160
Fiber analysis. *See also* Fabric analysis
automotive airbag forensics, 35–36,
47–54
Direct Analysis in Real Time (DART)
mass spectrometry, low-mass
fragments, 192–193
duct tape reinforcement, 300–304
glass cuts in:
associated glass fragments, 280
basic principles, 269–270
blade characteristics, 280–281
direction changes, 286
discontinuous cuts, 282–286
fabric properties, 280
hit and run accident case study,
274–276
homicide case study, 270–271
leather cuts, 287
parallel cuts, 286
recent or worn damage, 286–287
robbery case study, 271–274
slash cuts, 279–280
tearing *vs.* cutting, 276–279, 281–282
linear discriminant analysis, 349–354
multivariate statistical analysis, 339–
342, 362–364
principal component analysis,
dimensionality reduction,
343–348
surface-modified characterization:
distinguishing tests, 225
gas chromatgraphy/mass
spectometry, 231–234
preliminary examinations, 222–224
pyrolysis gas chromatograpy/mass
spectometry, 234–238
research backgroun, 221–222
scanning electron microscopy/
energy dispersive spectroscopy,
225–231
structural properties, 222
Filament tape, forensic analysis, 313–315
"Fingerprint" electropherograms, fiber
color analysis, 199–203
collision-induced dissociation, 209–210

Fingerprints:
  Direct Analysis in Real Time (DART)
    mass spectrometry, 180–181
  latent trace analysis:
    fluorescein testing, 119–125
    near-infrared dyes, 125–130
Firearms, smokeless powder, history,
    243–245
Fisher ratio plot, linear discriminant
    analysis, 352–354
Flake construction diagram, optically
    variable pigments, 376–379
Flattened ball smokeless powder,
    256–257
Flavorings, condom trace evidence, 87,
    95
Fluorescein testing, latent trace evidence,
    bloodstains, 117–125
Fluorescence. *See also* Luminescence
  defined, 143
  latent trace evidence, fluorescein
    interference, 124–125
Fluorescence microscopy:
  masking tape forensics, 316
  multivariate statistical analysis, fibers,
    362–364
Fly ash, cathodoluminescence, 168–169
Food products, authenticity and
    adulteration, stable isotope
    ratio analysis, 406–408
Fourier self-deconvolution (FSD),
    condom lubricant residue
    analysis, 104–105
Fourier transform infrared spectroscopy
    (FTIR):
  amphetamines, 362
  condom lubricant residue analysis, 92–
    96, 104–105
  case studies, 106–111
  document examination and currency
    analysis, 360–362
  fabric dye analysis, 214
  glitter characterization:
    criminal case studies, 30–31
    cross sectioning techniques, 15–19
  polypropylene packaging tape, 313
  pressure-sensitive tapes, 320
    duct tape, 305–307

electrical tape, 307–309
homicide case study, 327–329
masking tape, 316
sampling protocols, 317
smokeless powder characterization,
    260–262
surface-modified fibers, 223–224k
Fourier Transform mass spectometry
    (FTMS), ink analysis, 60–62
Fragment analysis, associated glass, with
    glass cuts, 280
Franck-Condon absorption bands, fiber
    dye analysis, 202–203
Fraud detection, ink analysis, 74–75
Frye v. United States, 335
F test, optically variable pigments, 395

Gamut of colors, optically variable
    pigments, creation of, 386–387
Gas chromatography (GC):
  multivariate statistical analysis,
    341–342
  smokeless powder identification, 242
  spltless injectors, 262–266
Gas chromatography/mass spectrometry
    (GC/MS):
  accelerant analysis, 359–360
  amphetamines, 362
  condom lubricant residue analysis, 93–
    96, 104–105
  case studies, 108–111
  document examination and currency
    analysis, 360–362
  fiber analysis, 363–364
  latent trace evidence, pepper spray
    analysis, 130–138
  smokeless powder identification,
    262–266
  surface-modified fibers, 231–234
Gas leakage, automotive airbags, 37–39,
    52–54
Geological samples,
    cathodoluminescence:
  provenance criteria, 166
  quartz, 162–163
  soil and sand characterization, 167
Geometric measurements, optically
    variable pigments, 390–391

Germany, condom market share in, 82–83

Glass:
cathodoluminescence, 169
cuts, as trace evidence:
associated glass fragments, 280
basic principles, 269–270
blade characteristics, 280–281
direction changes, 286
discontinuous cuts, 282–286
fabric properties, 280
hit and run accident case study, 274–276
homicide case study, 270–271
leather cuts, 287
parallel cuts, 286
recent or worn damage, 286–287
robbery case study, 271–274
slash cuts, 279–280
tearing *vs.* cutting, 276–279, 281–282
multivariate statistical analysis, 364

Glass transition temperature, polypropylene packaging tape, 310–313

Glitter:
characterization methods, 10–24
color, 11–12
cross sectioning, 15–16
infrared spectrosocopy, 16–19
morphology, 12–13
Raman microspectroscopy, 19–22
scanning electron microscopy/ energy dispersive spectroscopy, 22–24
shape, 13
size, 13–14
thickness, 14–1–5
components of, 1–2
contact trace properties, 210
collection, separation and concentration, 8–9
color, 5–6
computerized database capability, 9–10
cutting machine characteristics, 6–8
film and particle manufacturers, 6
individual characteristics, 3–8

invisibility, 2–3
layer characteristics, 4–5
morphology, 4
particle analysis, 9
size properties, 3
specific gravity, 4
structural characteristics, 3
transfer and retention, 3
transport vehicles for, 8
as criminal evidence, 24–31
ink analysis, 71–72

Glues, Direct Analysis in Real Time (DART) mass spectrometry, 191

Gonioappearance pigments, characteristics and function, 377–379

Graphical methods, optically variable pigment formulation, 385–386

Group membership classification, multivariate statistical analysis, 356–359

Guanidine nitrate (GuNi), automotive airbags, 35

Guncotton, smokeless powder history, 244–245

Gunpowder, history, 243–245

Guns. *See* Firearms

Gypsum, cathodoluminescence, 163–164

Hair analysis:
automotive airbag forensics, 45–54
stable isotope ratios, 413–414

Handling protocols, pressure-sensitive tape forensics, 316–317

Hematoxylene-eosine (HE) staining, condom powdering residue analysis, 88–91

Hercules Red Dot smokeless powders, 253–255

Hercules Reloder 12 tubular smokeless powder, 249–250

Heroin analysis, stable isotope ratio analysis, 408–411

Hexahydro-1,3-5-trinitro-1,3,5-triazine (RDX), Direct Analysis in Real Time (DART) mass spectrometry, 187–188

Hexamethylene triperoxide diamine (HMTD) explosive, Direct Analysis in Real Time (DART) mass spectrometry, 186–188
High-performance liquid chromatography (HPLC):
document examination and currency analysis, 360–362
fiber color analysis, 198–199, 203–204
trace quantity extraction, 213–214
oligonucleotide analysis, 76–77
Homicide cases:
glass cuts as evidence in, 270–271
glitter as evidence in, 24–31
pressure-sensitive tapes, 327–329
Hotelling's T^2 test, multivariate statistical analysis, 354–359
Human products, stable isotope ratio analysis, 411–414
Hydrocarbon contamination, cathodoluminescence spectral collection, 154
Hydrogen isotope ratios, animal and human products, 411–414
2-(4-Hydroxyphenylazo) benzoic acid (HABA), oligonucleotide analysis, 75–76

Identification techniques, cathodoluminescence, 165
Illuminants, optically variable pigments, color measurement, 380
IlluminatIR Infrared Microspectometer, glitter characterization, 17–19
Image collection, cathodoluminescence, 153
Improvised explosive devices (IEDs):
Direct Analysis in Real Time (DART) mass spectrometry, 186–188
pressure sensitive tape for, 291
smokeless powders, 241–242
brand identification, 242–243, 247–257
IMR 4198 smokeless powder, 250–252
Indigotin dyes, fiber color analysis, mass spectrometry analysis, 214
Individualistic properties of evidence, glitter, 3–8
Inductively couple plasma (ICP):

glass analysis, 364
pressure-sensitive tapes, 324–325
Infrared spectra:
glitter characterization, 16–19
surface-modified fibers, 223
Ink analysis:
analyte target molecules, 63–64
Direct Analysis in Real Time (DART) mass spectrometry, 193
laser desorption mass spectrometry (LDMS), 58–62
Inorganic pigments, ink analysis, 72–74
Insecticide-treated clothing, gas chromatrography/mass spectrometry analysis, 232–234
Interference pigments, measurement of, 394
Internal reflectance element (IRE), pressure-sensitive tapes, 320–324
International Atomic Energy Agency (IAEA) standard, stable isotope ratio analysis, 401
Intrinsic luminescence, cathodoluminescence and, 144–147
Invisibility properties, glitter, 2–3
Ionization techniques, fabric color analysis:
ESI-MS techniques, 204–207
liquid chromatography-mass spectrometry and, 211–214
Iron compounds, cathodoluminescence:
calcium carbonates, 157–158
feldspar group, 159–160
ISA/SPEX MiniCrime-Scope, latent trace analysis, fingerprints, 126–130
Isotope rationing mass spectrometry (IRMS):
manufactured items, 414–416
stable isotope ratio analysis, 403
"Isotopic peaks," ink analysis, 65–69

JASCO NRS-3100 Raman system, glitter characterization, 20–22

Kava starch grains, condom powdering residue analysis, 90–91

Kevlar fibers, multivariate statistical analysis, 339–342
Kidnapping cases, glitter as evidence in, 26–31
Kinetic energy:
    ink analysis, 61–62
    stable isotope ratio analysis, 401–402
Knife cuts, glass cuts compared with:
    homicide case studies, 270–271
    robbery case study, 271–274
    tearing patterns in fabric, 277–279
Known sample (K):
    glitter characterization and, 10–24
        color analysis, 11–12
        particle size, 13
    surface-modified fibers, 222–224
        pyrolysis gas chromatography/mass spectrometry, 234–238
Kumho Tire Co. v. Carmichael, 335–336

Lactones, latent bloodstain analysis, 118–125
Lamella powders, smokeless powder characterization, 248
Laser ablation inductively coupled plasma mass spectrometry (LA-ICP-MS), cathodoluminescence, glass materials, 169
Laser desorption mass spectrometry (LDMS):
    basic principles, 57–58
    failures of, 75–76
    ink analysis, 58
        analyte target molecules, 63–64
        fraud applications, 74–75
        pen ink dyes, 64–74
    instrumentation, 59–62
Latent invisible trace evidence:
    bloodstains, 117–125
        method overview, 123–125
    chemical detection principles, 115–117
    fingerprint detection, near-infrared dyes, 125–130
    pepper spray, 130–138
        chemical derivatization, 134–138
        near-infrared dyes, 131–133

Latex proteins, condom trace evidence, 85
    forensic analysis, 99–101
    residue analysis, 88
Layer thickness and numbers, of glitter, 4–5
Leather, glass cut analysis on, 287
Leave-one-out cross validation, multivariate statistical analysis, 357–359
Length measurements, tubular smokeless powder identification, 252–253
Leuco fluorescein:
    latent bloodstain analysis, 118–125
    latent trace evidence, disadvantages of, 124–125
Light-emitting diode (LED) lamps:
    latent trace analysis:
        bloodstains, 120–125
        fingerprints, 126–130
        pepper spray, 132–138
    optically variable pigments, color blending, 382–383
Light source stability, optically variable pigment measurements, 393
Linear discriminant analysis (LDA):
    accelerant forensics, 359–360
    development and application, 336
    document examination and currency analysis, 360–362
    drug analysis, 362
    fibers, 362–364
    glass analysis, 364
    group difference visualization, 348–354
    mineral trace evidence, 365–366
Linear measurements, smokeless powder, 259
Liquid chromatography (LC):
    fabric dye analysis, 203–204
    smokeless powder identification, 242, 262–266
Liquid chromatography/mass spectrometry (LC/MS):
    condom lubricant residue analysis, 93–96
    fabric color analysis, US-Vis spectrometry, 202–203
    fiber color analysis, 210–214
        generalized protocol, 216–217

latent trace analysis, pepper spray
analysis, 132–138
Liquid chromatography-tandem mass
spectrometry (LC-MS/MS),
fiber color analysis, generalized
protocol, 216–217
Locard's Principle of Exchange,
automotive airbags forensics,
33–34
Location characteristics, glitter traces, 9
Long tubular smokeless powders,
identification, 250–252
Low density polyethylene (LDPE), duct
tape analysis, 299–307
Lubricant coatings, condom trace
evidence:
Direct Analysis in Real Time (DART)
mass spectrometry, 182–184
forensic analysis, 104–105
materials characterization, 87
residue analysis, 92–96
Lugol's solution, condom powdering
residue analysis, 89–91
Luminescence:
basic technology, 143
cathodoluminescence:
anthropogenic materials, 168–171
cement and concrete, 168
duct tape, 170–171
glass, 169
paint, 170
slag, fly ash, and bottom ash,
168–169
camera equipment, 149–150
defined, 141–143
electron source, 143–144, 147–149
fading phenomena, 154
forensic applications, 164–166
authentication, 166
identification, 165
provenance, 166
screening and application, 165
geological soil and sand samples,
167
instrumentation, 147–150
limitations, 147
mechanisms, 144–147
microscope selection, 149
minerals characterization, 156–164

accessory minerals, 163–164
calcium carbonate group,
156–158
feldspar group, 159–160
quartz, 160–163
SEM-CL spectrometers, 150
spectrometer, 150
techniques and forensic analysis,
151–156
image collection, 153
instrumental conditions, 151–152
luminescence fading, 155
sample alteration, 155–156
sample preparation and
preservation, 152–153
spectral collection, 154–155
terminology, 143
theory, 143
gem authentication, 166
mechanisms, 144
Luminol, latent trace evidence,
bloodstains, 117–125
Luminoscope ELM-3R instrumentation,
cathodoluminescence, 151–152
Luster properties:
optically variable pigments, 376–379
smokeless powder, 258
Lycopodium spores, rough condom
powdering:
forensic analysis, 103
materials characterization, 86
residue analysis, 91–92

Machine edge characteristics,
polypropylene packaging tape,
313
Magnesium compounds,
cathodoluminescence, 146–147
calcium carbonates, 157–158
Magnification, optically variable
pigments, 391–392
Mahalanobis distance, multivariate
statistical analysis, 355–359
Manganese compounds,
cathodoluminescence, 146–147
calcium carbonates, 157–158
feldspar group, 159–160
Manufacturers' sources:
glitter, 6

pressure-sensitive tapes:
 equipment markings, 297
 identification, 326–327
 stable isotope ratios, 414–416
Marijuana, Direct Analysis in Real Time
 (DART) mass spectrometry,
 178–179
Masking tape, forensic analysis, 315–316
Mass measurements, smokeless powder,
 259–260
Mass spectrometry (MS):
 fabric color analysis, 202–204
 electrospray ionization and,
 204–208
 generalized protocol, 216–217
 glass analysis, 364
 instrumentation, 58–62
 latent trace analysis, pepper spray
 analysis, 132–138
 pepper spray detection, 133–138
 smokeless powder identification, 242,
 262–266
 stable isotope ratio analysis, 402–403
Material inhomogeneity, optically
 variable pigment
 measurements, 393–394
Matrix-assisted laser desorption time-of-
 flight (MALDI-TOF) mass
 spectrometry:
 condom trace evidence:
 case studies, 108–111
 spermicide residue analysis, 95–96
 ink analysis, 57–62
 oligonucleotide analysis, 75–76
MCS-400 MiniCrimeScope light source,
 latent trace analysis, 122–125
 pepper spray analysis, 132–138
Mean centering:
 multivariate statistical analysis,
 341–342
 principal component analysis, 343–348
Measurement techniques:
 cathodoluminescence spectra, 153–154
 color measurement, optically variable
 pigments, 379–382
 optically variable pigments,
 uncertainty levels, 392–393
 pressure-sensitive tape forensics, 319
 smokless powder, 252–253, 259–260

Mercury isotopes, ink analysis, 73–74
Metallic colors, additive color theory,
 383–385
1-Methylaminoanthraquinone (MAAQ),
 Direct Analysis in Real Time
 (DART) mass spectrometry,
 185
Methyl violet dyes, ink analysis, 66–69
Mica-based pigments, characteristics and
 function, 376–379
Micellar electrokinetic capillary
 chromatography (MECC),
 condom lubricant residue
 analysis, 93–96
Micrometry techniques, smokeless
 powder, 258–259
Micromorphology, smokeless powder,
 247–257
Microscopic equipment:
 cathodoluminescence, 149
 fabric dye analysis, 214
Microspectrophotometry:
 fiber color analysis, 198–199
 multivariate statistical analysis,
 334–336
 optically variable pigments, 389–390
Mineral sources:
 luminescence, 156–164
 accessory minerals, 163–164
 calcium carbonate group, 156–158
 feldspar group, 159–160
 quartz, 160–163
 multivariate statistical analysis,
 364–366
MK4 First Defense pepper spray,
 detection and analysis,
 130–138
Molecular weights (MWs), laser
 desorption mass spectrometry,
 60–62
Monazite, cathodoluminescence, 163
Monoaxially oriented polypropylene
 (MOPP), polypropylene
 packaging tape, 310–313
Morphine, Direct Analysis in Real Time
 (DART) mass spectrometry,
 178–179
Morphology characteristics:
 of glitter, 4, 12–13

smokeless powder, 246–247
micrometry techniques, 258–259
Multivariate analysis of variance
(MANOVA), separation,
classification, and outlier
detection, 355–359
Multivariate statistical analysis, chemical
characteristics:
accelerants, 359–360
data patterns, 333–336
document examination and currency
analysis, 360–362
drug analysis, 362
experimental design and
preprocessing, 336–342
fibers, 362–364
glass, 364
group separation, classification
accuracy, and outlier detection,
354–359
linear discriminant analysis, 348–354
principal component analysis
visualization, 342–348
trace minerals, 364–366
Muzzle-to-target distance, smokeless
powder identification and,
245–246

Nanotechnology, surface-modified fibers,
224
Near-infrared fluorescent (NIRF) dyes,
latent trace analysis,
fingerprints, 126–130
Near-infrared (NIR) dyes, latent trace
evidence:
fingerprint detection, 125–130
pepper spray detection, 130–138
Negative ion spectrum, ink analysis, 69
Neutron activation analysis, mineral
trace evidence, 364–366
Night vision equipment, latent trace
analysis, fingerprints, 126–130
Nikon D1X camera, pepper spray
detection, 133–138
Nikon microscopes,
cathodoluminescence, 151–152
Nitroaromatic explosives, Direct
Analysis in Real Time (DART)
mass spectrometry, 186–188

Nitrocellulose, smokeless powder,
243–245
Nitrogen isotope ratios:
drugs of abuse, 408–411
human bone, hair, and teeth, 413–414
Nitroglycerin (NG):
Direct Analysis in Real Time (DART)
mass spectrometry, 187–188
smokeless powder history, 245
Nitrosamines, condom trace evidence, 86
forensic protocols, 101
Noise-free values, multivariate statistical
analysis, 338–342
Nonoxynol-9, condom trace evidence:
Direct Analysis in Real Time (DART)
mass spectrometry, 182–184
forensics applications, 105
spermicide residue analysis, 95–96
Nonparametic sampling, multivariate
statistical analysis, 357–359
Normalization:
multivariate statistical analysis,
338–342
principal component analysis, 346–348
Norma 203 smokeless powder, 251–252
Numerical aperture:
cathodoluminescence microscopy, 149
optically variable pigments, 390–391
interference measurements, 394
Nylon windings, electrospray ionization-
mass spectrometry analysis,
207–208

Ocean Optics HR2000,
cathodoluminescence, 152
Octahydro-1,3,5,7-tetranitro-1,3,5,7-
tetrazocine (HMX), Direct
Analysis in Real Time (DART)
mass spectrometry, 187–188
Oleoresin capiscum (OC), pepper spray
detection, 130–138
Oligonucleotide analysis, laser
desorption mass spectrometry,
75–76
Opaque pigments, geometric
measurements, 390–391
Opiate alkaloids, Direct Analysis in Real
Time (DART) mass
spectrometry, 178–179

Optically variable pigments (OPVs):
  additive color theory, 383–385
  basic principles, 375–376
  blending protocols, 388
  color blending, 382–383
  color gamut creation, 386–387
  color measurement, 379–382
  form, characteristics, and function,
    376–379
  future research issues, 395–396
  geometric measurement, 390–391
  graphical methods, 385–386
  magnification switching, 391–392
  measurement uncertainty, 392–393
  microspectrophotometry, 389–390
  sample preparation and measurement,
    393–394
  sample size, 392
  spectral profiling, 394–395
  statistical analysis, 395
  weighted color blending, 387
Oriented film materials, polypropylene
    packaging tape, 310–313
Outlier detection, multivariate statistical
    analysis, 354–359
Oxygen isotope ratios, animal and
    human products, 411–414

Packaging characteristics, condom trace
    evidence, case studies, 106–111
Paint, cathodoluminescence, 170
Parallel cut patterns, glass cut analysis,
    286
Particle beam techniques, fabric color
    analysis, 211–214
Particle characterization:
  automotive airbag forensics, 36–39
  glitter, 9
  size and shape, 13
p-dimensional patterns:
  multivariate statistical analysis,
    337–342
  principal component analysis, 346–348
Peak broadening, cathodoluminescence
    and, 145–147
Peak wavelength, optically variable
    pigments, 378–379
Pearl luster pigments, characteristics and
    functions, 376–379

Peltier cooling, cathodoluminescence
    cameras, 150
Pen inks:
  analyte target molecules, 63–64
  laser desorption mass spectrometry,
    64–74
2,4,6-N-tetranitro-N-methylaniline
    (tetryl), Pentaerythritol
    tetranitrate (PETN), Direct
    Analysis in Real Time (DART)
    mass spectrometry, 187–188
Pepper spray:
  Direct Analysis in Real Time (DART)
    mass spectrometry, 184–185
  latent invisible trace evidence, 130–138
    chemical derivatization, 134–138
    near-infrared dyes, 131–133
Perception principles, color
    characterization, glitter traces,
    5–6, 11–12
Perfloration characteristics, tubular
    smokeless powders, 249–250
Permethrin-treated fabrics, gas
    chromatrography/mass
    spectrometry analysis, 232–234
Phlegmatization, smokeless powder
    history, 244–245
Phosphorescence, defined, 143
Photoluminescence, defined, 143
Photomultiplier tube (PMT),
    cathodoluminescence, 150
Photosynthetic pathways, stable isotope
    ratio analysis, 406–408
Phthalates, pressure-sensitive tapes:
  backing materials, 297
  electrical tape, 307–309
Pigments:
  ink analysis, 72–74
  optically variable pigments, color
    determination:
    additive color theory, 383–385
    basic principles, 375–376
    blending protocols, 388
    color blending, 382–383
    color gamut creation, 386–387
    color measurement, 379–382
    form, characteristics, and function,
      376–379
    future research issues, 395–396

geometric measurement, 390–391
graphical methods, 385–386
magnification switching, 391–392
measurement uncertainty, 392–393
microspectrophotometry, 389–390
sample preparation and
measurement, 393–394
sample size, 392
spectral profiling, 394–395
statistical analysis, 395
weighted color blending, 387
Pipe bombs, smokeless powders for,
241–242
P2i plasma process, surface-modified
fibers, scanning electron
microscopy/energy dispersive
spectroscopy, 225–228
Plastics:
Direct Analysis in Real Time (DART)
mass spectrometry, 191–192
pressure-sensitive tapes:
backing materials, 297
pyrolysis gas chromatography/mass
spectrometry, 326
Polarized cathodoluminescence
microscopy, 149
Polarized light microscopy (PLM):
automotive airbag forensics, 39–54
pressure-sensitive tapes:
component identification, 325–326
duct tape, 306–307
sampling protocols, 317
strapping/filament tape, 314–315
surface-modified fibers, 223
Polyanionic dyes, laser desorption mass
spectrometry, 75–76
Polydimethylsiloxane (PDMS):
condom lubricant residue analysis,
case studies, 106–111
condom trace evidence, 87
lubricant coating forensics,
104–105
residue analysis, 92–96
Polyethylene, rough condom powdering,
86
forensic analysis, 102–103
Polyethylene terephthalate (PET),
glitter from, characterization of,
18–24

Polymers:
elevated-temperature Direct Analysis
in Real Time (DART) mass
spectrometry, 190–191
pressure-sensitive tapes:
adhesive formulas, 297–299
backing materials, 297–298
duct tape reinforcement, 303–307
Fourier transform infrared analysis,
320–324
Poly(methyl methacrylate) (PMMA),
glitter from, 20–24
Polypropylene packaging tape, forensice
analysis, 309–313
Poly(vinyl chloride) (PVC), pressure-
sensitive tapes:
backing materials, 297
electrical tape, 307–309
Pooled standard deviation, multivariate
statistical analysis, 354–359
Positive ion spectrum, ink analysis, 64–69
Postburn powder characterization,
smokeless powder
identification, 246
Post-it® notes, glass fragment analysis,
280
Potassium Bromide, smokeless powder
characterization, Transmission
micro-Fourier transform
infrared spectroscopy, 260–261
Poudre B, smokeless powder history,
245
Powder particles:
condom trace evidence:
forensic protocols, 101–104
materials characteristics, 86–87
residue analysis, 88–91
smokeless powders:
ATR-FTIR spectroscopy, 261–262
basic properties and classification,
241–242
brand determination, 242–243,
246–257
ball-shaped powders, 248–249
disc shaped powders, 253–256
flattened ball powders, 256–257
lamella shaped powders, 258
tubular powders, 249–253
color, 257–258

Fourier Transform infrared
spectroscopy, 260–262
gas chromatograph/mass
spectrometry, 262–265
historical background, 243–245
identification, 242, 245–246
kernel/dot configuration, 258
liquid chromatography, 265–266
luster, 258
mass, 259–260
micrometry, 258–259
micromorphology, 247–257
morphology, 246–247
transmission micro-FTIR, 260–261
Preservatives, condom trace evidence, 86
Pressure sensitive tape:
adhesive formulas, 297–299
backing, reinforcement and adhesive
separation, 319
backing materials, 296–297
construction, 293–296
duct tape, 299–307
electrical tape, 307–309
elemental analysis, 320, 324–325
end matching techniques, 318–319
forensic analysis of, 291–292
case studies, 327–329
Fourier transform infrared analysis,
320–324
initial examination protocols, 316–317
interdepartmental protocols, 317
manufacturing source identification,
326–327
masking tape, 315–316
physical characterization, 319
polarized light microscopy, 325–326
polypropylene packaging tape,
309–313
machine edge offset, 313
orientation marks, 311
oriented films, 310
polarized ligh microscopy analysis,
310
thickness, 312
product variability, 292–293
pyrolysis gas chromatography/mass
spectrometry, 326
reinforcement fabrics, 299
strapping/filament tapes, 313–315

trace evidence recovery and
untangling of, 318
Presumptive testing:
latent bloodstains, 117–125
pepper spray detection, 134–138
trace evidence, 116–117
Primary colors, additive color theory,
383–385
Principal component analysis (PCA):
accelerant applications, 359–360
development and application, 336
dimensionality reduction, 342–348
document examination and currency
analysis, 360–362
drug analysis, 362
fibers, 362–364
glass analysis, 364
linear discriminant analysis, 349–354
mineral trace evidence, 364–366
Property damage accidents, automotive
airbag forensics, 50–54
Protonated molecules, drug/
pharmaceutical analysis, Direct
Analysis in Real Time (DART)
mass spectrometry, 177–180
Protonation detection technique, pepper
spray analysis, 131–138
Proton nuclear magnetic resonance,
condom lubricant residue
analysis, 92–96
Provenance, cathodoluminescence, 166
Pulsed-discharge electron capture
detector (PDECD), surface-
modified fibers, 238
Pyrolysis gas chromatography/mass
spectrometry (PGC/MS):
condom lubricant residue analysis,
92–96
document examination and currency
analysis, 360–362
pressure-sensitive tapes, 326
surface-modified fiber analysis,
234–238
Pythagorean theorem, multivariate
statistical analysis, 355–359

Quartz, cathodoluminescence, 160–163
Quencher atoms, cathodoluminescence,
146–147

Questioned sample (Q):
glitter characterization and, 10–24
color analysis, 11–12
particle size, 13
surface-modified fibers, 222–224
pyrolysis gas chromatography/mass
spectrometry, 234–238

Raman microspectroscopy:
glitter characterization, 19–22
multivariate statistical analysis,
339–342
surface-modified fibers, 224
Rare earth elements (REEs):
cathodoluminescence, 146–147
accessory minerals, 163–164
calcium carbonates, 158
feldspar group, 160
stable isotope ratios, 413–414
Rayon fibers, duct tape reinforcement,
303–307
Reflectance infrared spectroscopy:
ink analysis, 57–58
optically variable pigments,
376–379
Reflecting pigments, geometric
measurements, 390–391
Reflection measurements, optically
variable pigments, 389
Refractive index (RI):
glass analysis, 364
pressure-sensitive tapes, 326
Reinforcement materials, pressure-
sensitive tapes, 296
duct tapes, 299–307
Release coats, pressure-sensitive tape,
293–294
Retention properties, glitter, 3
Rhodamine dyes:
ink analysis, 69–71
latent bloodstain analysis, 121–125
Robbery cases, glass cuts as evidence in,
271–274
Rubber materials, pressure-sensitive
tapes:
adhesive formulas, 297–299
duct tape, 300–307
Rutile, cathodoluminescence, 168–171

Sample alteration, cathodoluminescence,
155
Sample-electron interactions,
cathodoluminescence,
141–142
Sample preparation and preservation:
cathodoluminescence, 152–153
multivariate statistical analysis,
336–342
randomization and classification,
357–359
optically variable pigment
measurements, 393–394
Sample size, optically variable pigments,
392
Savitzky-Golay algorithm, multivariate
statistical analysis, 338–342
Scanning electron microscopy/energy
dispersive
spectroscopy (SEM/EDS):
pressure-sensitive tapes, 324–325
surface-modified fibers, 225–231
low-voltage techniques,
230–231
3M protective finish, 229–230
P2i-treated samples, 225–228
Scanning electron microscopy (SEM):
cathodoluminescence, 149
instrumentation, 150
quartz, 163
document examination and currency
analysis, 360–362
glitter analysis, 22–23
glitter characterization, criminal case
studies of, 25–31
optically variable pigments, 376–379,
395
Scanning electron microscopy/
wavelength dispersive X-ray
spectroscopy (SEM-WDS),
pressure-sensitive tapes,
324–325
Scissors cuts, glass cuts compared with,
tearing patterns in fabric,
277–279
Screening techniques,
cathodoluminescence, 165
Scrim patterns, duct tapes, 300–307

Selected ion monitoring (SIM), surface-modified fibers, pyrolysis gas chromatography/mass spectrometry, 235–238
Self-defense sprays, Direct Analysis in Real Time (DART) mass spectrometry, 184–185
Sensitizer atoms, cathodoluminescence, 146–147
Separation techniques:
  glitter traces, 8–9
  multivariate statistical analysis, 354–359
  pressure-sensitive tapes, 318–319
Sexual assault crimes, glitter as evidence in, 26–31
Shape properties, of glitter, 3, 13
Shard glass, analysis of, 281–282
"Shimmer" materials, 2
Short tube smokeless powders, 252–253
Silica particles:
  cathodoluminescence, quartz, 160–163
  condom trace evidence, 86–87
  forensic analysis, 103–104
Silicon compounds, cathodoluminescence, feldspar group, 159–160
Silicone, condom lubricants:
  materials characterization, 86
  residue analysis, 92–96
Singe patterns, automotive airbags, 37–54
Singular value decomposition, principal component analysis, 343–348
Size properties, of glitter, 3, 13
Slag, cathodoluminescence, 168–169
Slash cut patterns, glass cuts and, 279–280
Smiths Detection Application Brief, glitter thickness characterization, 14–15
Smokeless powders:
  ATR-FTIR spectroscopy, 261–262
  basic properties and classification, 241–242
  brand determination, 242–243, 246–257
  ball-shaped powders, 248–249
  disc shaped powders, 253–256

flattened ball powders, 256–257
lamella shaped powders, 258
tubular powders, 249–253
color, 257–258
Fourier Transform infrared spectroscopy, 260–262
gas chromatograph/mass spectrometry, 262–265
historical background, 243–245
identification, 242, 245–246
kernel/dot configuration, 258
liquid chromatography, 265–266
luster, 258
mass, 259–260
micrometry, 258–259
micromorphology, 247–257
morphology, 246–247
transmission micro-FTIR, 260–261
Smoothing, multivariate statistical analysis, 338–342
Sodium azide gas, automotive airbags, 34–35, 37–39
Sodium hydroxide, automotive airbags, 35
Soft independent modeling of class analogy (SIMCA):
  amphetamines, 362
  document examination and currency analysis, 360–362
  multivariate statistical analysis, 356–359
Soil and sand characterization, cathodoluminescence, 167
Solid propellants, automotive airbags, 34–35
Source locations, stable isotope ratio analysis, 404–405
Specific gravity, of glitter, 4, 13–14
Spectral data collection:
  cathodoluminescence, 153–154
  latent trace analysis, fingerprints, 126–130
  multivariate statistical analysis, 334–336
  optically variable pigments, 394–395
Spectrometry equipment, cathodoluminescence, 150

Spermicides, condom trace evidence:
forensics applications, 105
materials characterization, 87
residue analysis, 95–96
Spherical® powders, smokeless powder
characterization, 249
Spinel, cathodoluminescence, 163–164
glass materials, 169
Sports doping, stable isotope ratio
analysis, 410–411
Stable isotope ratio analysis:
abundance variations, 401–402
basic principles, 399–400
delta notation, 401
forensics applications, 404–416
bones, hair, and teeth, 413–414
doping and drug abuse cases,
408–411
food products, authenticity, and
adulteration, 406–408
human/animal products and
sourcing, 411–413
manufactured items, 414–416
instrumentation, 402–403
isotope distribution and properties,
400–401
Standard Mean Ocean Water (SMOW),
stable isotope ratio analysis,
401
Starch particles:
automotive airbag forensics, 36–37,
41–54
smokeless powder history, 244
Statistical testing. *See also* Multivariate
statistical analysis
optically variable pigments, 395
Stereo light microscope (SLM):
smokeless powder characterization,
246–247
smokless powder identification, 242
Strapping/filament tapes, forensic
analysis, 313–315
Strontium, stable isotope ratios,
413–414
Student's *t*-test:
multivariate statistical analysis,
354–359
optically variable pigments, 395
Sulfur isotopes, ink analysis, 72–74

Surface anesthetics, condom trace
evidence, 87
Surface properties:
cathodoluminescence, calcium
carbonate, 158
drug/pharmaceutical analysis, Direct
Analysis in Real Time (DART)
mass spectrometry, 179–180
fiber analysis:
distinguishing tests, 225
gas chromatgraphy/mass
spectometry, 231–234
preliminary examinations, 222–224
pyrolysis gas chromatograpy/mass
spectometry, 234–238
research background, 221–222
scanning electron microscopy/
energy dispersive spectroscopy,
225–231
structural properties, 222
latent trace analysis, bloodstains,
122–125

Tackifiers, pressure-sensitive tapes,
297–299
Talcum powder particles:
automotive airbag forensics, 37–39
cathodoluminescence, 168–171
condom trace evidence, 86
forensic analysis, 103
Tandem mass spectrometry (MS/MS),
fabric dye analysis, 202–203
direct infusion protocol, 214–216
electrospray ionization and, 204–207
structural elucidation, 208–210
thermospray techniques and, 212–214
Tape products. *See* Pressure sensitive
tape
Tearing patterns in fabric, glass cuts *vs.*,
276–279, 281–282
Teeth, stable isotope ratios, 413–414
Tetrahydrofuran (THF), electrical tape,
307–309
Thermospray interfaces, fabric color
analysis, 211–214
Thickness properties:
of glitter, 4, 14–15
polypropylene packaging tape,
312–313

Thin film interference, optically variable pigments, 378–379
Thin-layer chromatography (TLC):
  fiber color analysis, 198–199
  ink analysis, 69
3M protective finish, surface-modified fibers, scanning electron microscopy/energy dispersive spectroscopy, 229–230
Time-of-flight (ToF) mass spectrometry, ink analysis, 60–62
Tin compounds, cathodoluminescence, 169
Titanium dioxide, optically variable pigments, 376–379
Total ion current (TIC) analysis:
  BUZZ OFF™ fabrics, 232–234
  fabric dyes, 202–203
    trace quantity extraction, 213–214
  pepper spray detection, 136–138
Total reflection X-ray fluorescence spectrometry (TXRF), cathodoluminescence, 169
Traffic violations, automotive airbag forensics, 50–54
Transfer properties, glitter, 3
Translucent pigments, geometric measurements, 390–391
Transmission measurements:
  cathodoluminescence spectral collection, 153–154
  optically variable pigments, 389
Transmission micro-Fourier transform infrared spectroscopy, smokeless powder characterization, 260–261
Transparent/semitransparent pigment blending:
  geometric measurements, 390–391
  optically variable pigments, 388
Transport medium, glitter, 8
Trapezoid kernels, smokeless powder characterization, 248
Triacetone triperoxide (TATP):
  Direct Analysis in Real Time (DART) mass spectrometry, explosive materials, 186–188

fingerprint analysis, 181
Triethylamine (TEA), fabric dye analysis, mass spectrometry analysis, 214
2,4,6-Trinitrotoluene (TNT), Direct Analysis in Real Time (DART) mass spectrometry:
  explosives, 186–188
  fingerprint analysis, 180–181
Tristimulus values:
  additive color theory, 384–385
  optically variable pigments, color matching functions, 380–382
Tubular powders, smokeless powder characterization, 249–253

Ultraviolet-visible spectrum:
  fiber color analysis, 199–203
  ink analysis, 57–58
    cationic dyes, 67–69
  latent trace evidence, pepper spray analysis, 130–138
  linear discriminant analysis, 351–354
  multivariate statistical analysis, 337–342
    dimensionality reduction, 342–348
    fibers, 362–364
  pepper spray detection, 133–138
UMPLFL lens, glitter characterization, 20–22
Univariate data analysis, group separation, classification accuracy, and outlier detection, 354–359
Urine, Direct Analysis in Real Time (DART) mass spectrometry, 181–182

Variable pressure (VP) scanning electron microscopy, cathodoluminescence, 165
Vehicle characteristics. *See also* Automotive airbags
  glitter traces, 8
Video microscopy, glitter particle size and, 13
Vilsmeier-Haack reagent, latent trace analysis, fingerprints, 129–130

Visible spectrum, optically variable
  pigments, 379–382
Visualization technology:
  latent trace analysis:
    chemical characterization,
      116–117
    fingerprints, 126–130
    linear discriminant analysis, 348–354
    masking tape forensics, 316
    pepper spray detection, 134–138
Vulcanization process, condom trace
  evidence, 86
VX agent ([*O*-ethyl *S*-(2-
    diisopropylaminoethyl)
    methylphosphonothioate]),
    Direct Analysis in Real Time
    (DART) mass spectrometry,
    189–190

Wavelength accuracy,
  cathodoluminescence spectral
  collection, 153–154
Wearing and worn damage, glass cut
  analysis, 286–287

Weight properties, optically variable
  pigment blending, 387
Winding samples:
  fiber color analysis, 200–203
  negative ion ESI-MS analysis, nylon
    winding, 207–208

Xlyoidine, smokeless powder history, 244
X-ray fluorescence (XRF):
  document examination and currency
    analysis, 360–362
  pressure-sensitive tapes, 324–325
X-ray powder diffraction (XRD),
  pressure-sensitive tapes,
  324–325

ZAF-related corrections,
  cathodoluminescence, 147
Zincite, cathodoluminescence, 168–171
Zinc oxide, pressure-sensitive tapes, duct
  tapes, 300–307
Zircon, cathodoluminescence, 163
Zoning, cathodoluminescence, calcium
  carbonates, 158